Esko Heikkonen

REAPING THE BOUNTY: McCORMICK HARVESTING MACHINE COMPANY TURNS ABROAD, 1878–1902

Esko Heikkonen

Reaping the Bounty: McCormick Harvesting Machine Company Turns Abroad, 1878–1902

Finnish Historical Society ■ Helsinki

Cover: The McCormick reaper and mower from 1883. The McCormick
Collection. The State Historical Society of Wisconsin.

ISSN 1238-3503
ISBN 951-710-022-1

Contents

■ Acknowledgements

Writing history is always a complicated process, where the researcher not only discusses with his sources but also with himself. A historian does not, fortunately, live in a vacuum, but stands on the shoulders of previous generations. How broad the shoulders have been that I have been able to lean on during my career, I have understood only at the final stages of this work.

My thanks can no longer reach Professor Matti Lauerma, who put a start to my research by giving me a seminar topic on American agricultural history. His patriarchal figure and courteous approach towards the novices on the vast field of history gave an unforgettable stimulus to my career as a historian.

During the period of his successor, Professor Kalervo Hovi, something of the same atmosphere has remained in the faculty of General History. As the supervisor of my doctoral dissertation, he has always found time for my questions. Without his advice and guidance this work would never have been accomplished. His guidance has also been invaluable in ensuring financial resources for this work. For all his support, Professor Hovi deserves my warmest thanks.

Associate Professor Reino Kero at the University of Turku took a risk when he appointed me as his research assistant after only one and a half years of studies. The time spent under his supervision taught me the true value and use of primary sources and research. Dr. Kero has never lost his trust in the fulfilment of this work and has given me much advice during our long discussions.

Dr. Keijo Virtanen, Professor in Cultural History at the University of Turku, has from the very beginning, in his American history seminar group, been a central figure in accomplishing this dissertation. He has provided me with many key ideas as well as methodological approaches. I wish to thank Dr. Virtanen for the much intellectual discussions on the critical points of this work. Of crucial importance has been his contribution to obtaining the publication of my dissertation in a highly respected series.

I owe warm thank to my brother Dr.Pol.Sc. Eero Heikkonen, who has with his brotherly hand guided my research. With special gratitude I remember the fishing trips during which my ideas of the meaning and logic of research have been greatly widened.

My understanding of American history has been deepened during my visits and studies in America. I wish to thank Alan L. Olmstead, Professor in Economics at the University of California for his contributions in the early stages of my dissertation. Special and warm thanks I owe to Professor Emeritus Allan G. Bogue at the University of Wisconsin, who always found time for his Finnish student, and has through the years provided me with critical comments. To Associate Professor Colleen A. Dunlavy at the University of Madison I want to express my gratitude for showing me what the history of technology could be. Lastly I want to thank Professor Morton Rothstein at the University of California for his advice in the shaping of this work.

Dr. Keith Battarbee, who has revised the manuscript, deserves warm thanks for his skilful work. Mrs. Maija Räisänen has taken care of the layout of my dissertation. The Executive Director of the Finnish Historical Society, Mr Rauno Endén, MA, has once again shown his professional skills in handling the publication of the text. I wish to thank him for his smooth and friendly co-operation.

For the provision of financial support for my research I direct my warmest thanks to the ASLA-Fulbright commission for the possibility to study and write my dissertation in America. In addition, I want to thank the following organization for their important support: the Aaltonen Foundation; the Finnish Cultural Fund and its Varsinais-Suomi and Satakunta Funds; the Öflund Foundation; the Federation of Academic Technicians; the University of Turku Foundation and University of Turku; the City Government of Turku; and the Finnish Historical Society, in whose series, Bibliotheca Historica, this work is being published.

Finally, I would like to express my great gratitude to my wife, Terttu, who has not only patiently listened to and followed me through the various stages of this study, but has also provided me with many invaluable comments on it. To our children Elina and Matti, the McCormick Co. has meant a demanding adaptation to many new conditions, not least school in America, and has taken up more than would have been appropriate of my time from them. The joint contribution of my family, in counterbalancing the research, has been invaluable.

Motala, October 1995
Esko Heikkonen

I

■ Defining the Problem

1.1. The state of research

The role of business has been central in the transformation of American society. Since the very beginnings of the nation, it has raised emotions, either in support or opposition. To Thomas Jefferson, the future of the United States was agricultural: "for the general operations of manufacture, let our work-shops remain in Europe". Similarly, John Adams declared: "...America will be the country to produce raw materials for manufacture; but Europe will be the country of manufactures." In the same spirit, too, Benjamin Franklin observed, "The great business of the continent is agriculture. For one artizan, or merchant, I suppose we have at least a hundred farmers."[1]

The history of the United States has proved these prophecies of the founding fathers only partly true. For the most part of the 19th century, the country remained agricultural; but alongside agriculture, the new technology, and industry based on it, developed rapidly.

Essential for the development of the U.S. economy were the British investments that flowed into the country. Britain was a major capital exporter, while the United States was the world's largest market. Sterling found its way into the various applications of steam to land and sea transport. In the long run, railroads lowered transport costs and integrated the vast country economically and politically into one entity. Although not as dramatic and rapid, there was a similarly far-reaching change in sea transport when steam replaced sail, at first on short routes in the 1850s and 1860s, and thereafter gradually on the long routes towards the end of the century. For business, one of the effects of this development was that it became possible to have meaningful coordination, control, and influence over long distances and even over international frontiers.[2]

The second great change in American life and society was the rapid growth of manufacturing industry. Machines replaced

---

1 *Kasson* 1988, 14-17.

2 *Foreman-Peck* 1983, 33-35; *Harley* 1970, 216-217, 219, 222-223, 227; *Wilkins* 1986, 86-87; *Wilkins* 1988, 8-9.

handicraft labor, new raw materials and sources of power gained importance, and workers became concentrated in factories. Alongside this process, there began to develop a new approach to the entire manufacturing process, the American System of Manufacturing, part of which was the concept of interchangeability of parts. Big business developed in American industry when the new national market created by railroads and improvements in technology combined with the impact of mass production and mass distribution after the Civil War.[3] The American economy expanded beyond its borders. During the period extending from 1871 to 1895 the expansion of American exports to Europe consisted mainly of foodstuffs, wheat being the major article. From 1895 to World War I, the paramount feature was the spectacular rise of the exports of semifinished and finished manufactures. By the last third of the century Britain was faced with a new situation; its industrial monopoly had given way to competition.[4]

The third big revolution of the period occurred in agriculture. Taking the prairies of the Mid-West and great plains under cultivation more than doubled wheat output alone from 1859 to 1880. This dramatic growth was based on new technology and increased acreage. The advance of refrigeration into railroad cars and steamships in the 1880s gave a further impetus to agriculture, but also caused a major crisis for European farmers. During the 1870s and 1880s, the cost of transporting wheat from New York to Liverpool was cut by about half, and consequently American grains and other foodstuffs began to flood into Europe. The collapse of wheat prices drove some Continental countries to protect their agriculture, while others changed their production structure.[5]

These changes, in connection with social changes that were occurring simultaneously on both sides of the Atlantic Ocean, are a necessary precondition for an understanding of the growth of American big enterprise. The focus of the present study is on the McCormick Harvesting Machine Company, as an example of a large company.

In Cyrus Hall McCormick's personality merge many fundamental features of the American nineteenth-century revolutions. He was an inventor, entrepreneur and a hardboiled business man. During his life time he turned his tiny blacksmith shop in Walnut Grove, Virginia, into the leading harvesting machine company in the world

---

3 *Blackford-Kerr* 1986, 84, 98, 102, 105, 152-153, 164.

4 *Simon-Novack* 1964, 592-593, 599-602; *Landes* 1989, 239.

5 *Hughes* 1987, 276-277; *Tracy* 1966, 100-103.

at the time of his death in 1884.[6] McCormick's reaper opened up one of the bottlenecks in American agriculture. After purchasing the reaper, farmers were able to extend their acreage and their dependence on casual labor was reduced.[7]

The most comprehensive study on McCormick has been produced by William T. HUTCHINSON.[8] Hutchinson was able to spend eight years writing his detailed, exactly documented work, over 1,200 pages long, on the invention of the reaper, its further development, the formation of big business, and the personal and family history of Cyrus McCormick. It is no wonder that all later researchers in this field have used his research as a cornerstone in their own work.

Before Hutchinson's monumental study, the question of the role of Cyrus McCormick as the inventor of the reaper had aroused hot debate. Relations between Cyrus McCormick and his younger brother Leander J. McCormick had gradually deteriorated from the 1860s until the final breakdown in 1880, which led to the resignation of Leander as the vicepresident of the McCormick Company. In the aftermath of the controversy, Leander tried to prove his brother's claims to the invention of the reaper false, and instead attributed the honor to their father, Robert McCormick.[9]

Herbert F. CASSON's "The Romance of the Reaper"[10] is a representative model of older business history in its romantic and heroic description of the McCormick and Deering Companies. It belongs to a series of romances written about numerous persons and subjects at the turn of the century in the United States. Casson's approach remains the same in his "Cyrus Hall McCormick. His Life

---

6 *Hutchinson* 1935, passim; *Carstensen* 1984, 107-118.

7 *Rogin* 1931, 79-82, 91-94, 125-141.

8 HUTCHINSON: Cyrus Hall McCormick. Seed time, 1809-1856. New York 1930; HUTCHINSON: Cyrus Hall McCormick. Harvest, 1856-1884. New York 1935.

9 *Hutchinson* 1930, 101-108; *Directors meeting* 6.4.1880. Mss M/I, box 17.
Neither the original book of 1885 nor the reprint of 1898 of "Memorial of Robert McCormick, Being a Brief History of His Life, Character and Inventions, Including the Early History of the McCormick Reaper" have been available for the writer. Information on this part is thus based on Hutchinson's research. *Hutchinson* 1930, 8-9 footnote 14, 99 footnote 1.
The second round of the game began in 1910, when the son of Leander McCormick, Robert Hall McCORMICK, and his nephew James Hall SHIELDS published their compilation of statements in homage to Robert McCormick: The Life and Works of Robert McCormick, Including His Invention of the Reaper. Chicago 1910. The question was reheated by John F. STEWARD: The Reaper. A history of the efforts of those who justly may be said to have made bread cheap. New York 1931, and N. LYONS: The McCormick Reaper Legend. The true story of a great invention. With a foreword by Robert Hall McCormick III. New York 1955.

10 New York 1908.

and Work."[11]

After Hutchinson, the first time the foreign business of the McCormick Harvesting Machine Company attracted attention was in the doctoral dissertation of George G. QUEEN.[12] Although he concentrates on economic relations between the U.S. and Russia, he takes harvesting machines as an example of American machinery in Russia. In this connection he has mostly used Hutchinson as his source, but he has also made use of some primary material in the McCormick Collection.

In the late 1950s and early 1960s, a couple of graduate theses examined McCormick's foreign activities.[13] A new approach to the

---

11 Chicago 1909. In reading both of Herbert Casson's books, there have to be kept in mind his contacts with the McCormick family, which render the validity of some of his statements suspect. Both Nettie F. McCormick, wife of Cyrus H. McCormick, and their son Harold F. McCormick made comments on Casson's manuscripts. Harold McCormick, for example, demanded "It is desired that all reference to Mr. Rockefeller be cut out". Besides, he made numerous corrections in details concerning the McCormick Company. In addition, he wanted Casson to contact his brother, Cyrus Jr., because "I would not like to pass on this unless it is necessary". Nettie McCormick for her part had very carefully read Casson's manuscript and dictated her suggestions for changes. "Page 36: Last six lines - Beginning with "Hussey's had a better cutting apparatus". Are these words historically true? I think not." Unfortunately it is not clear to which one of Casson's books these comments refer. *Criticism of H.F. McCormick to the third chapter of Mr. Casson's article. 9.26.1907.* Mss 6c, box 30; *Suggestions for changes in Mr. Casson's book, made by Mrs. McCormick, which she would like to have presented to Mr. Casson by Miss Smith.* No date. Mss 6c, box 30.

12 *QUEEN*: The United States and the Material Advance in Russia, 1881-1906. Ph.D. thesis, University of Illinois. Illinois 1942. All of the data concerning the amount of sales by the McCormick Co. in Russia are subject to inaccuracies. For example Queen has found in 1894 McCormick's sales to be only 132 machines, whereas in the accounts of George A. Freudenreich the figure given is 1050. *Queen* 1942, 140; *Recapitulation of all sales of machines made in Russia and Roumania during the season of 1894. Geo. A. Freudenreich. Genl. Agt.* Mss 2x, box 235; Material for the previous years shows a similar error in Queen's calculations. *Sales of the Odessa agency for the last 14 years.* Mss 2x, box 235.

13 All of these works are limited in their scale and scope. When foreign trade finally was explored as an independent topic, all the other aspects were left aside, and the necessary context for the whole phenomenon was thus forgotten.
The first in the series of three works was Eugene SHAPIRO: Expansion of the foreign market, 1898-1902. Inaugurating the Branch House System. Unpublished seminar paper, University of Wisconsin. Wisconsin 1958. He was followed by Eugene A. MANNING: Foreign Business of the McCormick Harvesting Machine Company, 1885-1902. Unpublished seminar paper, University of Wisconsin. Wisconsin 1961. Shapiro and Manning were the first who entirely concentrated in these papers on the McCormick Company's foreign business. Shapiro's aim was to figure out the main lines of the last phase of the McCormick Company's foreign organization. Manning, on the other hand, widened his scope also to include the earlier years of foreign sales. Both of them used Hutchinson and Queen as their sources, but also extended their research to the primary material. Besides, Manning owes a lot to Shapiro.
The next phase in the research was the graduate thesis of Howard Bernard SCHONBERGER: The Foreign Business of the McCormick Harvesting Machine Company. Unpublished MA thesis, University of Wisconsin. Wisconsin 1964. Schonberger also found and used the works of previous researchers, but fails to understand the continuity of the business from Crystal Palace. Besides, in spite of the use of original material, he tried to explain the operations of the McCormick Company

theme was found by Fred V. CARSTENSEN[14] in his doctoral dissertation, which was later published as "American Enterprise in Foreign Markets. Studies of Singer and International Harvester in Imperial Russia". Carstensen focuses his research on the Russian operations of the International Harvester Company, thus either omitting or only briefly mentioning the early phases of McCormick's foreign trade. His main emphasis is thus on the role of the International Harvester Company, and not on the McCormick Company.

The latest study that deals with the McCormick Company was first published in 1984 by David A. HOUNSHELL.[15] Hounshell concentrated in this thorough investigation on the technological and managerial sides of the central manufacturers in American business history. He has shown how primitive and simple methods continued to be used even in highly respected firms such as the McCormick Co. up to the beginning of the 1880s. He has also demonstrated the significance of organization and technology for the growth of an industrial enterprise.

The establishment of the International Harvester Company in 1902 ended the independent existence of the McCormick Company. That event has produced some seminar works and articles, too, but still awaits more comprehensive examination.[16]

---

by outside economic forces. While totally failing to formulate the crucial question, Schonberger frequently concentrates on details, like the Chicago World Fair or the debate on tariffs in the US Senate. Consequently, he has totally omitted effects arising within the company and the harvester business.

14 CARSTENSEN: American Multinational Corporations in Imperial Russia; Chapters on Foreign Enterprise and Russian Economic Development. Unpublished Ph.D. thesis, Yale University. Yale 1976; Chapel Hill 1984.

15 HOUNSHELL: From the American System to Mass Production, 1800-1932. The Development of Manufacturing Technology in the United States. Baltimore 1984.

16 In the research this event was first handled by Howard William BENSON: Organization and first years of the International Harvester Company. Unpublished MA thesis, the University of Chicago. Chicago 1936. Benson had to rely in his study on newspapers and other descriptive sources. The next development in the field was Helen M. KRAMER's master's thesis: Harvesters and High Finance: Formation of the International Harvester Company. Unpublished MA thesis, University of Wisconsin. SA. She had access to archival material, and was able to concentrate her interest on International Harvester's formation process. Unfortunately it has been impossible to find the original work, in spite of numerous efforts on the part of the writer, Wisconsin University Library, and the Wisconsin State Historical Society's personnel. Only the revised version of it has therefore been used in this study, together with the article published under the same title in Business History Review 3 (1964), 283-301. Barbara MARSH's A Corporate Tragedy. The Agony of International Harvester Company, New York 1985, deals more with the company's later development until its financial difficulties in the 1970s and 1980s.

The development of administration and professional managers as a decisive factor in the evolution of the modern firm was first demonstrated by Alfred D. CHANDLER, in his pathfinding studies "Strategy and Structure. Chapters in the History of the Industrial Enterprise",[17] and "The Visible Hand. The Managerial Revolution in American Business".[18] Chandler located the emergence of big American enterprise in the post bellum years as a result of integrated mass distribution and mass production. According to Chandler, "modern business enterprise took the place of market mechanisms in coordinating the activities of the economy and allocating its resources".[19] He argues that as modern business enterprise became the most powerful institution in the American economy, its managers became the most influential group of decision-makers. A chandlerian firm contains several distinct operating units, and is managed by a hierarchy of salaried executives. Chandler shows how the emergence of big business demanded new kinds of talents to manage it. This process gave birth to the professional, salaried manager, who was in contrast to the traditional single-unit firm operated by an individual or a small number of owners.[20] In "Scale and Scope. The Dynamics of Industrial Capitalism"[21] Chandler applied his model to

---

17 Cambridge, Massachusetts 1962.

18 Cambridge, Massachusetts 1977.

19 *Chandler* 1977, 1.

20 Chandler's thesis was soon attacked by a number of researchers. DUBOFF and HERMAN accept the merits of Chandler's approach to business history, but want to show its serious failings as well. The interplay between politics and business is almost totally ignored, as is also the social significance of oligopoly. The power of large firms is a result of their growth. The writers make the further criticism that Chandler's managers are presented as neutral, living in a vacuum, and that he thus fails to understand the goals these managers pursue. Duboff, Richard; Herman, Edward S.: Alfred Chandler's New Business History: Review. Politics and Society 10 (1980): 87-110.
Jeremy ATACK tests Chandler's basic concept of the evolution of the large business firm. By comparing companies before the Civil war and again at the turn of the century, he shows that there was no such radical change visible in the firms as Chandler states. The forces put in motion between 1850-1870 were rather an outcome of long-run developments. Atack, Jeremy: Industrial Structure and the Emergence of the Modern Industrial Corporation. Explorations in Economic History 22 (1985):29-52.
Naomi LAMOREAUX condemns the Cliometricians for passing over, with a couple of exceptions, the era of big business and the great merger movement. For the Cliometricians the merger movement was unimportant, since they believe that industrial concentration has not resulted in significant deviations from competitive pricing. An important point for the present study is her argument that mergers were a conjunction of particular circumstances. The simultaneous rapid expansion of many capital-intensive industries in the early 1890s, followed by the deep depression of 1893, gave rise to abnormally serious price wars and consequently to the great merger movement. Lamoreaux: The great merger movement in American business, 1895-1904. New York 1988.

21 Cambridge, Massachusetts, 1990.

European countries, by comparing US firms with British and German companies. Chandler introduced the idea of a crucial three-pronged investment - in machinery, management, and marketing. By utilizing these investments modern enterprises grew by taking advantage of the economies of scale, but also of those of scope by extending the range of products manufactured and marketed. In spite of great merits, Chandler's ideas have not met with unanimous acceptance, however. Although his basic concepts have survived under the critical eyes of economists, criticism has been raised because of the omission in analyses of such central factors as labor and industrial relations, business-state relationships, and legal and educational environment.[22]

The multinational enterprise can be seen as one phase in the development of big business. The most comprehensive picture of the American multinationals[23] before World War I can be found in Mira WILKINS' "The Emergence of Multinational Enterprise: American Business Abroad from the Colonial Era to 1914".[24] Wilkins only scrutinizes those companies that made direct investments[25] abroad. One of her objectives was to find out when, why, how, and where early American business went into direct foreign investments. In her later articles, Wilkins has broadened her field to embrace

---

22 Barry Supple: Scale and scope: Alfred Chandler and the dynamics of industrial capitalism. Economic History Review 44 (1991): 500-514 and William N. Parker: The Scale and Scope of Alfred D. Chandler, Jr. Journal of Economic History 51 (1991): 958-963.

23 The literature on multinational enterprises is manifold and it is beoynd the scope of the present study to discuss this in detail. Among the general works of the older generation, however, note should be taken of those of Frank A. SOUTHARD and Cleona LEWIS. Southard: American Industry in Europe. New York 1976. Originally published in 1907. Lewis: America's Stake in International Investments. New York 1976. Originally published in 1938.
On American firms, important information can also be obtained from reports by various authorities. These include the US Senate: American Branch Factories Abroad. Washington 1931. The US Department of Commerce: American Direct Investments in Foreign Countries. Washington 1930. The US Senate: Letter from the Secretary of Commerce. Washington 1933.
Lawrence G. FRANKO'S "The European Multinationals. A Renewed Challenge to American and British Big Business". London, New York 1976, was one of the first works on the development of the Continental multinationals. His work also helps to evaluate the role and size of the American multinational enterprise. His theories have been modified by many subsequent researchers. One of the most significant works is "The Rise of Multinationals in Continental Europe", edited by Geoffrey Jones and Harm G. Schröter. Bodmin 1993. Notice should be also taken of Mira Wilkins' "European and North American Multinationals, 1870-1914: Comparisons and Contrasts" in R.P.T. Davenport-Hines and Geoffrey Jones (eds.) "The End of Insularity, London 1988 and of Charles Wilson's "The Multinational in Historical Perspective" in Keiichiro Nakagawa: Strategy and structure of big business. Tokyo 1974.

24 Cambridge, Massachusetts 1970.

25 *Wilkins* 1970, ix-x, 19, 29.

Continental multinationals and their expansion in America as well, and she has contributed valuable new theoretical and methodical tools to research.[26]

1.2. The Central Questions

The main object of the present study is the foreign business of the McCormick Harvesting Machine Company during the period 1878-1902. This research is based on primary material collected in the McCormick Collection in the State Historical Society of Wisconsin Archives. Through this source material, for the most part previously unused, the present study offers new information as to why and how the McCormick Company extended its operations on the foreign field and turned into a multinational company. The primary information has been complemented by the existing research.

The foreign business of the McCormick Company has not previously been comprehensively examined, and the present study aims to fill this gap. This scrutiny of the McCormick Company has been broadened to include the whole harvesting machine sector in America and its expansion abroad, and the subsequent on-going competition in Europe. The importance and value of the agricultural machine trade becomes evident when it is compared with other sectors of industry. In 1917 the International Harvester Company, successor of the leading harvester companies, was in the U.S. ranked seventh by its assets ($m264.7).[27]

Comparison of the sales methods used in Europe with American business practises further widens the range of the present study to address the state of the European agricultural machine industry[28]

---

26 WILKINS, Mira: Defining a Firm: History and Theory (Peter Hertner and Geoffrey Jones (eds.): Multinationals: Theory and History). Shaftesbury 1986; WILKINS, Mira: European and North-American Multinationals, 1870-1914. Comparisons and Contrasts (R.P.T. Davenport-Hines and Geoffrey Jones (eds.): The End of Insularity. Essays in Comparative Business History). Chippenham 1988; WILKINS, Mira: Comparative Hosts. Business History 36 (1994): 18-50.

27 *Chandler* 1977, 510. Behind the International Harvester were such companies as E.I. du Pont de Nemours & Co. (8./$m263.3), U.S. Rubber Co. (9./$m257.5), General Electric Co. (11./$m231.6), Singer Mfg.Co. (15./$m 192.9), and Ford Motor Co. (16./$m165.9). Ibid. 503-512.

28 In Europe, mechanization of agriculture has attracted some interest among agricultural historians, but less interest has been shown in the agricultural machine industry. Among the basic works are G.E. FUSSEL's "The Farmer's Tools. The history of British farm implements, tools and machinery before the tractor came". London 1952, and Albert Eskeröd's "Jordbruken under femtusen år. Redskapen och maskinerna." Borås 1973.

seen through the material found in the McCormick collection. One of the key problems of the present work is also how the McCormick and other American companies were successful in conquering the European harvesting machine markets.

The present study is not the last word on this complex matter, but it does offer material, ideas and new facts for future studies. It should also be recognized that the approach chosen concentrates on only one aspect of the many activities of a large company, and other aspects have been taken into account only insofar as was necessary for clarification of the foreign business.

In the present study, many strong personalities emerge and are discussed in detail. It must be borne in mind that, although in business the actor is nominally the firm, behind corporate decisions there are always individual minds. Consequently, the meaning of family and company archives such as the McCormick Collection should be underlined.

Although the focus of the study is mostly on economic considerations, this is, nevertheless, historical research, carried out with the tools and methods of historiography. The new basic material and conclusions do, however, also offer possibilities for use by economists too.

The subject matter has been approached from four dimensions. On the personal level, Cyrus McCormick, his family and his company are at the center of the research: they are the actors in this game. When the investigator widens his scope from that level to business transactions, he ends up in the field of business history. The rapid expansion of American business led to the birth of giant corporations that expanded their operations abroad. The third starting point for the present study is therefore the multinational enterprise, and agriculture on two continents provides the necessary context for the study.

Foreign markets were opened to Cyrus McCormick after the Great Exhibition of the Works of All Nations in Crystal Palace in 1851. McCormick was, nevertheless, unable to maintain his dominant position in the European markets, and his name almost faded away from European farmers' awareness until in 1878 he displayed his new machine, the wirebinder, at the Royal Agricultural Society Show in Bristol. From this point on, foreign business began to

---

Paul A. DAVID's "The landscape and the machine: technical interrelatedness, land tenure and the mechanization of the corn harvest in Victorian Britain" (Donald N. McClosky: Essays on a Mature Economy: Britain after 1840. Papers and Proceedings of the Mathematical Social Science Board Conference on the New Economic History of Britain, 1840-1930, held at Eliot House, Harvard University, 1-3 September 1970), London 1971, is an attempt to describe adaptation of machines in England to the nature of the fields.

develop rapidly, and was made an integral part of the McCormick Company's activities.[29] This event has been taken as a starting point for the present study. Although the emphasis is on the McCormick Company stage, the foreign activities preceeding 1878 have been included as a necessary background.

The merger of the large harvesting machine companies to form the International Harvester Company in 1902 ended the history of the McCormick Company as an independent enterprise. In the new company, the role of foreign business was re-evaluated. Some of the former structures and strategies were continued in International Harvester, including the development of European branch houses and sales companies; but there were also drastic changes. The most crucial departure from the earlier period was the extension of manufacture to Europe, when International Harvester's first foreign factory was established in 1905 in Sweden; and in 1908, the board of directors of International Harvester approved plans to establish factories in France, Germany and Russia. These were in operation by 1910. In addition, International Harvester rapidly added new production lines to its structure, and was developed into a full-range agricultural machine manufacturer.[30] Consequently, the formation of International Harvester is a logical end for a study on the McCormick Harvesting Machine Company.

BLACKFORD and KERR explain that, although business history is closely related to economic history, economic historians concern themselves with larger forces that shape the formation of material wealth. "Those forces are the settings in which business firms arise, mature, and decline." By contrast, business historians examine individual institutions, not larger economic forces.[31] Consequently, these larger forces have been examined in this study only as creators of the context for individuals and firms. Behind all business transactions, there is always the driving force of an inventive mind; and that is why this work also deals with the history of entrepreneurs and entrepreneurship.

When the focus of the research is on the development of American big enterprise, Alfred D. Chandler's name towers over all others. In the present study, Chandler's approach has been enlarged to incorporate its possible implications for foreign business. An important, but too often neglected side of this process is the

---

29 *Hutchinson* 1930, 380-405; *Hutchinson* 1935, passim.

30 *Carstensen* 1984, 134-135, 145-146; *The International Harvester Co.* 1913, 141-147,

31 *Blackford and Kerr* 1986, 2.

evolution of production technology and improvements in the factory set-up. Both of these factors have here been taken into account.

The abundance of theories on multinational enterprises offers the researcher many competing approaches to select from. Alfred Chandler's extension of his theory of big companies to multinational enterprises has offered the necessary concepts for the present study.[32] For Chandler, one defining characteristic of a multinational is direct investment abroad. The foreign facilities of a multinational were administered from the central home-office and were administratively integral parts of the managerial hierarchy. Instead of economic analyses of the various theories of multinational enterprises, Chandler suggests a historical framework. He especially stresses the significance of long-term profit gained by reduction of unit costs, achieved by improving both technology and organization in the units of the multinational enterprise. The decisive factor in determining costs and profits in his analyses is throughput, the amount of output processed during a single day or other unit of time.[33]

To Chandler, multinationals were one stage in the development of big enterprises. They were born in the 1880s and 1890s, continued to grow in industries with similar technologies of production, and finally expanded their activities in the same manner. In the first stage, they replaced the wholesaler by investing in marketing and distribution, and purchased facilities and personnel. In the second phase, they acquired units producing raw and semifinished materials, and invested in research and development. Many of these companies became multinationals by investing abroad first in marketing and then in production.[34]

In addition to the outlines offered by Alfred Chandler, the works of Mira Wilkins have been of importance for the present study by offering central concepts for handling multinational companies.[35]

---

32 Use of the name 'multinational enterprise' can be traced to David E. Lilienthal's article "Management and corporations" in 1960. Since then, the theory of multinationals has passed through numerous stages, but has finally ended up with a conclusion that the concept of the multinational company itself has proved stimulating but in many respects misleading. "Multinational is merely shorthand for a wide range of capitalist enterprises which share only one common and non-definitive feature: that beyond there is fundamental diversity camouflaged under an umbrella term." In the current situation the conclusion to be drawn is that no single theory of international production can be found. What remains is that a multinational enterprise is a business enterprise which owns and controls income-generating assets in more than one country. *Fieldhouse* 1988, 1-26.

33 *Chandler* 1988, 30-31, 38-39.

34 *Chandler* 1988, 31-34.

35 The selection of Chandler's and Wilkins' theories as basis for the present study

Although foreign trade is not a new field of research, as Wilkins notes, "...studies of U.S. foreign investment ignore the evolution of business abroad and neglect the growth of the sales establishments... it was from such initial forays that there grew the large contemporary international investments."[36]

Wilkins included in her work only those companies that made direct investments[37] abroad. Direct investment is also to her one of the basic characteristics of a multinational company. Besides, she defined an American multinational enterprise as a US-based company that does business in two or more foreign countries.[38] This definition also embraces the first small-scale enterprises that involved only a minor amount of American capital.

Wilkins has shown that a feature common to most American multinationals was that the sales of goods was based on new technology, and that many of them were also trademarked. Many of these firms expanded abroad either to reach new markets or to obtain sources of supply. Besides, American companies substituted capital for labor, which enabled the use of unskilled labor. Wilkins also stresses the pressures of domestic economic and legal conditions forcing firms into foreign expansion. Trade rules encouraged mergers, and integrated business abroad was therefore simply a natural extension. Nevertheless, foreign business was in many cases only complementary to domestic business, and, crucially, all companies with business investments abroad are and have been shaped by economic and other conditions in their homeland.[39]

Wilkins has also formulated five parameters relating to multinational enterprise decision-making. The opportunity parameter

---

obtains futher justification from the latest works on multinational companies. According to JONES and SCHRÖTER, "the development of such organizational capabilities, which are stressed by Chandler as the core issue for international competition, were more important than lowering transaction costs by internalization." *Jones-Schröter* 1993, 17.

36 *Wilkins* 1970, 46.

37 *Wilkins* 1970, ix-x, 19, 29.
John H. DUNNING has found two distinctive features in direct foreign investment. Firstly, it embraces the transfer of equity capital, knowledge, entrepreneurship and sometimes goods as well. Secondly, the resources which are transfered between countries are not traded, they are simply moved from one part of the investing enterprise to another. *Dunning* 1971, 15-17.

38 In her early models, the distinction between a multinational enterprise and direct foreign investment was not yet clear, but in her later articles she explains the relationship as follows: "multinational enterprise make foreign direct investments - and carry on other tasks as well". *Wilkins* 1970, x; *Wilkins* 1988, 8; *Wilkins* 1994, 18-19.

39 *Wilkins* 1988, 8, 22-25, 31.

concerns the various prospects in markets, raw materials etc. The political parameter takes into account the political and legal environment both at the domestic and international levels. The third parameter is familiarity: a multinational is more likely to decide to invest in familiar conditions. The third-country parameter means that multinationals not only see in a host country opportunities, but also recognize third-country conditions. Finally, the corporate parameter relates to companies' internal decisions and experiences.[40] In the present work the definition of multinational enterprise has been understand in the way formulated by Chandler and Wilkins.

The agricultural implement industry can also be approached through innovation theories, as Jan KUUSE[41] has done. The typical central point in traditional innovation theories is the diffusion of ideas or implements in a society, and Kuuse's main interest lies, accordingly, in the development of the agricultural machine industry and its dependence on the development of agriculture. This basic idea is mounted in the process of diffusion of innovations. Kuuse has classified farmers who bought machines as pioneers, early adopters, early majority, late majority and laggards.[42] In the diffusion process he separates the following stages: information - awareness - interest - deliberation - attempt - adaptation. Using this division, he produced for the United States a model of adaptation of various harvesting machines, shown in Figure 1. Kuuse's research has been taken as an example here because of his connection with the

---

40 *Wilkins* 1994, 25-27. Multinational enterprise has also been explained by an eclectic paradigm. Behind this model is the idea that a multinational needs an advantage to compete and produce in an unfamiliar environment. There have been defined three conditions under which foreign companies would engage in foreign operations: There must be a location advantage compared with exporting to the host country; Multinationals have also ownership advantages compared to other firms supplying the same market; Finally there are internalization advantages which refer to economies of integration within the firm. *Dunning-Cantwell-Corley* 1986, 19-20; *Jones-Schröter* 1993, 5-6.

41 *KUUSE*: Interaction between agriculture and industry. Case studies of farm mechanisation and industrialisation in Sweden and the United States 1830-1930. Publications of the Institute of Economic History of Gothenburg University, 34. Göteborg 1974.
Everett M. Rogers, in his Diffusion of Innovations. New York 1962, together with F. Floyd Shoemaker, Communication of Innovations. A Cross-cultural Approach. New York 1971, has compiled and modified the main innovation theories. Although these works are certainly not the first volumes, they have been models for many later publications.

42 Gould P. Colman did not accept the traditional classification by the sociologists. He argues that in the evaluation of technological change there should also be taken into account, in addition to economic, social, and psychological chararacteristics, factors such as the soil, climatic conditions, and accessibility to markets. *Colman* 1968, 174.
See also David Grigg: The Dynamics of Agricultural Change. The Historical experience. Tiptree 1982.

McCormick Company and the agricultural machine industry. In contrast to traditional diffusion studies, he also takes into account the supply side of the phenomenon, although only implicitly.

Figure 1. Adoption of harvesting machines in America, 1840-1900.[43]

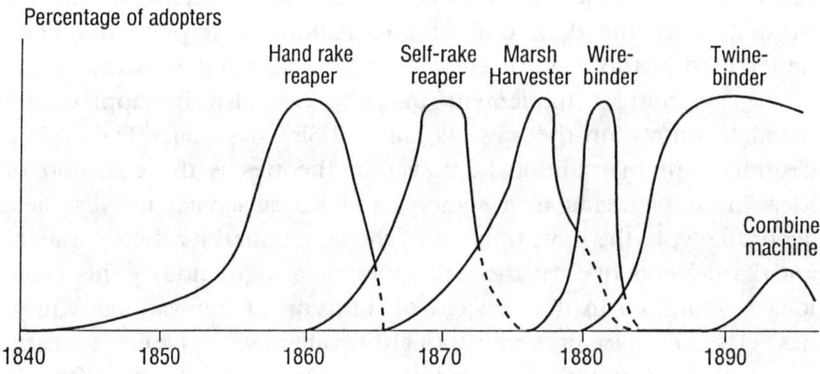

Source: Kuuse 1974, 256.

Johan GALTUNG[44] introduced a new approach with his theory of foreign policy attitudes and social position. He divided society into the center, periphery and extreme periphery. In the center, information, new ideas and attitudes spread rapidly, in contrast to the periphery, where the population wish to preserve the status quo. In other words, the center is a reformer and innovator, while the periphery tends toward preserving and defending the old. Consequently, information flows from the center to periphery. Galtung's theory has been transferred outside his own discipline, especially among ethnologists, and among the numerous researchers, Veikko ANTTILA[45] has modified Galtung's model in innovation studies. In his study on the diffusion of agricultural innovations in Finland, he shows how innovations spread sooner in western and southern than in eastern and northern Finland, and that the areas especially receptive to changes were those surrounding towns in southern Finland.

---

43 Unfortunately, Kuuse has omitted the quantity scale from his figure.

44 GALTUNG: Foreign policy opinion as a function of social position. Journal of peace research 1 (1964): 206-231.

45 *ANTTILA*: Talonpojasta tuottajaksi. Suomen maatalouden uudenaikaistaminen 1800-luvun lopulla ja 1900-luvun alussa. Helsinki 1974.

Albeit the meaning of the present study is not to evaluate various innovation theories, this situation tempts the researcher to survey these ideas from the perspective of a large enterprise expanding and seeking for new markets. How relevant are Galtung's and Anttila's models in such an environment, where competing companies resort to aggressive marketing to increase and create demand where it does not yet exist?

The concept of the company leads the researcher in the footsteps of Joseph SCHUMPETER,[46] who stressed the impact on the development of the national economy of the entrepreneur. Schumpeter's entrepreneur was a creator of new things, an innovator who with his creative energy pushed ahead the whole society. Was Cyrus McCormick the man who saved the world from starvation, as was stated?[47]

As a harvesting machine manufacturer, the McCormick Company was closely tied to the development of agriculture. In this study, agriculture is, however, merely the context for a case study of the operations of one company. Consequently, it has been possible to pay attention to agricultural phenomena only in such places where they were of importance for the spread and marketing of harvesting machinery, which is the focus of the present study.

1.3. The use of original material

The cornerstone of the present work is the McCormick Collection at the State Historical Society of Wisconsin. Any historian who works with primary material is of course interested in the origin and formation of his sources, because it decisively conditions the results of the research. This question is closely connected to the reliability of the material, but is at the same time a methodological question also.

The foundations for the McCormick Collection were laid already after the death of Cyrus McCormick Sr. in 1884, when his widow and children employed personnel to collect and organize material relating to him. This material was placed in 1951 in the hands of the Wisconsin State Historical Society, and was later enlarged with the papers of Cyrus' widow, Nettie Fowler McCormick, and his

---

46 *SCHUMPETER*: The Creative Response in Economic History. Journal of Economic History 2 (1947): 149-159.

47 *Casson* 1909, 46-47. To Casson, McCormick was a hero who "did more than any other member of the human race to abolish the famine of the cities and the drudgery of the farm - to feed the hungry and straighten the bent backs of the world."

children. The last additions were deposited in 1971. The International Harvester Company, successor of the McCormick Co., began in 1959 to deposit its material in the McCormick Collection, and continued to do so up to 1991, when the company's Russian collection was also sent to Wisconsin.[48]

The McCormick Collection consists of numerous overlapping segments which contain hundreds of archive boxes. The first shipment from International Harvester, for instance, included nineteen tons of financial ledgers.[49] Consequently, the collection covers a wide range of subjects. For the researcher, problems arise from the fact that the material has been acquired over a long period of time, and has been processed by many persons, often obscuring the provenance and leaving the organization of the material in some parts imperfect. This situation raises questions concerning possible gaps in the material. Had the company or some individuals any reasons to keep material out of the hands of outsiders? Is it certain that no material was lost during the formation of the collection?

The McCormick Collection is divided into five main segments: the McCormick Family Papers, the McCormick Company Papers, the McCormick Historical Records, Collateral Papers and Non-Manuscript Materials. Each segment is further divided into smaller parts by the name of family members and other agents. Basically letters are filed chronologically by year and alphabetically within the chronology by name of author. A guide, edited by Margaret R. HAFSTAD, gives an overview of the collection and facilitates the understanding of at first sight an endless mass of letters.

Sections of the collection can be identified by a number and letter code. The Cyrus Hall McCormick Papers form the A series, which comprises eight segments. For example, the series Mss 1a covers the years from 1788 to 1939 and comprises the main section of the Cyrus H. McCormick Papers. It consists of 125 boxes and contains correspondence received by Cyrus McCormick, his wife, his son Cyrus Jr., and other persons associated with the McCormick Company. The subjects covered are numerous, ranging from the development of the reaper to politics and Presbyterian Church activities. Series Mss 2a is the subject file in the Cyrus H. McCormick Papers.[50] The Nettie Fowler McCormick Papers[51] form the B series,

---

48 *Hafstad* 1973, v-vi.

49 Ibid, vi.

50 It consists of 57 boxes and 5 volumes and covers the years 1811-1884. In its chronological part, the emphasis is on business and economic affairs. Essential for the understanding of the integrity of the Cyrus H. McCormick Papers are the Mss 3a, which contain 24 volumes and 6 boxes of letterpress copy books (hereafter in this

which contains eight segments. The C series comprises the main section of Cyrus Hall McCormick Jr.'s correspondence.[52] In addition to the McCormick Family Papers, the McCormick Company Papers[53] have been essential for this research. The McCormick Estate Papers,[54] series Mss M/I, provided a positive surprise during the collection of material.

Although all of the family series are divided between outgoing and incoming letters, the division is not watertight. In many cases it has been impossible to find a reply to a letter, or an answer has been found but not the original to which it was a response. In some cases the reply has been found in a totally different series.

The McCormick Collection is a large archive, and the question therefore remains whether all the necessary material has been found and all the relevant sources examined. In the present study, the emphasis is on the McCormick Company's foreign business and related questions; consequently, the collection of source material has been based on that assumption. Forces have been concentrated on those parts of the McCormick Collection where foreign affairs are dealt with. For the remaining parts of the Collection, a sampling method has been used. Samples have been taken on the basis of the name of actors at critical points of the McCormick Company's

---

study: LPCB) of letters sent by Cyrus McCormick, his wife, Cyrus Jr., and the company personnel. This series has been used also to find replies to letters in other series.

51 The Mss 1b consist of 26 of boxes letters sent by Nettie McCormick. The Mss 2b contain 202 boxes of letters received by her. Equally important has been Mss 3b, which are in a similar way arranged chronologically by years and alphabetically by subject or author within the chronology. Altogether the Nettie Fowler McCormick Papers contain 453 boxes and 44 volumes of papers.

52 Series 1c consists of both incoming letters and outgoing responses and other letters. In series Mss 2c, there can be found part of the financial records of the McCormick Company, as well as papers concerning the foundation of the International Harvester Company. The Cyrus Jr. Papers contain 599 boxes and 268 volumes of letters and other material.

53 Series 1x contains 454 copies of letter press copybooks on domestic correspondence, covering the years 1856 to 1902. Sixteen more deal entirely with foreign trade from 1879 to 1902. Information on foreign business is supplemented with series 2x, which contains the McCormick Company's incoming letters. The series are not, however, overlapping and it is not possible in every case to find a reply in incoming letters to one in the outgoing communications, or vice versa. Series 2x consists of 420 boxes, in which the order of the letters is difficult to work with. The material is not in true chronological order, but is grouped by years. Besides, foreign correspondence is not stored separately, which is of course confusing for the researcher and entails a lot of extra work.

54 The 130 boxes and 87 volumes have offered new information not only on foreign business but also reports to general agents, the President's annual reports to stockholders, and material concerning the economic state of the company: profits, revenues, sales, costs, dividends and investments.

foreign operations as defined in the other material.

Entrance into a private archive opens huge opportunities for researchers, but on the other hand, they face equally great demands. They must be extremely careful with source criticism, both internal and external.

The correspondence of the McCormick Company contains messages sent and preserved by the company officials. Some of them are very confidential discussions between officials and family members; in other cases, they deal with daily company affairs or are reports of sales, accounts or other analyses. A material of especial interest and importance for this study is the correspondence between company officials such as the general manager E.K. Butler, and the European agents. There is hardly any reason to doubt the authenticity of this material. Where grounds for suspicion arise, however, material can to some extent be cross-checked, although the outgoing and incoming correspondence is far from overlapping.

A more problematic question for the present study has proved to be the internal criticism of sources. Are the sources reliable? Do they speak the truth in a certain question? Has the author had any reason to change the truth?

The situation is especially complex in the foreign material. At first, Cyrus McCormick Sr., and subsequently Cyrus Jr. were in charge of the foreign business. Consequently, they also took care of the correspondence. A major change in the Company's organization was made in 1886, when the General Manager, E.K. Butler took foreign business under his control.[55] Thereafter he or his subordinates took care of the ordinary daily foreign letters.

Correspondence between the two Cyruses and their representatives was confidential and private. With the takeover by a professional manager, there can be sensed a change in the way the agents are approached. At certain critical points in the history of the McCormick Company, such as the first attempt to merge the interests of the harvester manufacturers in 1890,[56] it was far from self-evident that all details were explained to agents. Such behavior can be explained in many ways, but leaves behind uncertainty on the reliability of the content of the letters. Can this material be used in explaining activities of the McCormick Company, or do they tell about activities of the General Manager? On the other hand, the function of the letters should also be remembered, since this explains their contents too.

---

55 *Cyrus Jr. to E.K. Butler* 8.24.1888 Mss 2c, box 112; *Butler to Cyrus Jr.* 8.28.1888. Mss 3b, box 9.

56 *Benson* 1936, 3-7; *The International Harvester Co. 1913*, 57-58.

First, it has to be noted that E.K. Butler was a manager in a family firm, the leading owner of which was also the President of the Company. This situation does not leave much room for personal operations, and consequently decisions of the General Manager were also decisions of the McCormick Company. Secondly, if something was omitted from letters to foreign agents, this merely illustrates the ways foreign business was dealt with, and is therefore also useful for the completion of this study.

In the McCormick Collection, as is often the case in the use of primary material, the information in the letters or other documents is not readily accessible. The researcher often has to read numerous documents, and assemble information, in order to obtain an overall view. On the other hand, within one letter European agents or a European manager might handle many different matters. The numerical material produced by the McCormick Co. personnel on sales, economic state and other related matters has also generated difficulties during this research. The statistical systems changed during the lifespan of the company and it has been difficult to render the surviving data commensurable. For this reason, the numerical material gives a good basis for qualitative examination, but is not reliable enough to give a reliable timeseries of the McCormick Company's sales and exports.

There still remains the possibility that the available material was originally collected and organized with future publicity in mind. In the case of the McCormick Company Papers, however, this is very unlikely; on the contrary, these papers represent the Company's private, confidential exchange of information. Nonetheless, it is of course possible that aspects prejudicial to the Company's interest may have been excluded from the archival material.

Perhaps the greatest risk for the reliability of the results of the present study lies in the one-sidedness of the material. The risk of bias has been taken into account during the various stages of the research; sources in the McCormick Collection have been compared for verification with other materials whenever that has been possible, e.g. with US consular reports, which have been of great value in estimating the state of business and competition in Europe as well as tariff and other restrictive methods used against American exports.

The formation of the International Harvester Company in 1902 soon raised the question of whether this was a monopolistic, unlawful trust in the sense of the Sherman Antitrust Act of 1890. International Harvester was the world's largest farm machinery maker, with its 85 percent share of the harvesting machine markets. The United States Bureau of Corporations launched in 1906 an investigation of the new company, which led in 1912 to the filing

of a suit against International Harvester by the Justice Department.[57]
On the basis of its investigation, the Bureau of Corporations[58]
published a report which has offered valuable information and
comparative material for this study.

The lawsuit against International Harvester also produced
hundreds of pages of legal material[59] which was later bound and
preserved in the Company archives. This material has been an
invaluable source and has also offered new information.

Complimentary material has been difficult to find in printed
sources. One of the few such examples are the sales catalogs, which
are preserved in Series 4z. Until now this material has been mostly
unused. From these catalogs, it has been possible to scrutinize the
distribution of agents and machine prices. Many catalogs also contain
both extensive descriptions of the history of the Company, and
numerous technical details. The Farm Implement News, a magazine
for dealers and other agricultural implement specialists, also provides
important information, but has to be approached with caution, since
there remains always a chance of biased articles.

Former research has not been of great value in the verification of
the source material here. All the earlier historians have used the
McCormick Collection without questioning its reliability.
Furthermore, location of the sources that have been used has proved
to be problematic because of inadequate references.[60]

The best comparative source for this study would be the archives
of another, competing company. Acquisition of such material has
been attempted, but the results have been discouraging. Most of
McCormick's competitors are no longer in existence, and have left
behind them no archives; on the other hand, the surviving
competitors were not engaged in the same line of business during
the period under investigation. Some material was found, however,
in the archives of Historical Societies in the Mid-West. Especially
interesting sources for future study could be the Historical Society
of Minnesota, and the catalog collections at the Agricultural History
Center in Davis, California.

---

57 *Marsh* 1985, 47-49.

58 The United States: Department of Commerce and Labor. Bureau of Corporations.
The International Harvester Co. 1913. Washington 1913.

59 No.624. In the District Court of the United States for the District of Minnesota.
The United States of America, Petitioner, vs. International Harvester Company and
others, Defendants. s.a., s.l.

60 This comment applies to the works of Eugene Manning, Eugene Shapiro and
Howard Schonberger.

1.4. Cyrus McCormick and his family

While it is not the intention of this study to write a McCormick
family history, it is essential for an understanding of the development
of the harvester business also to know the main actors in the play.

Time normally glorifies deeds of great men, but contemporaries
are usually inclined to remember things in a different light. In the
mid-19th century Mid-West there was hardly a person who denied
the achievements of Cyrus McCormick. With the invention of the
reaper, he had raised himself in two decades from a blacksmith and
a farmer to one of the most prominent businessmen of his time.
Such a career demands special skills combined with unyielding
character to win the battles in business. The winner normally has
few friends. Such was Cyrus McCormick, too: widely respected, but
loved by few.[61]

Cyrus McCormick was a hardworking man, who expected the
same from his associates. He was constantly traveling from city to
city, managing his businesses. The everyday functions of the
Company he left to his brothers Leander and William. Cyrus made
decisions on Company policy, which his brothers then implemented.
He had more important tasks in mind than the details of production.
Cyrus understood early in his career that great wars in business were
won in courtrooms and in the chambers of Washington. He fought
numerous patent wars against his competitors, in order to establish
an adequate basis for future production and new models, and
acquired money from East Coast bankers to accomplish his plans.
For his brothers, these operations were difficult to understand and
accept.[62]

McCormick's hard entrepreneurship was fused with deeply
religious convictions. During his early years he had no time to waste
on social life, but even in the smallest cities he had time for a
sermon. On the day of defeat he silently accepted the ruling of
Providence, but more important is that his mounting success made
him convinced that the same Providence was on his side in the long
run.[63]

At the age of 49 years, fame and prosperity had made McCormick
one of the most eligible bachelors in the country, who nevertheless
had no time to waste on the gentler sex. In 1857, however, at a
party in Leander McCormick's home, he met Nancy Fowler, who

---

61 *Hutchinson* 1930, passim.

62 *Casson* 1909, 56-57, 179-182; *Hutchinson* 1930, 453-454, 456-457.

63 *Casson* 1909, 158-165, 178-181.

sang in the choir at the Presbyterian Church. A year later, the couple was married. In Nancy (Nettie) Fowler, Cyrus gained not only a reliable companion, but also a business counselor who took an active part in crucial decisions.[64]

Nettie broadened Cyrus' otherwise narrow range of interests. She opened Cyrus' eyes to recognize that he had something to say about things unrelated to his business. Besides his church activities and generous philanthropy, Cyrus took active part in politics as a stout Democrat. During his political career Cyrus contested the offices of mayor, governor, congressman, senator, vice-president, and ambassador, but never held an office as the result of an election. As Hutchinson noted, for Cyrus life without competition would merely be existence. This Hutchinson saw as giving a singular unity to his career.[65]

By the 1860s, Cyrus began to complain of rheumatism, the attacks of which grew stronger over the years. He was concerned about the state of his health, and every year spent weeks at minerals springs throughout the country. In the same proportion as Cyrus's energy weakened, the position of Nettie McCormick at the stern of the business strengthened. Her responsibilities grew even larger after 1878, when a malignant carbuncle came near to killing Cyrus at the moment of one of his greatest victories, his promotion in Paris to the rank of Officer of the Legion of Honor.[66]

Nettie Fowler McCormick took an active part in high society and also introduced her husband to its members; but even these moves were only to promote the glory of the great invention. Nettie learned to take the case of the reaper as her own. She shared Cyrus' conservatism, which was to mark all their actions both at home and in business. The family was the other pillar of Nettie's life. By 1874, they already had five children, two others having died in infancy; yet it was not until 1876, when Cyrus Jr. was already seventeen years old, that they began to build their first own home. During all those years, they had lived in a hotel or in a rented house.[67]

If Nettie's position was decisive in business matters, it was pre-eminent at home and in the question of the children's education. From the very beginning the children were taught to save time. Nettie urged them to write down how much time they used to brush their teeth or put on their clothes. During the construction of their

---

64 *Hutchinson* 1930, 458-461.

65 *Hutchinson* 1935, 3-41.

66 Ibid. 671-673, 763-768.

67 Ibid. 3-5, 736-741.

house in Chicago, the eldest son Cyrus Jr. visited the site each day and kept his parents informed of the progress. The children had to learn to work and understand that the comfort and wealth that surrounded them could not be taken for granted, but had to be earned every day.[68]

Especially Cyrus' McCormick's eldest child and namesake, Cyrus Hall McCormick Jr.,[69] who was born on May 16, 1859, learned these facts from his birth. As the first child, he was sent to a normal city high school, while the other children were instructed by private tutors and in private schools. After long discussions with Nettie, however, his father let Cyrus Jr. continue his studies at Princeton. The ageing inventor had begun to feel the need of a reliable assistant, and was obviously ready to turn over some of his workload to his son. Cyrus Jr. had no formal business training, but instead a traditional classical education. Nevertheless, he was born to an atmosphere where life was business, and he gained his experience in action. After the firing of Leander McCormick and his son Hall from the McCormick Co. in 1880, Cyrus Jr.' responsibilities grew year by year, until after his father's death in 1884 he was elected as the new President of the Company.[70] When the International Harvester Company was formed in 1902, Cyrus was made its President until 1918, when he turned the Presidency over to his brother Harold.[71]

Although the McCormick family managed their business in a marvelous way, in personal life they suffered many ordeals. The oldest daughter of Cyrus and Nettie McCormick, Mary Virginia, was found to be mentally retarded at the age of nineteen. Their second daughter, Anita, married an attorney, Emmons Blaine who, however, died after three years of marriage and left her with one son. Mary Virginia and Anita McCormick did not take an active part in the business, although each had an interest in the company.[72]

Harold F. McCormick, the second son of the inventor, joined the McCormick Company in 1895 following his graduation from Princeton University. In 1898 he was made Vice-President of the

---

68 Ibid. 737-761.

69 Originally younger Cyrus's second name was Rice, but it was changed to Hall about 1870. *Hutchinson* 1935, 110 note 39.

70 *Directors meeting* 6.4.1880. Mss M/I, box 17; *Extract from the minutes of Board of the Directors.* 4.5.1880; *Hounshell* 1987, 178-180; *Special meeting of the Board.* 5.22.1884. Mss 1b, box 25, vol.1; *Harvester World* 7/1936. In memory of Cyrus H. McCormick 1859-1936.

71 *Hafstad* 1973, 20.

72 Ibid. 27-29.

Company, and later succeeded his brother at the head of the International Harvester Co. Although Harold McCormick left only a few business records, his marriage with Edith Rockefeller connected the McCormicks to another mighty family.[73]

The youngest son, Stanley R. McCormick, was like his brothers educated at Princeton, and entered the McCormick Company first as a salesman, then as Comptroller and finally as a Director and Assistant Secretary of the company. Stanley McCormick represented the Company at the Paris World Fair in 1900, and his trips to Europe and reports therefrom are a valuable addition to this work. In 1906, however, in the prime of his manhood at the age of thirty-two, Stanley McCormick broke down and became mentally incompetent.[74]

Another person who must be counted in the inner circles of the McCormick family was Eldridge M. Fowler, brother of Nettie Fowler McCormick, who became Vice-President of the McCormick Company in 1890.[75]

---

73 *President's annual report to the stockholders*, 7.13.1899. Mss M/I, box 18; *Hafstad* 1973, 38-42.

74 *President's annual report to the stockholders.* 7.13.1899. Mss M/I, box 18.

75 *Hafstad* 1973, 10-12.

II

■ The Entrepreneur Explores the Field

2.1. The farmer's options

In 1831, when Cyrus Hall McCormick made his first experiments with the reaper, agriculture in America was still practised with traditional methods. Work was hard, carried out using hand tools, and farms were short of labor. However, the frontier moved steadily towards the Mississippi River valley, and by the 1830s was on the brink of seemingly endless prairie. It took nearly fifty years to substitute wheat and corn for prairie grass. Simultaneously, settlers pushed ahead into Kansas, Nebraska, Wisconsin, Minnesota and the Dakotas.[1]

The further west the wagon trains slowly moved, opening new territories for agriculture, the more expensive and difficult became transportation. Farmers were neither able to sell their farm products in the markets, nor haul consumer goods through the muddy roads back to their homesteads. The opening of the Erie Canal in 1825, and during the next decades of the other canals that connected the Ohio River to Lake Erie, however, made commercial farming possible in the Middle West.[2] In time, canals were replaced by more flexible and faster railroads, although no uniform route system was completed before the Civil War. Railroads were private enterprise, and cargo often had to be transhipped and reloaded several times before it reached its final destination, because of the differences in rail gauges between the different railroad companies.[3] Whereas in 1835 Ohio was the only state in the region of the Great Lakes from which grain was transported to the east, three years later the first carloads of grain took their course east from Chicago. In 1860 rails were extended to the western parts of Iowa and thus to the outer

---

1 Bogue 1968, 8-12.

2 Taylor 1951, 15-27, 33-35, 43-48, 54; Schlebecker 1975, 89-91; van Metre 1939, 20-21.

3 Schlebecker 1975, 93.

limits of permanent farming.[4]

The story of free American land was a mere illusion up to the Homestead Act in 1862, and in practise even after it. The would-be farmer had first to acquire title to his land. The principles for the distribution of land were laid down in the Land Ordinance of 1785, which set the minimum size of the farm at 640 acres, and thus opened the way for speculators, since only the wealthiest proprietors could afford to buy such a lot at the required price of one dollar per acre. In the Homestead Act, the standard farmsize was set at 160 acres on terms of occupation and improvement only. This, however promising it seemed, did not help the newcomers as much as had been expected. Establishing a homestead was an expensive business, and in spite of the availability of free government land, many a farmer preferred to buy land from the railroad companies, which could provide financing and help with settlement; moreover, their land was often in better regions.[5]

A considerable number of land seekers had to give up their expectations for the time being, however, and to be content with the life of a tenant: they could not afford either to buy a farm outright, or to meet the requirements of breaking up prairie land and equiping a farm. For many, tenancy was a good way to begin, and it can be seen as a ladder toward a homestead of one's own. According to Allan G. BOGUE, between 1860 and 1880, the number of tenants in Clarion township in Bureau County, Illinois, rose, while "would-be farmers saw the cost of land rise and the amounts of machinery needed for farming increase".[6]

---

4 Railroads did not necessarily mean lower freight costs than the riverboats. SCHLEBECKER states that by 1850 it cost 25 cents per ton and mile to transport goods by train. By 1860 rates had dropped to 4 cents, in comparison with 2 cents by boat. Schlebecker 1975, 93. SHANNON on the other hand mentions that in 1868 to carry a ton of wheat by rail from Chicago to New York cost 42.5 cents and by boat 25.3 cents. In 1882 the figures were 14.6 cents and 8.7 cents respectively. Shannon 1945, 177. Transportation costs were of great importance to the implement industry also, in which Cyrus McCormick tried to exert influence. Hutchinson 1933, 10-12. McCormick Collection, special reports file, box 5.

5 *Schlebecker* 1975, 21, 47, 61-67; *Gates* 1960, 89-91; *Hughes* 1987, 90-91, 94-95, 274.
The 1785 Land Ord'nance opened up the way to speculators. The parcel size and price were too high for a normal farmer, who throughout his lifetime on average earned only $50 in cash. In 1800 the size was reduced to 320 acres, but the price was raised to two dollars. In 1841 Congress accepted a general pre-emption law for squatters, and limited pre-emption to 160 acres. Pre-emption laws were required because thousands of land-seekers flooded into the opening lands of the west and could not understand the government's claims to the land. *Bidwell-Falconer* 1941, 151-155; *Hughes* 1987, 94; *Bogue* 1968, 29-30.

6 *Bogue* 1968, 56-66.

The real life of a settler began only after the land purchase. He had to plow tough prairie land, build a house and a barn, make fences and buy some machines for his homestead. Breaking the tough prairie land was a hard and expensive task to accomplish. Custom brakers charged from $2.00 to $4.00 per acre for their services during the 1860s and 1870s. After preparing the seedbed, the farmer had to surround his new fields with good fences against free-running hogs. Before the time of barbed-wire, this demanded much, both in time and dollars.[7]

Most of the would-be farmers were so young that they could not rely on their children's help in daily work on the farm. By the age of fifteen, at the latest, a farm boy was generally performing a man's work. Before that time, and after his son's marriage, the farmer had to rely on hired hands. In the 1860s, farmers had to pay for day laborers as much as $3 per day, compared to 35 to 50 cents per day in the 1840s.[8]

In addition to these costs, farmer had to add the burdens generated by the rapidly expanding mechanization of agriculture. A settler could not grow more than he was able to take under cultivation. On the other hand, it was a waste of time to plow more than one could cope with at harvest time. Harvest time was short, normally about two weeks, and free hands were especially short at that time.[9] With mechanization, the farmer could save both labor and time, and even expand his farming land or cultivated area. Table 1 shows how much more time was required for harvest than for preparing the seed bed.[10]

---

7 Ibid. 70-81.

8 Ibid. 182-184.

9 *Gates* 1960, 279; *Danhof* 1969, 181-182.

10 Similar development can be found in Europe too. In France after the introduction of the combine harvester and mower in the 1880s the harvest time per hectare was reduced from 16 man-days to five or six. *Kindleberger* 1964, 213 note 18.
Alan Olmstead and Paul Rhode have, however, in their latest article noticed that the efficiency of the reaper declined unless it was followed by a proper number of workers; at least a driver, a raker, and four to six followers to bind and shock. Consequently they argue that the view that the reaper eased demand for outside harvest labor is incorrect. *Olmstead-Rhode* 1995, 41-42.

Table 1. Man-hours needed to grow an acre of wheat.

Year	Before harvest	Harvest	Total
1800	16	40	56
1840	12	23	35
1880	8	12	20
1900	7	8	15

Source: Cooper-Barton-Brodell 1947, 3.

A pioneer farmer's equipment usually consisted of a wagon, plow, axe, shovel, scythe, cradle, fork and rake. In a couple of decades, horse tools were substituted for hand tools. Farmers began to buy John Deere's steel plows,[11] spring tooth and disk harrows and seed drills.[12] As Table 1 shows, harvesting machinery was of great importance to agriculture. Even in the 1880s, more time was spent on the harvest itself than on the work before it. Before the time of the reaper, the harvest work consisted of the cutting, binding and shocking of the grain. Thereafter, the harvest had to be threshed.[13]

According to Bogue, the first settlers in the prairies area in the 1830s had to pay about $300 for their machines and draught animals. Horses and other draught animals increased costs considerably, and in 1870 the Iowa Railroad Land Company estimated the cost of a team (horse or oxen) and outfit as between $515 and $847. Included in this outfit were wagon and harness, two plows, cultivator and harrow, and a combination of reaper and mower. In the nineteenth century, nevertheless, investments in machinery and implements never constituted more than six percent of the farm's total investments.[14]

The transition from handtools to horsedrawn implements was closely connected with developments in metallurgy and engineering. New machines required better raw materials and also standardization of production. This was evident, for example, in the development

---

11 Normal wooden or even cast iron plows were inefficient in tough prairie lands full of intermingled roots. John Deere was certainly not the first to experiment with steel plows, but he was the first to make a profitable industry out of it. He made the first model in 1837 and in 1856 turned out 14 000 plows in a year from his factory in Molin, Illinois. *Broehl* 1984, 43-50.

12 *Rogin* 1931, 59-64; *Danhof* 1969, 206-207; *Schlebecker* 1975, 107-108.

13 *Rogin* 1931, 125.

14 *Bogue* 1968, 148, 169-170, 286.

of plows. Especially during the Civil War, when farms were short of labor, mechanization took rapid steps forward. When reaping machines became a common part of a farmer's implements, it also meant new requirements for draught animals. Oxen were too slow to keep reaping machines in proper motion and were changed for horses. Wayne D. RASMUSSEN has even called this change the first revolution in American agriculture, the impetus for it being the Civil War.[15] Machine power had been used for plowing since the 1830s and 1840s. The steam engine found its proper place in the farm as a stationary source of power. and was used mostly for threshing only. Steam engines reached their heyday at the turn of the century, when the first combustion engines were already being tried out for traction purposes.[16]

2.2. From reaper to harvester

Reaping has been for centuries one of the most important stages of agriculture, and numerous rites and traditions have been associated with it. It is therefore no wonder that the question as to who was the inventor of the reaper[17] in the late nineteenth century has excited the minds of countless researchers and manufacturers.

Sickle, scythe and cradle were in common use in Europe throughout the nineteenth century. Nonetheless, it was in England that the first attempts to mechanize the laborios work of harvesting were undertaken. Capel Lofft was probably the first to suggest the idea of mechanical reaping. Lofft's idea, published in 1785, was put

15 *Rasmussen* 1962, 578-582; *Rasmussen* 1965, 193-195. Thomas C. Cochran, Stanley Engerman and Alan Olmstead do not accept Rasmussen's arguments and state that the agricultural machine industry had already begun to expand before the Civil War and coincided with it only accidentally. Olmstead, furthermore, shows how labor costs did not rise faster in agriculture than in other sectors. Besides, agriculture was short of horses during the hostilities: after the War the number of horses had fallen by eight percent. *Cochran* 1961, 170-171; *Engerman* 1966, 195-198; *Olmstead* 1976, 35-50.

16 *Danhof* 1969, 181-182; *Hounshell* 1987, 3-8; *Danhof* 1972, 81; *Spence* 1959, 107-108, 112-115; *Wik* 1951, 182-186; *Wik* 1953, 208-211.
The first tractor that used a combustion engine was patented already in 1889, but not until the Hart-Parr Company began to sell its tractions in 1901 did it become a success. Actually Hart-Parr was the first to call this monster a "tractor". The name was then commonly used to mean all machines that used their own power for moving. *McKibben-Griffin* 1938, 3-8; *Wik* 1964, 80-84; *Owings* 1911, 170-174; *Gray* 1954, 15, 19-21.

17 The word reaper is normally used to mean a machine which during the first phases of its development only cut the grain; a raker raked the cut grain into sheaves from the cutting table. In the later models, a self-raking mechanism freed farmers from a separate raker. *Ardrey 1894*, 40-48; *Miller 1902*, 11-19.

into practice by William Pitt, who in 1786 or 1787 constructed the first model of a reaper. However, it was not until 1799 that the first machine was patented. Among the most important developers of the reaping machine were Henry Ogle and Patrick Bell. The reaper was seen as a machine of the future, and it inspired numerous inventors; by 1831 there are records of 2 French, 1 German, 33 English and 22 American reapers. Although Cyrus McCormick and Obed Hussey are the most famous inventors of reaping machines, they did not live in a vacuum. Before their reapers, the British had invented the first reciprocating cutter and divider, the platform, the reel and side draft features.[18] How much McCormick and Hussey were able to get information on these earlier inventions is impossible to establish, but it is obvious that at least news of these inventions also spread to America.

In 1833, Obed Hussey patented the first practical American reaper, which also sold in fair numbers: about 45 machines during the 1830s. Cyrus McCormick inherited his interest in reapers from his father, Robert McCormick, who had been making unsuccessful experiments with his reaper since 1816. In 1831 the machine fell into Cyrus McCormick's hands. According to his own claims, Cyrus made the necessary improvements, and was ready to introduce his first reaping machine during the same year. It was then remodeled, and tried in the field in 1833, but patented only in 1834. Thereafter, McCormick put aside his invention for nearly a decade and together with his father embarked on ruinous efforts with an iron-smelting furnace, which ended in 1837 in financial disaster. Two years later he devoted more of his time to the reaper, but it was not until 1840 that he sold his first two machines.[19] McCormick's reaper raised wide interest and enthusiasm among farmers. In 1842 it was tried in field

---

18 The model which Ogle finished in 1822 already incorporated the cutter and reel, with a reciprocating knife over stationary fingers and a platform behind the cutter. In spite of the fact that Ogle's machine thus already contained some of the main elements of a successful reaper, it was never very popular. Patrick Bell, on the other hand, constructed a reaper which found some popularity among British farmers, and a couple of machines were also brought to America. *Miller* 1902, 7-10, 12-13, 15-18; *Ardrey* 1894, 42-44; *Jones* 1979, 117-126; *Fowler* 1895, 354; *Oliver* 1956, 226-227.

19 *Ardrey* 1894, 45-46; *Miller* 1902, 24-25; *Rogin* 1931, 73-74; *Jones* 1929, 127-135; *Oliver* 1956, 226; *Thwaites* 1909, 235-237, 242-243.
Thwaites gives the credit to McCormick for introducing and connecting for the first time in one machine what he calls "four vital elements": a platform, reciprocating knife, horizontal and adjustable reel and a divider. *Thwaites* 1909, 240-241. These were, however, already used in the English machines, as well as in Hussey's reaper. The main difference between the Hussey and the McCormick was that the latter had a reel and a divider. The former had, nevertheless, a unique cutter. It consisted of "a series of slotted iron fingers through which vibrated a number of triangular knives fixed to a flat bar." In 1847 Hussey modified his cutter by introducing a new guide, that was open at the back and an upper part. Hussey's cutter mechanism was later used as a model for mowers. *Miller* 1902, 24, 26.

tests and, according to testimonials, made a favorable impression.[20]

In spite of favorable opinions among the public, McCormick had difficulties in finding customers. The cost of the reaper, in 1844 $100 in cash or $106 in four monthly payments, was a major investment even for the wealthiest proprietors. Besides, he was not alone in the field. In addition to Hussey and McCormick, there were numerous other inventors who were trying to get their share of the promising markets. It was therefore no wonder that altogether 138 patents were granted to reaping machines in 1834-1854.[21] The two foremost competitors, though, were Obed Hussey and Cyrus McCormick. Their first encounter in 1843, from which McCormick emerged victorious, was the first of its kind, and was a model for the machine trials and contests so typical later.[22]

Cyrus McCormick was also able to expand his production geographically. Cyrus hammered together his first reapers in his father's blacksmith's shop in Walnut Grove, Virginia. As the demand for reapers began to grow, in 1844 Cyrus sold about fifty and a year later about 150 machines; he now also introduced his reaper in New York, Illinois, Wisconsin and Missouri. Cyrus understood the potential of the new western states, and especially the prairie, which at that time was opening up to farmers. Meanwhile, Cyrus

---

20 Farmers Register Nov. 30, 1842 in W.J. Hanna: Collection of Historical Material. 1885. S.L., 3-5.

21 Petition to the Legislature of Wisconsin. 1854. 2-5.

22 *Hutchinson* 1930, 187-193.

McCormick continued to hammer out his machines in the home shop; but by 1845, he had contracted with various machine shops in Cincinnati, Ohio, to produce his reaper. Manufacture of a reaper was technically demanding, and not all the licensee manufacturers could turn out the quality that McCormick expected. Broken machines were bad advertisements; eventually, to avoid more confusion, in 1847 McCormick establised his factory together with Gray, in Chicago. The following year this first reaper factory in the west was able to turn out 500 reapers, and 300 more were produced by licensees.[23]

During most of the 1840s, besides McCormick and Hussey, there was only the Easterly Harvester worth mentioning; but the situation changed dramatically in 1848, when McCormick's orginal patents expired. In a couple of years a number of strong competitors, some of them Cyrus's old partners or licensees, founded their own factories. Especially Seymour & Morgan was active, and by 1849 had begun to manufacture its New York reaper. Production rose to about five hundred machines in 1851, with additional production by Warder and Brokaw at Springfield, Ohio.[24]

---

23 Cyrus himself moved to Brockport, New York, where in 1846 he supervised the manufacture of two hundred machines at the works of Seymour and Morgan, his later competitors. From Brockport, the machines could be transported both westwards and eastwards through the Erie Canal. *Fagan* 1931, 1-3; *Rogin* 1931, 72-75; *Thwaites* 1909, 244-245. *The invention of the reaper* 1931, passim.

24 *Rogin* 1931, 74-77; *Fagan* 1931, 3-4; *Hutchinson* 1930, 278-279, 292. Construction of the reaper developed considerably during the 1840s, when McCormick, too, modified, completed and patented his versions. In 1845 McCormick received a patent on a divider and two years later on a raker's seat. *Rogin* 1931, 87-89; *Miller* 1902, 28. Obed Hussey took out a patent in 1847 for improvements in guards, but he could not compete with McCormick or any other manufacturer in volume. In 1840-1849, according to his own statement, he had sold altogether ca. 358 reapers, and during 1851-1854 Hussey himself made about eleven hundred machines. To this number should be added reapers made under royalties or territorial licenses. Many a manufacturer copied Hussey's and McCormick's principles, but there were also imaginative inventors like Nelson Platt, who took a patent in 1849 for a self-acting rake. It was later assigned to Seymour and Morgan, who in the 1850s improved it, added to it another sweep-rake patent, and began to make their famous "New Yorker" self-rake reapers. John H. Manny, on the other hand, developed a mower attachment to the old reaper construction so that the same machine could be used both for reaping and mowing. This opened the way for combined machines. The first attempts to bind grain by machine were also made in the 1850s. Although none of these yet achieved great success, the Marsh brothers developed a successful model which was improved during the 1860s, and the later binders were based on its principles. Researchers usually give to C.W. and W.W. Marsh the honor of the development of the first practical harvester. Harvesters include self-binders and combine harvesters and threshers: in other words a machine that not only cuts the grain, but either binds it into sheaves or also simultaneously threshes it. *Ardrey* 1894, 53; *Miller* 1902, 33-35. Obed Hussey is normally regarded as the father of the mower. Later inventors followed his ideas of a reciprocating knife and slotted guards. In the development of a mower, two names tower over the rest, namely William F. Ketchum and Cyrenus Wheeler. Ketchum was the first to put mowers on the market as distinct machines. His mower was of a rigid bar type with a single drive wheel. Wheeler's machine was fundamen-

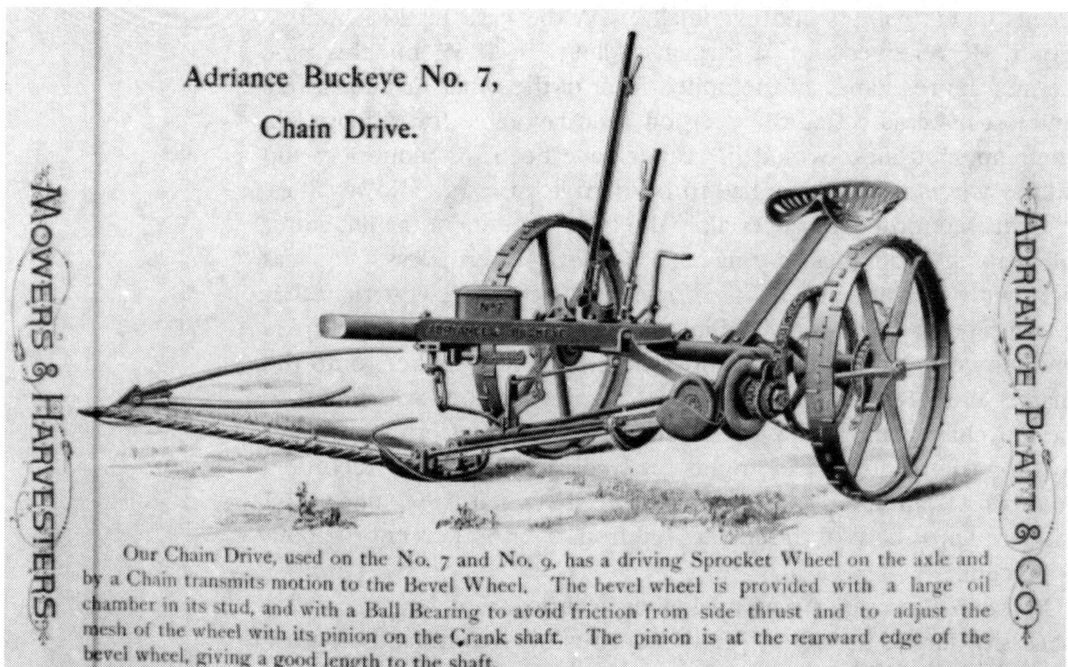

Adriance, Platt & Co.'s mower from 1900 is of a flexible bar type. (McCormick Collection. State Historical Society of Wisconsin).

The manufacture and trade of the reapers was, as noticed, subject to tough competition. Competition increased not only because of the expiry of Hussey's and McCormick's original patents, a situation which their former licensees naturally exploited; in addition, countless improvements and totally new inventions had been added to the machines. Furthermore, mowers and reapers began to follow their own paths of development. Besides, there was also a new field of contest: continuous warfare between the manufacturers with patents, which in the 1860s led to the formation of the first patent pools.

Numerous new manufacturers sprang up around the beginning of the 1850s, McCormick's patents having expired in 1848. Inevitably, after this setback, he decided to fight his competitors on every front.[25] In 1851 McCormick sued Seymour & Morgan for infringe-

---

tally different, with two drive wheels and a flexible bar, thus laying the basis for the subsequent division between the two types. *Miller* 1902, 30-31, 34-37, 39-40; *Deering Harvester Co.* 1900, 13-14, 42, 48-49; *Ardrey* 1894, 47-49, 58-60, 78-82; *Fagan* 1931, 4-6; *Humphries-Gray* 1949, 6-11; *Fowler* 1895, 355.

25 The most notable of the other manufacturers were two already mentioned: Seymour & Morgan of Brockport, New York, and Warder & Brokaw at Springfield, Ohio, who had been making the "New Yorker" reapers under their license. Perhaps the toughest competitor, however, was John H. Manny of Waddams Grove, Illinois,

ments of his patents and eventually won the case in the Supreme Court. He renewed the case against Manny in 1855, but this time he met the resistance of the united front of the other manufacturers and lost his case.[26] Had the decision been favorable for McCormick, the reaper business would thereafter have been his monopoly and all the others would have had to pay him license fees and royalties.

Cyrus McCormick was certainly the most prominent manufacturer of reaping machines during the 1850s. Nevertheless, he was constantly losing his relative advantage compared with other entrepreneurs who were beginning to manufacture self-rake reapers and mowers. McCormick did not bring his first self-rakers onto the market until 1862, and when the McCormick reaper was found to be too clumsy and heavy for efficient mowing, McCormick bought in 1865 the manufacturing rights for Hubbard's mower.[27] According to Alan OLMSTEAD and Paul RHODE, between 1850 and 1865 McCormick's share of the market fell from over 50 percent to about 5 percent.[28]

By 1855 the reaper was no longer a curiosity: it had become a necessity for farmers, and production numbers continued to grow for several decades. Most of the reaper factories were short-lived, and so it is difficult to obtain even approximate estimates of production during the 1850s and 1860s. Table 2 shows the steady rise of McCormick's output from 1853 onwards.[29]

---

who later moved his factory to Rockford, Illinois. He obtained a patent for his reaper in 1851, and began to manufacture them in fairly large quantities, in 1854 about eleven hundred a year. *Rogin* 1931, 77; *Ardrey* 1894, 48; *Deering Harvester Co.* 1900, 51-52.

26 *Deering Harvester Co.* 1900, 51-52; *Hutchinson* 1930, 432-443.

27 *Hutchinson* 1930, 374, 379-382, 386, 390; *Heikkonen* 1989, 69, 71; *Fagan* 1931, 7.

28 *Olmstead-Rhode* 1995, 28.

29 *Incomplete list of sales of reapers and mowers manufactured by reaper and mower companies other than McCormick Company.* Mss Special Reports File. Box 14. The source of this information is unknown and may therefore be in some measure unreliable. However, the information concerning the Walter A. Wood Co. and its sales seems to be correct. It is, nevertheless, possible that Wood's sales have been collected from the company catalogs which were also used in Table 2.
According to Rogin, in 1854 John H. Manny built at least 2500 reapers and Atkins 1200 of his 'automatons'. When Wood's 600 machines are added to this number, the total known output for 1854 is about 6000 reapers. The material in the McCormick Collection indicates, nevertheless, an even larger volume than Rogin's calculation. Material found in the Special Reports File records that John S. Wright, who made Atkin's Automaton in Chicago, sold 1200 machines in 1855, and 5000 the next year. In addition, there were 10 036 machines built by Manny in 1853-1855 and 25 089 machines in 1856-1858. Although there were also numerous other smaller manufacturers (Hussey for example continued to work and had sold production rights to several blacksmiths), it is obvious that McCormick had by this time cornered at

Table 2. Number of harvesting machines produced in 1850-1864 by C.H. McCormick & Bros. and the Walter A. Wood Co.

| | McCormick Co. | | Walter A. Wood Co.[a] |
Year	Reapers	Self-rakers	Reapers
1850	1603	-	..
1851	1004	-	..
1852	1011	-	..
1853	1101	-	500
1854	1558	-	600
1855	2534	-	1300
1856	4076	-	2500
1857	4065	-	3800
1858	4565	-	4500
1859	5118	-	5500
1860	4083	-	6000
1861	5491	-	6500
1862	4965	203	5500
1863	2259	2053	6500
1864	2027	4063	7500

Source: "McCormick Machines Built since 1841", McCormick Estate Papers, M/I, Box 18; Walter A. Wood Company, catalog for 1873. Mss 4z, box 25.

a) Walter A. Wood's figures include only sold machines and should be approached with caution in view of the origin of the information.

These figures confirm that while in the 1850s McCormick was the market leader, he subsequently gradually lost his position to his competitors. As one reason for this, David HOUNHSHELL points toward conflicts between Cyrus McCormick and his two brothers Leander and William. Leander was the superintendent of the factory and responsible for production. William, on the other hand, took care of the management and book-keeping. Both of the younger

---

least one fourth of the total market. This becomes even more evident in 1856, when Atkins made 2800 machines, Wood 2500, and McCormick 4076. ROGIN, however, has estimated that by 1858 there were in America altogether 73 000 reapers, of which McCormick had produced 23 200, which makes about one third of the total production. It must be taken into account that not all these machines were reapers; at least 20 000 of them were mowers. *Rogin* 1931, 78, 96 note 151; *Olmstead* 1975, 347-348. Rogin estimates that in 1862 the total output was 33 000 reapers and a year later 40 000 reapers and mowers together. In 1864 production was probably 85 000-90 000 mowers and reapers. *Rogin* 1931, 91, 93. Rogin's calculations are based on newspaper material and other secondary sources and have to be considered with some caution.

brothers hesitated to expand production to such an extent and at such a pace as their elder brother, the owner of the enterprise, demanded. HOUNSHELL has demonstrated that the question was in part one of two totally different kinds of business ideologies, and in part of inferior and inflexible production technology.[30]

What Hounshell presents is undoubtedly true, and he opens up a totally new approach by stressing the impact of production technology. Besides these admittedly important internal difficulties, McCormick's competitors were also pressing him harder year by year. John H. Manny was one of the most aggressive entrepreneurs, who not only manufactured himself, but also sold production rights to other makers. One of the latter was Walter A. Wood at Hoosick Falls, New York, who became involved in the business about 1852. This firm was incorporated as the Walter A. Wood Mowing and Reaping Machine Company in 1865 and began to produce in 1873 a wire binder patented by Sylvanus D. Locke.[31] The Wood Company was an expansive one-man concern in the same manner as McCormick's, and despite their rivalry, the two men seemingly understood each other and during the 1870s even discussed co-operation.[32]

Cyrenus Wheeler obtained the first of his seventeen patents on mowing machines in 1854. He supplemented these with sixty-seven patents bought from other inventors and in 1859 incorporated two smaller mowing machine companies and one machine tool company. The Wheeler Association began to make the famous "Cayuga Chief" mowers, and tried to prevent competition with numerous patents. Another outstanding concern developed in Greentown, Ohio, around John Miller, Cornelius Aultman and Ephraim Bell, who started a reaper business by manufacturing Hussey's reapers under license.

---

30 *Hounshell* 1987, 156-157, 159-172.

31 *International Harvester Co.* 1913, 49; *Miller* 1902, 36; *Ardrey* 1894, 60; *Farm Implement News* 3.21.1892,vol.XIII, no.3. p.19. Although International Harvester Co. 1913, a report from the Deparment of Commerce and Labor is regarded as reliable, there seem to be minor flaws in it. The report states that Wood began to make Locke's wirebinder in 1870; nevertheless both Miller and Ardrey mention the year 1873. Ibid.

32 Seymour & Morgan had developed the famous New York reaper and put it on the market in 1854. Their company was later bought by D.S. Morgan, and was finally incorporated in 1894 into Adriance, Platt & Co. In 1854 B.H. Warder from Springfield, Ohio, also began to make the New Yorker under Seymour & Morgan's license. The name of the company was changed when new partners were taken into the concern, and finally the firm took the name Warder, Bushnell & Glessner. The Warder group pooled its business in 1867 with Whiteley, Fassler & Kelly, also from Springfield: the outcome of this pooling of interests was the Champion Machine Co. When Whiteley, Fassler & Kelly failed in 1887, the partnership was changed to a corporation, the Warder, Bushnell & Glessner Co., which thereafter controlled the Champion lines. *Hutchinson* 1935, 542-543, 546-548; *International Harvester Co.* 1913, 47-50.

With their own patents and patent purchases, the partners developed in 1856 the "Buckeye" line of mowers. Adriance, Platt & Co of Poughkeepsie, New York, which was incorporated in the 1850s, began to make Buckeye mowers under license and later became part of the Buckeye line. By 1860 Wheeler, Aultman & Miller and Adriance, Platt & Co. owned the central mower patents. Since the patents overlapped, they became convinced that if either of them attempted to assert rights over the other, the result would be an endless list of litigations. All of them were experienced reaper men and understood the problematic situation, but could also see its possibilities. After negotiations they decided to pool their patents. The country was divided into selling areas, the mower was standardized, and license fees and royalties were imposed on other manufacturers. This hinged-bar pool proved to be profitable. In royalties and license fees alone, by 1868 it had gained more than $530 000 from more than twenty-five producers. The position of the pool was further reinforced, when it reached agreement on production rights with Walter A. Wood and Moses Hubbard.[33]

Patents and lawsuits were from the very beginning an integral part of the reaper business. Every substantial reaper and mower company had both lawyers and inventors on their payrolls. "Lawsuits for infringements, and purchase of likely patents, either for use or to prevent competition, were as normal parts of a year's business as the appointment of agents and the shipment of machines".[34] Cyrus McCormick, too, had brought numerous charges against his rivals. Now he himself had his back against the wall. The McCormick hand-raker was a combine machine and could be used as a mower, too, but it was already totally outmoded. McCormick had developed a new self-rake reaper, but was not able to introduce a combined self-raker until 1869.[35]

Since every company maintained a battery of lawyers, lawsuits were prolonged, and finally only the most prominent manufacturers

---

33 *Deering Harvester Co.* 1900, 49; *Hutchinson* 1935, 370-376; *International Harvester Co.* 1913, 47-50.

34 *Hutchinson* 1933, 2.

35 Members of the mower pool had reserved for themselves some of the patents, and began to sell them to outsiders. McCormick tried to use this loophole to his advantage by buying in 1865 production rights for Moses Hubbard's patent for five dollars per produced machine. The agreed selling area was too limited for McCormick, and he had to make a contract with the pool in the same year. He could now make and sell an efficient mower in the entire Mid-West, except in Ohio. As a hardboiled entrepreneur McCormick bluntly declined to pay the royalties. Finally Wheeler in 1869 charged him for breach of patent. Two years later, the mower pool was demolished, when Wheeler sold his patents to Aultman & Miller; but the case dragged on, and was settled only at the end of the 1870s when McCormick agreed to pay Miller $25 000 in unpaid royalties. *Hutchinson* 1930, 374, 379-382, 386, 390.

THE McCORMICK IMPERIAL SELF-RAKE REAPER.

survived, in a field of contantly intensifying competition. The patent case of the mowers was repeated with self-rake reapers and later with self-binders. By around 1860 Cyrus McCormick, finally acknowledged the necessity to develop a self-raker to replace the old hand-raker. He was, nevertheless, a latecomer. Other manufacturers had already been experimenting, developing, producing and patenting self-rakers for a decade. Therefore, McCormick had to reconnoitre and protect his moves very carefully. First, in 1861, he bought patents from Benjamin Fitzburgh and McClintock Young, and supplemented them with patents bought from Isaac and Henry Russel. He further acquired production rights to Seymour & Morgan's self-raker. Finally, in 1862, McCormick's began to make a self-raker of their own. As in the case of the mower, patent lawsuits and accusations of breach of patent followed. The self-rake reaper was already a fairly complicated machine, and its construction was covered by hundreds of patents. It was just a question of time who would sue whom. Moses Hubbard began in 1865 to gather various patent rights and finally constructed the Harvester Rake Pool. In one way or another, the pool involved Adriance, Aultman, Warder and

In the 1860s, the combined reaping and mowing machine was a favorite machine of many farmers. It was also sold in Europe by James T. Griffin. (McCormick Collection. State Historical Society of Wisconsin).

Whitley. Two years later Samuel Johnston assured himself a dominating position in the pool with his rake patent, which became a standard in the pool. Pool raised a litigation case against McCormick in 1870 but it was once again delayed, and after a lawsuit lasting nearly a decade, McCormick and Johnston finally agreed on terms.[36]

Simultaneously with the development of the self-rake reaper the real harvester also took its first steps. According to C.W. MARSH, he and his brother W.W. Marsh of DeKalb County, Indiana, experimented with the Mann machine (probably J.J. and H.F. Mann, of Clinton, Indiana) in 1856 and 1857, and finally next year patented a harvester of their own.[37] The introduction of the harvester was laborios. It had an energetic advocate in J.D. Easter, who became interested in it in 1864 in a trial near De Kalb, and acquired in the same year, with his partner E.H. Gammon, a license for the machine. By 1879, various companies had produced about 100 000 Marsh Harvesters.[38]

---

36 *Ardrey* 1894, 51; *Hutchinson* 1930, 395-404; *Deering Harvester Co.* 1900, 71.

37 The Marsh brothers' invention was one of the breakthroughs in the history of the reaping machine. It cut the grain and then a continuous canvas elevated the grain into a receiving box to the waiting hands of normally two binders. This basic idea was later used by all the leading self-binder manufacturers. In 1894 Ardrey noted: "it (Marsh harvester) has never changed materially, in principle or form, since; and if the same old machine as used in 1858, and painted as others now are, were seen standing to-day in any field in America, Europe or Australia, with binder's table off, one familiar with such machines would wonder, as he came forward for a closer inspection, whether it was McCormick's, Wood's or Deering's, Samuelson's or Hornsby's harvester and binder, and why the binder was not in place." *Ardrey* 1894, 58-59.

38 *C.W. Marsh to J.D. Easter* 12.7.1900. Mss 1a, box 116. *Statement by Mr. J.D. Easter*, 6.4.1900. Mss 1a, box 115. Interviews of J.D. Easter give a lively picture of the reaper business and offers many important details of the entrepreneurs and of their relations. Easter, who once was one of the big names, also underlines the importance of patents and how easy it was to infringe someone's rights. Easter and his pool had to pay to Aultman & Miller $36 000 for their claims and to Mann $25 000 and still $10 000 to E.S. Morgan Brockport, New York. Unfortunately Easter does not mention when these events happened. *Statement by J.D. Easter* 6.4.1900. Mss 1a, box 115. Easter also tells of a patent case against Cyrus McCormick in the interest of the Marsh patents. McCormick offered to pay $25 000 on 5000 machines already sold and was ready to pay a $5.00 royalty on each machine he would make in the future. Easter would not accept the offer and lost everything when McCormick won the case. *Reminiscences of Hon. J.D. Easter* 11.16.1900. Mss 1A Box 113. Of the Marsh brothers Easter says that they "were very poor men as manufacturers. I never could get them to build a good machine. They would not build it according to sample". *Reminiscences of Hon. J.D. Easter.* Mss 1a, box 113. His other business partners do not get a better evaluation. "In getting interested with Gammon & Deering and Emerson in the harvester business, I got in with a d--- set of sharpers. An old methodist preacher for the first one was as sly as a snake... Emerson was one of the sharpest old rascals that ever lived in this country". *Reminiscences of Hon. J.D. Easter.* 11.22.1900. Mss 1a, box 113. Easter's outburst is understandable, for his business had failed in 1878 and was taken over by Gammon & Deering. *Reminiscences of Hon. J.D. Easter* 11.27.1900. 1a, box 113. While Easter's information is interesting, it should be approached with some caution because of the intimate relations of the parties.

The Marsh harvester showed the way for the future development of harvesters, and its idea and construction were also used in selfbinders. Sylvanus Locke was the first inventor to attach a binding mechanism to a Marsh harvester. The first samples came out in 1873, from Walter A. Wood's factory. Two years later the experimental phase was over, and Wood manufactured about 300 of them. Locke was not, however, the only inventor to work with binding mechanisms: Charles B. Whittington, of Janesville, Wisconsin, developed perhaps the best and most advanced binding attachment, which was attached to a McCormick harvester. In 1875 and 1876, McCormick's made their first experimental machines, and began large-scale production in 1877, when 1040 of them were put on the market. McCormick's continued to expand production of wire-binders until 1884, when it was finally replaced by the twine-binder.[39]

The development of harvesters and wire-binders, and the opportunities to make a profit, attracted William Deering to enter the field as well. In 1869, Deering acquired an interest in the factory of Marsh & Steward at Plano, Illinois, and formed a partnership in 1874 with Gammon.[40] In 1879, the partnership between Deering & Gammon was dissolved and Deering moved his factory to Chicago. The concern was incorporated in 1883 as William Deering & Co., and renamed as the Deering Harvester Co. in 1894. Deering introduced the twine-binder with the Appleby knotter in 1880, and some 3000 machines[41] were put on the market.[42]

---

39 *Miller* 1902, 36-37; *Ardrey* 1894, 74-77; *Rogin* 1931, 110-112; *McCormick Machines Built since* 1841. Mss M/I Box 18.

40 Deering & Gammon manufactured at their Plano factory, for the most part, wire-binders of the Gordon type, as did D.M. Osborne & Co. Deering earned his reputation, however, as the maker and developer of the twine-binder.
In his bitter statement J.D. Easter somewhat strips Deering's fame. Easter describes Deering as a tough and quarrelsome person. Easter had met Gammon in Europe and Gammon had been in 1873 "about half crazy" and told how Deering was everlastingly going for him. On the development of the Marsh harvester Easter maintains that Deering had nothing to do with it, and had wrongly inherited the name and merits of the Marsh Company. Furthermore Easter gives new evidence on the development of the wire-binder, when he explains that McCormick was in the 1870s clearly the market leader in wirebinders. That is why Deering was compelled to make something else, and the answer was the twinebinder. *Reminiscences of Hon. J.D. Easter*, 11.22.1900. Mss 1a, box 113.

41 Competition was also growing on the Canadian side of the border, where the Massey Manufacturing Company claimed to make 1000 binders for the 1884 harvest. Similarly A. Harris, Son & Company reported 1000 for the same season. *Denison* 1949, 82-83.

42 With the exception of Walter A. Wood Co., all the prominent firms, including McCormick, Champion and Osborne, secured production rights to Appleby patents. From 1882, therefore, the leading companies were all producing practically the same machine, with only minor improvements, for example, in bearings and substituting steel for wood; the basic construction and binding mechanism remained unaltered.

After the emergence of the self-binder, only minor improvements were made in the harvesting machines. (McCormick Collection. State Historical Society of Wisconsin).

Competition continued to intensify up to the end of the century. Large companies could invest in research and develop new models year after year, and were able to use money freely for buying promising new patents. As the construction of the machines became more complex, the threshold to enter the business grew higher. In addition to this, only large companies could build a nation-wide sales and service system. To this must be added the problems of unsold machine stock and part payment, which naturally strained companies' liquidity.[43]

International Harvester Co. 1913, 47; *Fagan* 1931, 8-10; *Rogin* 1931, 110-112; *Miller* 1902, 36-37; *Ardrey* 1894, 74-77; *Kuznets* 1937, 558; Forty harvest seasons. Deering 1898, 1.

43 *International Harvester Co.* 1913, 56-57.

Table 3. Number of manufacturers and the value of production in the US agricultural machine industry, 1849-1909.

Year	Manufacturers	Production ($)
1849	1333	6 842 611
1859	2116	20 831 904
1869	2076	52 066 875
1879	1943	68 640 486
1889	910	81 271 651
1899	715	101 207 428
1904	648	112 007 344
1909	640	146 329 268

Source: Cencus of the U.S., 1910. Manufacturers; Heikkonen 1989, 88.

From Table 3 it can be seen how in the 1850s the agricultural machine industry expanded and attracted new entrepreneurs, but from the 1860s a declining trend in the number of manufacturers is visible. Alan Olmstead and Paul Rhode have estimated the number of harvester firms to be approximately 200 in 1864 with a total output of about 90 000 machines. In 1890 there any more were only 21 harvester and mower makers.[44]

In 1883, severe competition forced the leading companies into negotiations on evolving some kind of common policy. For two days in Chicago, the binder manufacturers discussed the price question, and decided to continue their meeting at Niagara Falls. A special committee of five manufacturers, Osborne, Deering, Walter A. Wood, Buckeye and McCormick, was appointed to prepare a plan for consolidation.[45] At the Niagara Falls session, the companies tried to find solutions to the price question, discounts and payments, commissions and especially the total number of binders to be built. The companies were, however, unable to reach a mutual understanding. The most complex question turned out to be production quotas. The small, owner-managed companies, in particular, such as Easterly and Whitley, opposed any kind of interference in their business.[46]

---

44 *Olmstead-Rhode* 1995, 27; *Clark* 1929, 8.

45 *Cyrus Jr. to Cyrus H. McCormick* 8.11.1883. Mss 1a, box 90.

46 *Cyrus Jr. to Cyrus H. McCormick* 8.31.1883. Mss 1a, box 90.
The committee of the manufacturers recommendeed the retail price for 5 ft. harvesters and binders to be $230 and for 6 ft. machines $240, with a five percent discount on cash sales. Payments that would have been made before November 1st were considered as cash sales. The maximum commission to local agents was planned

Despite the initally promising prospects, therefore, this collective undertaking by the harvester companies did not materialize. The idea of some kind of combination was not buried, however, and in 1884 D.M. Osborne held conversations with the General Manager of the McCormick Company, E.K. Butler, over a merger of the largest companies. Osborne suggested that the large companies could then buy out the remaining 15 manufacturers.[47]

In 1890, the next time the leading companies seriously moved to prevent competition, they finally agreed on a merger, and the American Harvester Company was incorporated, with the same twenty-one companies still in the game.[48]

2.3. Machines had to be sold

Invention and development of an implement was only part of the tasks of an entrepreneur. If he wanted to make a living out of his invention, he had to find ways and means of selling his products. For most of them, this turned out to be a complicated task. Similarly, the reaper manufacturers found it a difficult task to convince poor and reluctant farmers of the benefits of their machines.

In his works, Alfred D. Chandler has defined the outlines of this process. Mass distribution of products was one of the central elements in Chandler's concept of the emergence of big enterprise. In the selling of their products, manufacturers traditionally used territorial agents, who initially hired subagents or dealers, but in later phases changed to using jobbing houses.[49] The modern industrial enterprise, the birth of which Chandler dated to the 1880s, integrating mass production with mass distribution, could no longer

---

at 20 percent. The most difficult part was the settlement over the volume of production, which was suggested to be 65 000 binders for 1884. There was also agreement on the uniform cutter bar lengths. The Committee was also unanimous that "the combination should take the form of a contract between the manufacturers; with a commissioner appointed by the association who should have full power to assess damages and forfeits, investigate all charges for accusation of violation of contract, and whose decision upon any point should be final".
Report of Cyrus Jr. shows that the total number of binders built during 1883 was about 77 000. All the main companies agreed that output was too high and Cyrus Jr. expressed his opinion that prices should be kept as high as possible.

47 *Diary of Cyrus McCormick Jr.* 4.7.1884. Mss 4c, vol. 18.
The big companies were McCormick, Wood, Osborne, Deering, Champion and Buckeye. Cyrus Jr. considered such an idea utopian.

48 *International Harvester Co.* 1913, 57.

49 A jobbing house or jobber is a wholesaler who sells to dealers. *Leigh* 1924, 54, 80, 85-86.

tolerate intermediaries in the trade. Consequently, distribution too was transferred to the manufacturer. The reasons for this transformation were, according to Chandler, the inability of the marketers to distribute products at the volume and speed now required. Besides, some items required specialized distribution and marketing services, which wholesalers, retailers, and other middlemen were unable to provide.[50]

Markets for the reaper were affected by numerous factors. There was evidently a shortage of labor but, as Allan Bogue has shown, farmers tried to overcome this, as far as possible, by using their families' own resources. Where these were not sufficient, they hired hands for harvest.[51] Before a farmer decided to buy a machine to lighten his burden, he had to be sure of the markets for his own products. This in turn was at best uncertain before the construction of transportation facilities and growth of demand for agricultural products. Furthermore, manufacturers had to publicize their ideas. For this purpose they used from the very beginning farm magazines such as The American Farmer, The Genesee Farmer or The New England Farmer.

In the 1830s, and even in the 1840s, the marketing and distribution of commodities was still undeveloped. However, in the larger cities some implement and seed dealers had already established their warehouses, and also had shops, where simple implements were manufactured. These repositories could also buy patent rights to some patented articles and hammer them in their own shops. There is some evidence of the first jobbing business too.[52]

In 1842 Cyrus McCormick succeeded in selling six of his reapers, and as his trade began to grow, he decided to benefit from the existing marketing channels; the next year, he sold county rights both to manufacture and to sell the reapers to four persons in Virginia and Maryland. Besides, McCormick, who himself manufactured reapers at Walnut Grove, closed down the shop for harvest time and followed his machines into the field. From the very beginning this was one of his principles; a man had to stand behind his machines. In addition, McCormick announced that he would sell

---

50 *Chandler* 1977, 209-239, 285-288; *Chandler* 1987, 67-69; *Chandler* 1988, 41-42. Chandler's basic concept can be found in his famous "The Visible Hand" published in 1977. In his later articles Chandler adds new details to the basic idea and illustrates his concept of throughput.

51 *Bogue* 1968, 182-184.

52 Another possible channel for the distribution of machines was the sale of county or state rights. This could include either the right to manufacture and sell, or to manufacture only. *Leigh* 1924, 12-15.

his machines at fixed prices all over the vast country. This intention was made widely public; publicity was to be used in numerous variations in the war for farmers' souls and against other manufacturers.[53]

During his journey in 1844 to the new western states of Ohio, Illinois, Michigan and Wisconsin, McCormick realized the future potentials of the prairie area. He made a contract for the following year with some manufacturers in Ohio, New York, Wisconsin and Missouri; these small local shops both manufactured and sold the reapers, and paid $20 royalty per machine. When demand for the reapers grew, since McCormick could not guarantee the quality of the reapers made under license,[54] he decided to move his factory to the center of the wheat-growing territory.[55]

Expansion of business also meant new requirements for sales operations. When McCormick's own establishment in Chicago was finally able to meet the demand, he tried to free himself of the manufacturing agents. Since 1844, McCormick had been canvassing, and in 1845 he appointed an agent to collect orders in the eastern states. Thereafter he began to construct his new market organization. McCormick appointed general agents to specific territories. In 1848, McCormick seems to have had nineteen agencies. It was, nevertheless, too expensive a way of doing business to have agencies in all the new communities; therefore the number of the general agents was reduced, and they selected assistants or subagents to take care of local sales.[56] If local agents were successful, they could even

---

53 *Eckles* 1953, 41; *Leigh* 1924, 16.

54 Cyrus McCormick sent to his contractors a model of his reaper. It was not a normal size machine, but a miniature model. Besides he provided technical assistance if necessary. *Hounshell* 1987, 155.

55 *Leigh* 1924, 39-44; *Rogin* 1931, 74-75. Warren Leigh, basing his information on McCormick's statement on his extension case in 1845, separates four different kinds of contracts. McCormick could give the right to sell in a certain area or manufacture and sell for an agreed sum. McCormick used also contracts, where the royalty was bound on sales. It was not a fixed sum, but whatever he could get. On these bases he got $15 on reapers made in Virginia, in the West it was $20, and Gray & Warner and D.I. Townsend paid $30. Sometimes royalties were connected to the sales price. Thus Seymour & Morgan paid $22.50 on machines priced at $105. The great variety of contracts express McCormick's business talents and ambitions. In other cases he got $5 commission on every machine sold while he himself collected orders. *Leigh* 1924, 47-48.

56 Subagents were normally local farmers, blacksmiths or postmasters, usually well known persons in their vicinity; and even if they were unable to sell machines, they boosted the idea, and made it known to possible future customers. Allan G. Bogue quotes D.R.Burt's letter to Cyrus McCormick where he tells of the hard rivalry between the competitive companies' agents in 1854. Burt also describes how he succeeded to persuade the local agricultural society's president to change his Manny's reaper for a McCormick. This farmer was furthermore interested in acting as a local agent for five dollars for each machine he helped to sell. *Bogue* 1968, 208.

hold some machines in stock. General stores, too, might stock machines; but in the 1850s, the field was not yet ready for a retailer. Neither did the jobbing houses find a steady foothold in the western states.[57]

By the beginning of the 1860s, McCormick had brought his distribution system to its ultimate phase. The General Manager of Agencies at headquarters in Chicago controlled from ten to twelve general agents, who in turn had some 400 subagents of various kinds, covering over one thousand counties and thirty-one states. Other manufacturers of course tried to copy McCormick's ideas, but he was able to back his operations with better economic standing, and his organization reached practically to every farmer in the main grain growing area.[58]

During the 1860s, retailers and dealers entered in the picture. They developed from general store keepers into specialized implement dealers, at least to some extent, from the impetus of the Civil War, usually selling most of their machines on consignment. By the end of the 1870s, machines had developed considerably, and were fairly complicated. Retailers were not always able to set up or instruct the farmers carefully enough. Besides, they normally also had other machine lines on sale, as well as harvesting implements, and could not entirely concentrate on that trade: understandably, since the harvester trade was limited to a short period of the year. In addition, service problems and the question of spare parts and freights grew hand in hand with the growth of the trade.[59]

McCormick resolved the widening dilemma by establishing his own warehouses and general agents at key junction points.[60] McCormick's first branch house was opened in either 1877 or 1878

---

57 *Leigh* 1924, 49-56. *Eckles* 1953, 43. Subagents normally got from seven to ten percent of the collections of each reaper. Orders were normally taken in early spring and sent to the factory in Chicago. McCormick was not, however, satisfied with this order and adopted the system of sending machines to agents to keep them in stock ready for delivery. Machines were normally sent in lots to a certain point, wherefrom the purchaser paid the freight. *Leigh* 1924, 50-52.

58 *Leigh* 1924, 77-77; *International Harvester Co.* 1913, 56; *Eckles* 1953, 43-44.

59 The dealer obtained his stock from the manufacturer, who also set a fixed price for the machines sold to farmers. Dealers received a fixed percentage on each machine, and few of them were wealthy enough to buy machines directly from the manufacturer. Those who were able to do so, however, received an extra discount on their machines. Dealers opened their branches in new localities and began to supersede manufacturers' own canvassers and even the general agents. This of course freed the manufacturer from keeping his own sales force, but had also some drawbacks. *Leigh* 1924, 78; *Eckles* 1953, 43-44; *International Harvester Co.* 1913, 55-56. Leigh tells about a firm in Peoria, Illinois (Lukens & Branden), who normally contracted the farmer's payment out of increased yield from the same acreage resulting from the improved methods of harvesting. Ibid, 79.

60 J.J. Glessner of Champion line had, according to his interview, already established the first branches in 1870 in St. Louis, St. Paul, Omaha and Denver. *Leigh* 1924, 89.

at St. Louis. Branches supplied the retailers and also monitored the company's credit policy. McCormick also concentrated collection and credit departments in branch offices, keeping only supervisory power at the central office in Chicago. When the branch house organization was set up and elaborated during the 1890s, these dealers were the only independent factors in the field.[61]

The founding of the branch houses in the harvester business was no unique phenomenon: development was much the same among implement makers, and for example, the Moline Plow Company opened its first branch in 1878 in Kansas City. And on an even more general level, evolution of the agricultural machine industry followed the overall trends in American industry.[62]

A reliable and well-lubricated marketing organization was, nevertheless, only one aspect of successful commerce. As has been noted earlier, farmers did not hold large amounts of cash; moreover, farm incomes came in just once a year, at harvest time. This, combined with farmers' conservatism, led even well-established ones at first to ask a neighbor or a custom reaper to cut their grain.[63]

---

61 The Deering Company also had a similar system. General agents governed areas alotted to them with a sales force of about 3300 local agents. An undated and unnamed memorandum of the Deering Co, 1890. Mss w, box 2.
According to the Federal Trade Commission an unnamed but, however, one of the larger manufacturers had the following employees in a branch house: 1 manager, 1 assistant manager, 2 bookkeeppers, 1 cashier, 1 shipper, 5 clerks, 5 stenographers, 16 salesmen and 5 experts. *Report of the Federal Trade Commission on the Causes of High Prices of Farm Implements* 1920, 55

62 *Thomas* 1976, 40-41; *Chandler* 1977, 287-288.
The branch system was certainly more efficient than jobbers or independent general agents, but it also represented a heavy burden to the manufacturer. Reaper manufacturers normally produced just harvesters, mowers and reapers. All of these machines were used in the fall, and therefore production was also very seasonal. Factories began to warm up their engines and call in labor in the fall, and the production peak was at midwinter. The seasonal nature of the manufacturing was reflected in marketing, too, and general agents gathered their sales force in spring as the plants began to close their doors. The canvassers' and retailers' high noon lasted to the end of July or beginning of August, and they had to move with the harvest from south to north. Most of the sales were made within four months just prior to the harvest. After the harvest, the general agents visited the local agents to take stock of unsold machines off their hands and to approve their accounts. Seasonal fluctuation was trying for the manufacturers, who could not keep all the work force on the payroll for the whole year. To overcome this problem, some manufacturers began to add new production lines to their assortment. *Leigh* 1924, 82-92, 103-110; *International Harvester Co.* 1913, 55-56; Undated and unnamed memorandum of the Deering Co., 1890. Mss w, box 2.

63 According to Olmstead-Rhode the farms of reaper owners, on avarage, were larger than the farms of typical farmers. The number of reaper owners was, however, increased by machine sharing. In Illinois in 1859 about 18 to 22 percent of farmers owned a share of a reaper. *Olmstead-Rhode* 1995, 32, 34, 46-49.
Allan Bogue's carefully collected material shows that the expansion of reapers began during the Civil War and continued into the 1870s, when even the most prejudiced farmers began to acquire machines to cut their grain, either alone or in partnership with a neighbor. *Bogue* 1968, 261-262, 271-272.

Cyrus McCormick knew his customers, and developed an elastic credit system. In the 1850s, McCormick's normal policy was to acquire from the purchaser $50 in cash, plus freight costs of $5, on delivery of the machine. The remaining sum of $100 was to be paid on the first of December of the same year at six percent interest.[64]

A heavy burden was also laid on the shoulders of the agents, who were responsible for enforcing the credit terms set to farmers. They had to be sure of the reliability of their customers. The McCormick Company frequently stressed in its circulars to agents the importance of good sales. Renters should be avoided, except in cash sales, and all freights and charges must be paid in cash too.[65] Other manufacturers had also faced the same quandary. John J. Glessner asked his agents to examine the interior of the home of a possible part-payment purchaser; if the house was neat, the farmer was a good risk.[66]

Sales on credit were a stress not only on small manufacturers, but also on the larger ones, and it was no wonder that the McCormick Company emphasised the importance of cash collection whenever possible. It was not a complicated task to make a deal, but collecting claims was more intricate, as the Company frequently had to remind its agents. Intensified competition forced McCormick to extend his credit policy; by 1873, the purchaser had to pay half of the price during the fall of 1874 and the other half in fall 1875. In 1880, the McCormick Company allowed its agents to give farmers a discount of ten percent for any amount of cash down payment. It also further

---

64 *McCormick Reaper Order* 1850 (blank). Mss 5x, box 1; *McCormick Reaper Order* 7.14.1954. Mss 5x, box 1; *McCormick Reaper Order* 4.24.1958. Mss 5x, box 1. Normally the deal was made just before the harvest, but the money was collected only when the harvest was sold. Under the strain of growing competition, McCormick & Bros had to modify their credit terms. In 1862, the company promised to extend credit for two to five years, at twelve percent interest on the outstanding loan. Four years later, interest was six percent again, but with freight and other charges added. *Credit Policy of the McCormick Co.* 1839-1902. Mss. Special Reports File, box 2; *Machine order* 6.1.1866. Special Reports File, box 2. The information of the customers included exact co-ordinates of the farms under mortgages, their standing and amounts due and paid and when the payments should be made. Alan Olmstead also includes the interest freights and service charges and ends up with a nineteen percent interest rate. *Olmstead* 1975, 332-333.

65 *Private circular to agents*. March 1867. Mss 5x, box 1. Agents are asked to sell the old machines first, even at reduced prices. The freights were agents' own separate business and had to be kept out of the Company records. Furthermore McCormick required agents to keep uniform prices for all customers, whereas the terms of trade might vary. Agents were also asked to make intensive canvassing and even to haul machines to a reluctant farmer's field for a trial. "It is the only way to make numerous sales"..."Sales can be more quickly and pleasantly closed up by note or cash while the farmer is using the machine than afterwards.". An interesting note is the McCormick Company's strong comment on the banks which should be avoided in the transfer of money because they are unreliable and liable to go bankrupt.

66 *Leigh* 1924, 98 note 2.

widened the time limits for credit. The ultimate limits were reached for sales closed in 1880: one-third on January 1st 1881, one-third on December 1st 1881, and one-third on June 1st 1882. On the other hand, the Company required its agents to enforce the published prices; otherwise, the agents would bear the loss.[67]

The Deering Company's conditions at the end of the 1880s were very similar. Its credit provided that one third should be payable in October 1888, one third in January 1889 and one third in October 1889. In some territories, however, it was possible to extend credit up to October 1890. Interest on promissory notes given in payment for machines varied from six to even ten percent in those territories where there was no limit on the rate of interest; it was, however, always the highest rate that State law allowed.[68]

Table 4 explains McCormick's concern over bad notes and his repeated demands for cash payment. From the records, it can be seen how the amount of notes increased, both in relative and in absolute terms. This situation imposed a heavy burden on the Company's liquidity. To be able to meet production costs, McCormick had to finance factory operations with loans from East Coast bankers.[69]

---

67 The McCormick Co. recommended October or the weeks immediately thereafter as the best time for collections. *To the Agents for the McCormick Machines.* June 23, 1869. John Edgar, General Agent Mss 5x, box 1; *Private Instruction for Agents.* C.H. McCormick & Bro. March 1, 1869. Mss 5x, box 1; *Private Instruction to Agents.* July 31, 1873. John Edgar, Agent. Mss 5x, box 1; *For Agents Only.* May 1, 1874. C.H. & L.J. McCormick. Mss Special Reports File, box 2; *Private.-For Your Information Only.* May 3, 1880. E.C. Beardsley, General Agent. Mss 5x, box 4.
The McCormick Company included among those whom agents should avoid those who lived on homesteads, those on railroad and school land and those whose property is exempt by law. *To the Agents of the McCormick Machines.* Mss 5x, box 1.

68 An undated and unnamed memorandum of the Deering Co., 1890. Mss w, box 2.

69 Other manufacturers were, however, in the same situation, as was reflected in their business practices, which were similar to McCormick's. In 1856, Atkins sold his reaper and mower at $200 in three payments during the same year; $10 was added for longer credit, and ten percent interest if the purchaser failed to pay the amount due on the first of May. On the other hand, Manny promised cash buyers discounts as high as 25 percent. The Minneapolis Harvester Works demanded eight percent interest on time sales. It is, nevertheless, difficult to make a reliable comparison between McCormick and the other firms, because of the incomplete material. All the manufacturers warned against bad credits, wanted more cash payments, but year by year made terms more attractive to farmers. *Atkins Automaton* 1856, 2. 4z, box 1; *Minneapolis Harvester Works. Contract 1881.* Mss 4z, box 16; *Minneapolis Harvester Works. Order for Harvesting Machines 1882* Mss 4z, box 16; *D.M. Osborne. To Our Agents.* 1.16.1882 Mss 4z, box 17; *Aultman, Miller & Co. Binder order 1884.* Mss 4z, box 1; *Minneapolis Harvester Works. Contract of Robert Newton* 12.20.1881. Mss 4Z, box 16.

Table 4. Total annual sales and financial standing of the McCormick Company, 1848-1858.

Year	Annual sales of reapers	Total proceeds ($)	Proportion in cash ($)	Proportion on credit ($)	Notes as proportion of total proceeds (%)
1848	500	60 000	19 800	40 200	67
1849	1490	178 800	54 247	124 553	70
1850	1598	191 760	58 170	133 590	70
1851	999	119 880	36 362	83 518	70
1852	994	142 950	43 356	99 549	70
1853	1079	161 850	53 410	108 440	67
1854	1549	232 350	74 933	157 410	68
1855	2524	390 675	118 032	272 673	70
1856	4039	626 045	189 128	436 917	70
1857	3937	609 995	126 815	483 180	79
1858	3917	581 560	110 496	471 064	81

Source: Leigh 1924, 67.

The evolution of the McCormick Company's machine prices in Table 5 show a marked fall. When a new model was introduced, prices were at first high but came down in a couple of years, as was the case with the harvester. In most years, McCormick was not able to maintain the published rates; in 1883, for example, the actual rates were on average from five to ten dollars under the announced price level. The same phenomenon was repeated the following year, and for 1885 Cyrus McCormick Jr. ancipated still stiffer competition and therefore suggested to his mother they should begin with the prices of 1884.[70] In the price rates there were clear regional differences. All the states were classified into various rate categories.[71] The reason for this price

---

70 1882 seems to form an exception to this general rule: probably the question is of a special price for one territory only, in this case San Fransisco, and the figures possibly also included extra charges for freight. *Cyrus McCormick Jr. to Nettie Fowler McCormick* 1.11.1884. Mss 3b, box 3.

71 *McCorcmick Harvester Machine Co. Draft for 1891.* 12.5.1890. Mss W, Box 2.

Table 5. Unit prices of McCormick machines ($), 1860-1890.

Year	Reliable Advance Imperial[a]	Mowers	Harvesters	Daisy reaper	Dropper
1865	265/280[b]				
	215/230				
1872	175	120			
	200				
1874	195	115			175
	180				
1880	160	75	290/300[c]		140
	170				
1882	210	100/115	350		
1883	160	75	230	115	140
1884	160	75	225/235	125	
1888		50/55	150/155/160	80	
1889		50/55/65	155/160/165	80	
1890		50/55/65	150/155/160		

Source: C.H. McCormick & Bros. Circular 1865. Mss Special Reports File, Box 2; C.H. McCormick & Bro. Machine Order 1872. Mss Special Reports File, Box 2; C.H. & J.L. McCormick. For Agents Only, May 1st, 1874. Mss Special Reports File, Box 2; The McCormick Harvesting Machine Company. Private.- For your information only, May 3, 1880. Mss 5x, Box 4; N.B. Barnes to Geo. O. Bates & Co., Feb 20, 1882. Mss Special Reports File, Box 2; Records of the Director's Meetings. Secretary Records. Jan. 2nd. 1883. Mss 1A, Box 90; E.K. Butler to Ben Craycroft, Sept. 17, 1888. E.K. Butler to N.B. Fulmer, February 7, 1889. Mss Special Reports File, Box 1. E.K. Butler to W.P. Patch, Apr. 15, 1890. Mss Special Reports File, Box 2.

[a] The Reliable was a combined reaper and mower which was in production in the 1860s, replaced at the end of the decade by the Advance and in the late 1870s by the Imperial. They all were combined machines. In the figures for 1872 and 1874, the first row means the Reliable and second the Advance. For 1880, the first row means the Advance and the second the Imperial; thereafter, there is only the Imperial.
[b] The first figure means a hand-rake reaper and the second a self-rake reaper of various cutting widths.
[c] Parallel figures express various cutting widths of machines. The normal widths were 4, 5 and 6 feet.

elasticity was the stiff competition to which McCormick had to adjust its activities.[72]

According to the information collected from contracts, catalogs and price lists, on average, the McCormick Company held its prices higher than its competitors. The figures used are companies' list prices; actual deals in individual cases, both with the McCormick Co. and with other companies, were often made at lower rates. For example, in 1884, the McCormick Company gave a ten percent discount on cash sales on binders, and discounts on other machines varied between ten and twenty percent.[73]

The McCormick Company, like other manufacturers, still relied in the 1880s on jobbing houses; but once they established their own branch offices, most of the trade was done through retailers, who sold machines at commission. Commissions of course reduced manufacturers' profits. During the early years, McCormick allowed his agents a commission of between eleven and fifteen percent. About twenty years later, when competition was stiffer, the Minneapolis Harvester Works gave its agents twenty-five percent commission on all machines; on the other hand, the company set out to increase cash payments by promising dealers five percent commission on cash sales. At about the same time, McCormick Co. agreed to give eighteen percent commission for all kinds of machines and an extra commission of twenty percent in lieu of the said eighteen percent on cash sales. In addition, the Company allowed

---

72 Competition increased considerably during the 1880s. Nettie McCormick remarked in 1883 that the Deering Co. sold twine binders at $200, which was $30 under McCormick's list price. Nettie received confirmation for her doubts during the next fall, when Deering, Osborne and the Minneapolis Co. made contracts with some agents for rates as low as $135 for harvesters and binders. In 1884, Cyrus Jr. wrote to Nettie that "there is no doubt that prices may be very much demoralized for next year, as the small collections will tend to make farmers very backward about ordering machines next year, and there is liable to be an overproduction". Competition hit the smaller manufacturers harder than McCormicks' or Deerings'. D.M. Osborne & Co. required its agents to canvass every foot of their territory and offered help for closing deals. Osborne wanted its agents to dispose of everything still unsold, even at reduced prices and with longer credit. Walter A. Wood expressed his fears in a letter to Adriance, Platt & Co., stating "as long as the McCormick and Deering Companies keep the attitude they now have and have maintained towards each other, nothing can be done to advance prices in this country". Furthermore, Wood states that none of McCormick's sales of twine-binders were over $110 and most were as low as $95. *Nettie McCormick to Cyrus Jr.*, 3.21, 1883. Mss 1B, Box 25; *E.K. Butler to Nettie McCormick*, 10.13, 1883. Mss 3b, box 2; *Cyrus Jr. to Nettie McCormick* 1.11.1884. Mss 3B, Box 3; *D.M. Osborne & Co. To Our Agents*. 5.28, 1886. Mss 1a, Box 104; *D.M. Osborne*. 7.8.1886. Mss 1A, B 104; *Walter A. Wood to Adriance, Platt & Co.* 12.17.1889. Mss W, Box 1. The Deering Company's prices for agents in 1890 for the mowers was between $36 and $42.50, and for binders $105 and $130. These prices were well under McCormick's list prices for its agents. *An undated and unnamed memorandum of the Deering Co.*, 1890. Mss w, box 2.

73 List price refers to the prices published in the Company catalogs. *Cyrus Jr. to Nettie McCormick* 10.20.1884. Mss 2B, box 27.

a ten percent cash discount.[74] The Deering Company required its agents to guarantee promissory notes given by farmers. Their agent's net price was fixed, and everything he brought in above this price was his commission, normally fifteen percent.[75]

Without any doubt, the McCormick Company kept prices of its products on a higher level than its competitors, as David Hounshell has stated. This became possible because of its marketing strategy, which included retail dealers, service system, and an installment purchasing plan. Alfred Chandler had already earlier noted how by 1917 the large companies had become the price leaders, due to the same factors which Hounshell defined.[76] Although the McCormick Company was not initially in the 1880s a clear market leader, it possessed many such characteristics. These factors, combined with its good economic standing and reputation, allowed it to maintain its prices. The development of machine prices which has been shown in the present work concur with the official material produced by the US Department of Agriculture.[77]

Parallel to credit problems, reaper manufacturers encountered the service question. From today's perspective the first reapers were simple, and could be produced by almost any blacksmith; but to a farmer in the middle of the nineteenth century the case was different. The manufacturer had a two-sided dilemma to solve: to set up and keep the machines going and, on the other hand to take care of spare parts supply.

At first both Hussey and McCormick closed the doors of their shops and personally followed their machines to help to set them up. Often reapers needed some adjustment, which is no wonder in view of the crude production methods and the many producers to whom licenses were sold. The set-up expences significantly increased as distribution of the machines expanded. To get round these problems, McCormick began to publish special instruction

---

74 *Contract of A.R. Metcalf,* 1862. McCormick & Bros. Mss 5X, box 1; *Contract of L.G. Dudley* 1.11.1864. McCormick & Bros. Mss 5x, box 1; *Contract of Robert Newton* 12.20.1881. Minneapolis Harvester Works. Mss 4z, box 16; *Agency Contract of Wright & Co.* 12.1.1883. McCormick Harvester Machine Co. Mss 5x, box 2; *Cyrus Jr. to Nettie McCormick* 10.20.1884. Mss 2b, box 27.

75 An undated and unnamed memorandum of the Deering Co. 1890. Mss w, box 2.

76 *Hounshell* 1987, 5-6; *Chandler* 1977, 409.

77 *The course of prices of farm implements and machinery* 1901, 13, 19.

Table 6. Unit prices of the competing harvesters by manufacturer ($), 1855-1885.

Manufacturer	Year	Reaper	Combined	Mower	Binder
J.P. Manny	1855	135/145		120	
Atkins	1856	175	200		
Buffalo Agric. Machine Works	1858			120	100
Warder, Mitchell & Co.	1866	170	200	140/160	
Warder, Mitchell & Co.	1876	160	180/190	100/120/130	
1.Minneapolis Harvester Works	1881		150	80	300
2.Adriance, Platt & Co.		135	160	70/75/80	300/310/325
3.D.M.Osborne		135	160	80	
Minneapolis Harvester Works	1882		135		270
Aultman, Miller & Co.	1883		150	80/85	260
					275
Warder, Mitchell & Co.	1884	100/115/120	130/135	70/75	225
Aultman, Miller & Co.					190
1.Warder, Mitchell & Co.	1885				180
2.Aultman, Miller & Co.					160/180
3.D.M.Osborne					180/200

Source: J.H. Manny. Patent adjustable reaper and mower compined and single mower. 1855. Mss. 4Z, box 14; Atkins' Automaton. Self-Raking reaper and mower. 1856. Mss 4Z, box 1; Buffalo Agricultural Machine Works, 1858. Mss 4Z, box 2; Warder, Mitchell & Co. Price List No.1. 1866. Mss 4Z, box 24; Warder, Mitchell & Co. Price List No.1. 1876. Mss 4Z, box 24; Minneapolis Harvester Works. Contract 12.20.1881. Mss 4Z, box 16; Adriance, Platt & Co. Net cash prices for machines. 1881. Mss 4Z, box 1; D.M. Osborne & Co. Price card 1881. Mss 4Z, box 17; Minneapolis Harvester Works. Order for harvesting machines. ? May 1882. Mss 4Z, box 16; E. K. Butler to Nettie McCormick 10.18.1883. Mss 3B, box 2; Aultman, Miller & Co. Retail prices of machines. January 1883. Mss 4Z, box 1; Champion. Revised net cash price list 1884. Mss 4Z, box 3; Aultman, Miller & Co. Machine order 1884. Mss 4Z, box 1; Morgan Bros. to W.H. Hatch 3.7.1885. Mss 3B, box 5.

leaflets on how to set up and handle a reaper.[78]

The normal spare parts which a farmer was able to change himself, even a canvasser or a local agent could take care of. Salesmen simply carried with them the parts that normally first wore out, or a general agent had them in store. More problematic were parts that demanded special handling. A partial solution was found from the factory, which normally closed for harvest time; in this way, mechanics were freed for use as travelling experts. And experts were needed, for the McCormicks relied upon specialty contractors to supply their factory with some parts even up to the 1890s. During the early years, agents often complained about the quality of McCormick's machines and advised the Company to devote more time to improved workmanship. The use of castings and careful specifications for the contractors guaranteed uniformity in some measure in the same year's models. McCormick's also founded a special repair department which maintained a duplicate of every year's model and also the patterns. The customer needed to state the year, model and possibly the number of the part, and was responsible for the fitting of the new castings or wrought-iron parts.[79]

Development of the harvesting machines made it impossible for the farmer to set up a machine himself. A skilled mechanic was a necessity, but very costly to maintain. Even in the case of jobbing houses, the manufacturer had financially to carry the heaviest load. The poor mechanic had to carry the physical burden, traveling on wagons or on horseback hundreds of miles from one customer to another. His task was to keep the machines running and farmers satisfied. In a word, he was the show window for the manufacturer. Companies recognized this side of competition as extremely important, and farmers frequently

---

78 *Leigh* 1924, 36-38, 68; *Hounshell* 1987, 159; *Names of parts and directions for putting together & operating M'Cormick's patent Va. reaper*, 1850. Mss 5x, box 1. According to the above mentioned leaflet, machine parts were numbered and marked with paint. Besides, in the brochure there was also a picture of the reaper and directions were so detailed that they enabled the purchaser to set up his machine. The other manufacturers also adopted McCormick's methods. John H. Manny published in 1856 directions for using his reaper, where he had point by point minute instructions. *Directions for using Manny's patent adjustable reaper and mower!* Mss 4z, box 14.

79 *Hounshell* 1987, 157-160. The McCormicks relied on the old system but modified it in 1875 when they began to cast or stamp the numbers on the parts. Nonetheless, they had problems with frequent model changes. Agents should know that " a machine may be made in one year and sold in some future year, and in this way lead to mistakes. We may also add, that the year when patented gives no information when made". *Price list and catalogue of C.H. & L.J. McCormick*, 1875. Mss 5x, box 1. McCormick's system was obviously common among the other manufacturers, too. Cayuga Chief Manufacturing Company asked its customers in ordering extras to give the number of the part, of the machine and the name of the manufacturer. *Cayuga Chief Mower and Reaper. Cayuga Chief Manufacturing Company*, 1867. Mss 4z, box 3.

received their repairs totally free of charge. The importance even in
this question can be seen, for example, in the McCormick Company's circulars to agents, where it is stressed they should "spare no
time to look after them (machines) in the field" and furthermore
agents should be masters in handling the machines.[80]

Closely connected to the service question was the warranty that
a manufacturer gave to his machines. In the first years McCormick
guaranteed his reaper to cut a certain amount of acres in a day and
also to save grain normally scattered by cradling. He also gave a
warranty for the workmanship of his machines. The other manufacturers followed suit, and both their acreage and quality guarantees
were very similar.[81]

The same warranty policy was continued in the 1860s but in
McCormick's agent's contracts there was no mention of a warranty of
any kind. In the 1880s manufacturers set out exact acreage figures and
simply guaranteed that the machine is well made of good material, and
if in a fair test, it did not satisfy the customer, he could have a new
machine or have his money refunded. Aultman, Miller & Co., on the
other hand, included in their contracts a clause guaranting their
machine to do as good work as any of its size in America. Only in the
event that it failed to do so, would the machine be changed.[82]

---

80 *Private circular to agents.* McCormick Bros. 1867. Mss. Special reports file, box
2; *Private instruction to agents.* 7.31.1873. Mss 5x, box 1.
The importance of the service and repair question becomes evident in the comments
of John J. Glessner and L.J. Martin. Both mention service as one of the main reasons
for the establishment of the company branch house system. Another key question
was the credit problem. When service became part of the branch house, it was
separated to form a separate department which could take care of almost any work
that emerged. In this way, organizational power was transferred downwards and
nearer the customer. *Leigh* 1924, 98-100.
John J. Glessner was formerly president of Warder, Bushnell & Glessner and after the
great merger in 1902, director of the International Harvester Company. L.J. Martin
was a General Manager of Experimental Department at International Harvester
Company. Leigh has obtained this information by interviews with both of the persons
mentioned. Ibid. 149.

81 *McCormic's Patent Virginia Reaper.* 1848 (?). Mss 5x, box 2; *McCormick's Patent
Virginia Reaper.* 1850. Mss 5x, box 1; In 1848 McCormick promised his reaper could
cut one and a half acres in an hour and in 1850 two acres in an hour and to save
"at least three-fourths of all the wheat scattered by ordinary cradling."
J.H. Manny guaranteed his machine would "Mow as well as can be done with the
Scythe, and Reap as well as can be done with the Cradle." Furthermore he warranted
that his reaper would cut from ten to fifteen acres a day. *J.H. Manny's Patent
Adjustable Reaper and Mower Compined and Single Mower.* 1855. Mss 4z, Box 14.
Atkin's tried to convince purchasers by calculating the amount of money that a
self-raker saved compared to a hand-raker. *Atkin's Automaton,* 1855. Mss 4z, box 1.

82 *Agency contract.C.H. McCormick & Bros.* 1862. Mss 5x, box 1; *Agency contract.
C.H. McCormick & Bros.* 1864; *Sales contract. C.H. McCormick & Bros.* 1866. Mss
Special Reports File, box 2; *Cayuga Chief Manufacturing Company.* Catalog 1867.
Mss 4z, box 3; *Privat circular to agents.* C.H. McCormick & Bros. 1867. Mss Special
Reports File, box 12; *Minneapolis Harvester Works. Order for Harvesting Machines,*
5.?.1882. Mss 4z, box 16; *Aultman, Miller & Co. Sales contract.* 4.26 (?).1884. Mss
4z, box 1.

Both of the inventors of the reaper realized the value of publicity for the future of their business. Obed Hussey introduced his reaper to farmers for the first time in 1834, and continued his appearances during the following years. Cyrus McCormick made his first serious public appearance in 1839.[83] This manner was a preface for future field tests. Farmers wanted to see the reaper at work, and perhaps to drive it themselves, before they were convinced of its benefits. Some manufacturers, like McCormick, began even to lend machines for free trials. The agricultural societies also became important factors in the trade. They arranged tests and appointed committees to investigate competing machines, and eventually also published reports of these investigations. When new firms came into the field, competition became stiffer and mere cutting tests were no longer enough. To stand out from the others, one had to find something more attractive, until reapers and mowers were driven at full gallop against poles or horses drew them in opposite directions so that they broke apart.[84]

Cyrus McCormick opened a totally new field in trials in 1851, when he shipped his reaper to the first World Fair in Crystal Palace in London, where he won the Great Council Medal, beating his arch-rival Obed Hussey. After the London Fair, McCormick presented his reaper in all the major fairs all around Europe and reaped medals and honor with his machine. From the very beginning, this fame was extensively used on the homefront; but he could not keep this privilege for many years. Other producers followed the lead, went over to Europe and exploited their success in the Old World to boost their standing in America.[85]

Newspapers and magazines were another integral part of marketing efforts. Manufacturers could publish descriptions of new models, publish reports of trials and of course actual advertisements. An essential part of newspaper advertising, and later on of sales catalogs, consisted of written testimonials from those who already

---

83 *Greeno* 1912, 39; *Leigh* 1924, 10-11.

84 *Heikkonen* 1989, 81; *Warder, Glessner, Bushnell & Co.* 1900. Mss 4z, box 24; *Leigh* 1924, 26, 94. Importance of the trials and fairs becomes evident for example in D.M. Osborne's circular to agents, where the company tried to push its agents, by emphasising how in Mount Morris Fair twenty-nine binders were sold and for good prices. *D.M. Osborne & Co. Circular to agents* 5.28.1886. Mss 1a, box 104.

85 *Reaper contract of Henry Werner* 4.24.1858. C.H. McCormick. Mss 5x, box 1; *Walter A. Wood.* Catalog 1873, 1874. Mss 4z, box 25; *The Johnston Harvester Co.* Catalog 1876, Mss 4z, box 14; *Wm. Anson Wood.* Catalog 1878. Mss 4z, box 5; Walter A. Wood Company published in 1900 a sales catalog, which listed all the company's main achievements in field tests. Of the 35 pages of descriptions of the results fifteen were reserved for foreign awards. *Walter A. Wood Company*, 1900. Mss 4z, box 25.

had bought a machine.[86]

Sales catalogs were a direct channel to affect farmers' opinions. Together with handbills, flyers and showcards, they provided effective means to hit competitors. A common feature of marketing was the use of various ways to make competitors look ridiculous in the eyes of purchasers. In 1890, the Deering Company published a showcard where the Company's salesman introduced their latest reaper to its competitors, who were represented as rednecks with mucus running from their red noses: "Deering Binder carefully studied and imitated by all binder manufacturers". In another card, McCormick's harvester was displayed as a skunk which a farmer was carrying away holding his nose: "We hold our nose as we treat the subject (McCormick)".[87]

Catalogs are also a part of cultural history. They reflect the state and opinions of the latter part of the 19th century. During the first years catalogs were simple and usually illustrated only with black and white drawings. The same catalog contained information on the latest models, testimonials, sites of depots and offices, victories in trials and other achievements, a history of the factory, and instructions on how to order machines and extras. And of course, how supreme this reaper was compared with the others.[88] Scenes and pictures were simple, stoic and showed only the machine at work. In the 1860s, the number of pages began to grow and pictures were also added on inside pages, but the main line was still informative. Manufacturers wanted to send a message to purchasers and emphasized qualities of their products, changes in the factory, production numbers etc. A slow change became visible in the 1870s. On the covers some (pale) colors were used. Themes became romantic scenes with maidens holding scythes, and grain wreaths, encircling the main theme, the reaping or mowing machine. No drastic changes occurred in the contents, however, except that the

---

86 See for example *Atkin's Automaton* 1856. Mss 4z, box 1; *C.H. McCormick & Bros.'flyer* 1860. Mss 5x, box 1; *McCormick Harvesting Machine Co. Sales catalog 1882.* Mss 5x, box 1; *D.M. Osborne & Co. Circular to Agents 8.24.1886.* Mss 1A, box 104; *Heikkonen* 1989, 82-83.

87 In 1855 John A. Manny warned McCormick not to infringe his mower patent, and makers of Atkin's automaton called McCormick's hand-raker a back-breaker. Attacks continued with complaints of humbuggery and with a challenge to a contest during the next season. *J.H. Manny's Patent Adjustable Reaper and Mower Compined and Single Mower* 1855. Mss 4z, box 14; *Atkin's Automaton* 1856. Mss 4z, box 1; *Deering Harvester Co. Showcard* 1890, 1894. Mss 4z, box 4;

88 See for example *Atkin's Automaton* 1856. Mss 4z, box 1 ; *John P. Manny.* Catalog 1869. Mss 4z, box 14; *Walter A. Wood.* Catalog 1873. Mss 4z, box 25; *McCormick Harvester Machine Co.* Catalog 1884. Mss 5x, box 1; *William Deering and Co. Catalog* 1887. Mss 4z, box 4.

pictures, which were still hand-drawn, became more detailed.[89]

The general romantic atmosphere was preserved during the 1880s and the 1890s, with colorful and decorative covers, which might represent exotic themes like the first gallic header or oriental transportation methods. Orientalism was several times repeated, probably in order to show the universality of the products. Equally common were romantic harvest scenes, where happy children were playing in their Sunday clothes while their parents watched. The idea was to project an image of peace and comfort, which only the harvester made possible. There were also patriotic themes from the War of Independence, Congress and of course the American farmer; harvesters cutting wheat on the prairie while a remnant from the past, the buffalo bull, looks quietly from the mountains at his lost empire.[90]

---

89 *C.H. McCormick & Bros.* Catalog 1863. Mss 5x, box 1; *John P. Manny.* Catalog 1869. Mss 4z, box 14; *Walter A. Wood.* Catalog 1873, 1878, 1879. Mss 4z, box 25; *D.M. Osborne & Co.* Catalog 1876, 1879; *The Johnston Harvester Co.* Catalog 1876. Mss 4z, box 14; *Wm. Anson Wood.* Catalog 1878. Mss 4z, box 5.

90 *McCormick Harvester Machine Co.* Catalogs 1881, 1883, 1884, 1885, 1887, 1889, 1894. Mss 5x, box 1 and 2; *Wm. Deering Harvester Co.* Catalog 1887, 1895, 1897. Mss 4z, box 4; *Plano Manufacturing Co.* Catalog 1891. Mss 4z, box 20; *Adriance, Platt & Co.* Catalog 1896. Mss 4z, box 1; *Aultman, Miller & Co.* Catalog 1884. Mss. 4z, box 1;

What were the crucial factors behind the success of the McCormicks? Cyrus McCormick was one of the inventors of the reaper. It was a big advantage in the fight over the market, and Cyrus also understood its marketing value. As the inventor he was able to patent some critical parts of the machine and in that way either to collect royalties or prevent competitors reaching the market. This policy lead to continuous lawsuits with his rivals. In these he showed himself to be a hardboiled and impudent businessman. A great deal of McCormick's success was due to the marketing strategy Cyrus designed. It included warranty and service of the machines, well lubricated delivery systems and long credits to customers. That was made possible partly because of the accumulated wealth during the first decades of the business. McCormick was far from being a market leader in the 1860s, neither were his machines more intelligent than those of other makers. Therefore it is tempting to consider that due to his fame, combined with the rapidly growing market, he was able to stay in the game, in spite of his problems with patent pools, and with his aging machines. Nevertheless, McCormick's held their machine prices at a higher level than their competitors and had strict contract terms with its dealers.

The evolution of the firm of Cyrus McCormick followed surprisingly closely the outlines of the development of modern enterprise as defined by Alfred D. Chandler. By the beginning of the 1880s, it had replaced wholesalers, and built instead a network of its own branch houses. Distribution of high technology binders required investments in specialized, product-specific facilities and personnel, which independent agents were not able to offer. Thus, the harvesting machine industry was part of the modernization of American industry.

III

■ Changing Patterns of European Agriculture

3.1. Old fashions and new winds

The American harvesting machine companies that enlarged their operations to Europe soon noticed the great differences between one country and another. The underlying reasons for these are countless. Some researchers have pointed towards the varied ways in which different countries had shaken off the remnants of their feudal past. General economic progress had not been even, either, and its influence in the development of agriculture consequently differed from country to country.

In England, enclosure and consolidation of farms had led to large holdings. The old landed aristocracy were able to hold their position as landlords by letting out their land to tenant farmers; wealthy landowners therefore had both the means and an interest to invest in their farms. In France, on the other hand, the state of agriculture by the middle of the nineteenth century was the opposite of the dynamism and receptivity to new ideas found in Britain. After the Revolution, farms had been split into small units, and almost 40 percent of the farming land was in farms of less than 20 hectares. This problem was further aggravated by the division of land between the heirs on the death of the owner. In the western parts of Germany, farms were similarly small and fragmented. Peasant agriculture also had a long history in parts of southern Germany, Switzerland and the Tyrol. On the contrary, the old landed aristocracy (the Junkers) dominated the eastern parts of 'the new German Empire, and development was much the same also in Russia and Austria-Hungary.[1]

In most European countries, farms in the 1870s were still

---

1 *Tracy* 1964, 19-21; *Tracy* 1966, 98-99; For example, in Thuringia, where the land was subdivided into smallholdings, the owner of a farm of 12 hectares could barely support himself and his family. Those possessing less than 12 hectares had to work part time in the nearby industries for their living. *Reports of the consuls of the United States No.2, 1880,* 97-98; In Russia the United States consuls underlined the primitive character of agricultural pursuits in general. Russia with its 80 000 000 inhabitants used annually only about $3 000 000 on agricultural implements. The dearth of money was so great that an association had to be formed to enable a peasant to purchase a $30 plow. *Reports from the consuls of the United States No.20, 1882,* 233-234.

self-sufficient, producing only for the use of their own household, although there were big regional differences. Britain, as a result of its expanding industrial population, had become a large importer of food. France also imported some grains, while it exported others. Germany imported grain in its western regions, while exporting grain from the great estates in the east. The largest exporter of grain in the world was Russia. Nevertheless, agricultural trade was still largely an intra-European affair.[2]

The opening up of the lands of the American Mid-West, together with the rapid improvements in methods of transportation, produced a flood of cheap American wheat, which was imported in increasing measures at the end of the 1870s. Simultaneously, Russian grain began to flow in ever larger quantities to central Europe through the railroads, which now extended to the Baltic. The impact of these imports was multiplied by exceptionally good harvests in America at the end of the 1870s, while Europe was suffering from bad weather. Normally, bad harvests were accompanied by higher prices; but now, cheap exports from America forced prices even further down until 1896.[3]

Table 7. Area under wheat in the main grain growing countries, 1860-1900 (millions of hectares).

Year	U.S.A.	Russia	Canada	Argentina	Australia
1860	–	–	0.6	–	0.2
1870	8.4	–	0.6	0.1	0.5
1880	15.3	11.6	0.9	–	1.2
1890	14.7	13.7	1.1	1.0	1.3
1900	19.8	20.2	1.6	3.2	2.1

Source: Grigg 1974, 262.

Although Table 7 shows a continuous increase of the area under wheat cultivation, this was not directly reflected in world trade. Expansion of grain growing took place in very similar environments. The new lands were grasslands with progressively more arid climates as settlement moved out from the core areas. Farmers in the new agricultural lands were confronted in establishing farms by many

---

2 *Tracy* 1964, 21.

3 *Tracy* 1964, 22-24; *Tracy* 1966, 101-102; *Klein* 1973, 122-123; *Haushofer* 1972, 238-239; *Grigg* 1974, 174.

problems similar to those faced by the prairie farmers in America. Besides, there were questions over the systems of land disposal. The transportation of the harvest also had to be taken into account. In Canada, Argentina and Australia it was not thought possible for wheat production to be carried on profitably more than 25 kilometers from a railroad.[4]

The cheap grain pouring into Europe from overseas and from Russia caused problems for European farmers. Although the effects of grain imports did not hit all farmers as severely, the phenomenon was soon labeled an agricultural crisis. In its early stages, it mostly affected grain growers;[5] livestock producers actually benefited from the fall in grain prices, since they used grain for feeding their animals. There were also major geographical variations in the effects of the depression. In England the arable farms were mostly in East Anglia and the south of England. The plains of the Paris basin and Northern France were the central grain growing regions in France, while in Germany grain production was concentrated on the large Prussian estates.[6]

Responses to the agricultural crisis varied from one country to another. Throughout the Continent there had been a growing demand among industrialists to raise barriers against growing imports and to protect own production. This attitude gained new thrust from the nationalistic feelings aroused after the Franco-Prussian War of 1870. It was therefore no wonder that the farmers and industrialists in central Europe found a common ground to raise a protectionist wall against imports. In 1881, France raised its tariffs and its duties on livestock products considerably, followed by new tariffs in 1885 and 1887 whereby the duties on all agricultural products were raised. Germany imposed moderate duties in 1879 on agricultural products. Duties on grain were, however, raised in 1885 and 1887. After the dismissal of Bismarck in 1890, for a time Germany followed liberal tariffs as the country tried to obtain markets for its industrial products.[7]

---

4 *Grigg* 1974, 259-263.

5 As a consequence, prices of arable products fell and the wheat acreage in England contracted from 3.3 million acres in 1871 to 1.8 million acres in 1904. Over the same period the area of pasture rose from 11.2 million acres to 16.2 million acres. *Jewell* 1976, 126.

6 *Tracy* 1964, 25-26; *Klein* 1973, 125-126; *Kindleberger* 1964, 216-217. Kindleberger remarks that in France farms in the densely populated areas became smaller but were more efficient in the northern part of the country, the east-center area and around Paris. Ibid. 216.

7 *Tracy* 1964, 27-28, 65-70, 87-89; *Klein* 1973, 124; *Skalweit* 1937, 584-585; *Haushofer* 1972, 247-248; *Grigg* 1974, 174.

Britain took a totally opposite attitude towards cheap foreign grain. The government did not offer the slightest protection to domestic agriculture. Industrialists preferred cheap food and raw materials, and since the farmers were unable to form a united front to push their case, foreign competition hit British farms with its full strength. Good harvests during the 1880s brought some relief to the situation, but hopes of a better future were destroyed in the early 1890s by drought and cold summers in conjunction with further increased American competition. The decline of the agricultural population, which had continued from the middle of the 19th century, continued at an accelerating rate. Whereas in 1861 agriculture still occupied 18 percent of the total workforce, by 1901 the proportion of the active population in agriculture was reduced to a mere 8 percent of the total.[8]

In the small European countries, Denmark and Netherland held firmly to free trade. Both countries were so dependent on foreign trade that they could not even consider protectionism. Here the reaction was a fundamental transformation of agriculture from grain growing to dairy production. A distinctive feature in Danish farming was the development of co-operatives, which played a vital role in its transformation. Co-operatives offered a good solution for independent smallholders, who could not afford themselves to arrange processing and marketing of their livestock products. As a result, exports of Danish livestock products grew many times over in three decades.[9]

The development described above also had its impacts on sales of agricultural machinery. The expansion of agricultural land in America, together with increased export of agricultural products, undoubtedly promoted production and sales of machines. In Europe, where farmers now faced a totally new situation, corresponding changes in the demand for agricultural machines and implements might be predicted.

There were, however, many other factors affecting the use of machinery in agriculture, as Folke DOVRING has shown. All machines save labor, but where labor is abundant and capital scarce, there is no incentive for mechanization. Nevertheless, there are other factors promoting labor-saving machinery. New technology can produce higher output, improve the quality of the harvest or, as in

---

8 *Tracy* 1964, 30, 46-50.

9 *Tracy* 1964, 106-113. Exports of butter increased from 9 thousand tons in 1870-74 to 70 in 1901-05, of pork from 8 thousand tons in 1881-85 to 76 in 1901-05. Tracy 1964, table 33, page 112; *Skalweit* 1937, 584-585; *Larsen* 1895, 145-148, 161-166; *Fussel* 1966, 219-220.

the case of the harvesting machinery, rescue crops from spoiling.[10]

The labor question was, nevertheless, one of the key components affecting mechanization of agriculture in Europe. On the Continent, the main increase in population occurred in the countryside, and the change from a mainly agricultural to an urban population did not take place until the latter part of the nineteenth century. In Britain, however, the non-agricultural population exceeded the agricultural population well before 1800, and in Belgium, too, before the middle of the 19th century, when half of the population in France was still occupied in agriculture. The agricultural population remained very stable in actual size until World War I. In Germany, the agricultural population was 18.7 million in 1871 and 18.5 million in 1895, when it still made up 34.9 percent of the total population. In eastern Europe, on the other hand, the population increased considerably during the 19th century. In view of the lack of industry, most of this increase had to be absorbed by agriculture.[11]

In most west European countries, the field systems underwent a thorough change, and agricultural land was re-allocated in a more economic way. This movement was reinforced by the introduction of new agricultural techniques, especially new crops and improved crop rotation systems. Root crops, clover and other leguminous fodder crops were introduced into the rotation, intensifying tillage of the soil and making control of weeds more efficient. New plants and consequent rotations also intensified the use of the workforce. As can be seen in Table 1, harvesting was one of the bottlenecks in U.S. agriculture, and the same can be presumed to have been the case in Europe too. Besides, during the 19th century, corn, root and leguminous crops competed with each other for labor. E.J.T. COLLINS even states that grain production grew faster than the supply of harvest labor. Consequently, he argues that there existed growing disparities between harvest work demand and harvest labor supply. The farm labor market became increasingly ruled by the trade cycle. Supplies of labor were unpredictable and could fluctuate from season to season. In Britain, Collins has found that cyclical peaks coincide with shortage of farm labor and increased farm mechanization. Accordingly, there was a close correlation between the state of the labor market and the diffusion of labor-saving machinery. Agricultural machinery did not directly raise the level of output and saved more labor than land. A key problem in the

---

10 *Dovring* 1966, 645-646.

11 *Dovring* 1966, 604-608; *Klein* 1973, 121; *Krzymowski* 1939, 272-273. According to Krzymowski, the agricultural population was 42.5 percent of Germany's total population in 1882, and 28.6 percent in 1907; *Berend-Ranki* 1982, 16-21.

mechanization of British agriculture was the question of the poor.
Farmers preferred to give work rather than charity and consequently
opinion was against the use of machines as long as there were
available hands in the parish. Even when the machines had been
acquired they were not necessarily used to full advantage. The
reaping machine could be used only for wheat, leaving other grains
to be harvested by hand. The harvesting machine was welcome as
long as earnings and employment were unaffected.[12]

According to Collins, the drying up of the flow of migrant workers
in Britain and France accelerated the shift to reaping machines. On
the other hand, the small average size of farms in France and
Germany prevented mechanization of agriculture. In fact, in many
areas of western Europe, technological change took the form of a
switch from lower to higher capacity hand tools. Resistance to these
new more efficient tools was strongest where labor was abundant.
Those who supported the use of the sickle against the scythe argued
that the latter was wasteful of grain. In Scotland, there were still
doubts expressed in the 1860s whether the scythe could ever be
transformed into a successful grain harvesting tool. Up to the
mid-1860s in Britain and throughout Continental Europe the scythe
gained ground faster than the reaping machine between 1850 and
1880. In many areas farmers were reluctant to introduce technology
that could cause unemployment among resident workers. Besides,
only the largest farms could provide a sufficiently large seasonal
work-load to justify the adoption of harvesting machines. JEWELL
estimates that less than half the British cereals acreage was harvested
by machine in 1870.[13]

J.R. WALTON's findings confirm Collins' observations. He found
that in Oxfordshire in Britain, innovations that did not involve

---

12 *Dovring* 1966, 626-630, 636-640; *Collins* 1969, 61-65, 71, *Collins* 1989, 204-206, 208, 211-215.

13 *Collins* 1969, 71-72, 78-79, 83-86, 93-94; *Long* 1963, 22; *Kindleberger* 1964, 213; *Jewell* 1976, 127; *Long* 1963, 22; Grantham 1899, 17.
On the Thuringian smallholdings reapers were still rare in 1880. Most of the harvest was cut with scythe and sickle, which were preferred, as doing less damage to the grain. *Reports from the consuls of the United States No.2, 1880*, 99.
In Finland even the landed aristocracy resisted the introduction of hay in crop rotation. In the third general meeting of Finnish agriculturists in 1852, J.G. von Bonsdorff, announced that he would oppose growing of fodder on the fields until people learnt to eat grass. *Liakka* 1920, 129-130. A similar typical fate met Professor Lerche, who was obliged to leave his new swing plow and iron harrow untouched to rust. *Soininen* 1975, 103. This phenomenon was not unknown either in America, where the machine-breakers' movement expanded in 1878 from Ohio to the main Mid-West grain-growing states. It was aimed against farmers who had bought or intended to buy a binder and thereby diminish demand for labor. *Arger-singer-Argersinger* 1984, 397-409. As late as the 1880s in some parts of Hertfordshire in Britain it was no uncommon practice for the laborers to burn reapers and binders. *Collins* 1989, 207.

substantial reductions in labor were adopted at rates consistent with known improvements in the quality and reductions in the price of machinery from 1830. On the other hand, innovations that reduced labor were ignored, even though available at an early phase. Therefore, in 1850 there was in Oxfordshire no evidence of any use of mowing and reaping machines, whereas by 1880 both appeared in over 30 percent of the advertisements. In the five-year running means that Walton calculated, adoption of mowing and reaping machines increased rapidly after the 1860s. By 1880 the majority of barn operations and a large part of the hay and corn harvests had been mechanized.[14] Jewell adds an interesting point that competition between American harvester manufacturers accelerated the adoption of the harvesting machines in England.[15]

In an article published in 1966, Paul A. DAVID presented a new approach to the mechanization of agriculture in America. He calculated the threshold size for a farm where it would be profitable to buy a reaper. Factors affecting this threshold were the prices of labor and the cost of reaping machines. David set the threshold size at 46.5 acres in the beginning of the 1850s, but a couple of years later it was lowered to 35.1 acres, as wages rose faster than machine prices. As a result, reapers began to spread among the farmers.[16]

In his later study David has applied his method in Britain. He attacks the interpretations of Folke Dovring and other researchers, who had explained the slow mechanization of agriculture by labor abundance. Instead, David offered as an explanation the state of the farming landscape and the attendant expenses involved in rendering it suitable for mechanical farming. The greatest impediments to the reaper were the nature of the field surfaces across which the machines would have to be drawn, and the size of the farms. Often, the breadth of the swath cut by the generally available makes of reaping machines often turned out to be insufficient for the width of the cultivated tracks, causing loss of straw and leaving behind sheaves of varying lengths. Besides, fields were in need of drainage. Consequently, David argues that these factors made the use of

---

14 *Walton* 1973, 8-10; *Collins* 1989, 205.

15 *Jewell* 1976, 128.

16 *David* 1966, 13-16, 21-22; David's ideas have not met with unanimous acceptance. Alan L. OLMSTEAD showed David omitted from his calculations joint-ownership of machines and professional reaper men who circulated from farm to farm after the harvest. Besides, farmers began to increase their holdings only after the introduction of the reaper. In addition, the reaper went through a radical modification and evolution process which reduced the average draft alone over 40 percent. *Olmstead* 1975, 330-344. The latest turns in the discussion can be read in *Olmstead-Ankli* 1995, 27-57.

reapers more costly than the use of hand labor on all but very large farms. Grain growing in Britain was for the most part in the hands of tenant farmers, which further diminished the incentive for improvements or financial abilities thereto. On the other hand, rising labor costs in relation to the price of grain and price of machinery made it attractive to invest in machinery. David, however, argues that unfortunately wage costs for the improvement of land also rose which then began to hinder mechanization.[17]

S. TVEITE's observations in Norway follow the main ideas of David's original threshold model. In his study, Tveite found labor supply and wages to be the key factors affecting the mechanization of agriculture. Following David's example, he has calculated threshold acreages for Norway too, but has expanded his approach also to include the horses, which were initially too small to draw the first mowers, and had to be changed for larger ones. The question of labor costs also turn out to have been more complicated than had been estimated. Wages differed from work to work. Mowing was done normally with scythes by men, while grain was harvested with sickles. Since women were also used for this work, wages tended to be lower than in mowing. Besides, the calculations also have to include efficiency of both the hand labor and the machine. In this way, Tveite estimates that the decisive factors favoring purchase of the first reapers and mowers was a combination of higher wages in hand mowing with the development of better machines. He claims that in 1875 there were about 4000 farms that had adequate economic resources or for which it was economically reasonable to buy a harvesting machine. The acreage even on many of these farms was curtailed by natural hindrances like steep hills, which in many cases considerably cut savings obtained by the purchase of a reaper or mover.[18]

The size of the farm is not, however, the only factor determining investment in machinery, as B.H. SLICHER van BATH has noted; the kind of cultivation practised on the farm also has to be taken into account. Large machines could not be used in small-scale diversified farming. Instead, light, small and simple tools were used. Slicher van Bath especially underlines the influence on farmers of market prices in the purchase of machinery. More tools and machines are invented and applied in periods of high prices for farm produce than in periods of low prices. If dairy production is more profitable than

---

17 *David* 1971, 145-146, 148-151, 156, 158, 160-161, 173-175.

18 *Tveite* 1980, 1-18, 20, 24. Tveite estimates there must have been in 1890 about 12 000 harvesting and mowing machines in Norway and in 1907 49 190 machines on 225 795 farms. Ibid., 28-29.

grain, inventive activity moves towards machines for milking, churning, hay mowing, etc.[19]

In summary, it can be noted that Europe emerged very unevenly to the state where it was profitable to mechanize harvesting. In Britain, that point was reached early in the 1860s, when reapers and mowers began to expand in increasing numbers. On the Continent, the situation was more complex. In western Europe farms were smaller and land ownership scattered. However, there were regional differences, such as the Paris basin. As a whole, distribution of harvesting machines was restricted by the supply of cheap hand labor. On the other hand, the fluctuation of the agricultural product prices should also be taken into account. The price of the machines themselves was also an important factor, especially relative to the price of labor but also relative to the efficiency of the machine.

In relation to the labor question, it should be noted that although, for example, Germany lost through emigration in 1881-1890 some 1 342 400 people, and 529 000 more in 1891-1900, at the same time new people were arriving in the country. In 1880 there were about 300 000 foreigners in Germany, and twenty years later 1 260 000. Thousands of them were wandering farm workers, especially from Poland.[20]

In this light, the question of the effects of emigration on farm mechanization appears to be much more complicated than, for example, Jan Kuuse has explained[21]. Besides, more effective hand tools, such as the scythe, were still spreading at the same time as reapers and mowers emerged on the market. So when the American harvester companies began to compete in Europe, there were in Britain in 1874 about 80 000 reapers, in 1880 in Belgium about 1500, and in France in 1882 about 35 000. In Germany in 1882 there were 19 600 reapers and mowers.[22] In this sense, the American har-

---

19 *Slicher van Bath* 1960, 4-5, 12-14. Slicher van Bath also adds costs of labor, capital investment, size of the farms and the initiative of the farmer in his list of the determining factors conditioning the development of the agricultural tools. Ibid. 8.

20 *Klein*, 1973, 21.

21 *Kuuse* 1977, 274, 278-279.

22 *Collins* 1969, 75; *Haushofer* 1972, 226; The number of machines at a certain moment does not tell the whole story. It has to be compared with the number of farms before the real state of mechanization becomes evident. From France this kind of account is available. According to G.E. FUSSEL there were in France 52 000 mowers and 51 000 reapers in 1892 which means that in a country of 5 672 000 holdings less than one percent of farms had some kind of harvesting machine. On the other hand, 4 034 000 of the farms were smaller than five hectares. *Fussel* 1966, 199. There were 29 000 farms classed as big holdings possessing over 250 acres of which only 23 000 had reaping machines; that makes less than one reaper on each holding. *Clapham* 1951, 171.

vester-makers faced very demanding but also potentially very lucrative markets. On the other hand, expansion into European markets coincided with the agricultural crisis and consequently implement makers had to fight on two fronts: against competitors and against economic strains. One of the tasks of the present work is to figure out the factors that helped the American harvester companies to conquer the European market.

3.2. Competition from the European agricultural machine industry

When Cyrus McCormick and Obed Hussey brought their reapers for the first time to Europe in 1851, harvesting machines already had a long history behind them in Europe. However, none of the European machines had become popular among the farming community before the introduction of the American machines.[23] The reason for this was apparently that the early models were insufficiently well constructed, and the time was not yet ripe for their expansion.

After Crystal Palace both Hussey and McCormick made extensive agreements with European machine companies for the production of their reapers.[24] The effects of the success of the American reapers were also reflected in patents. From 1849 to 1854 thirty-two patents were issued on reapers in England. Impacts were felt also on the Continent. French inventors had made some experiments with reapers prior to 1851, and even presented a French reaper at Crystal Palace. Only after the International Exposition in Paris in 1855 did the reaper gain success in France, when McCormick contracted with Fleischmann for fifty machines.[25]

Although the American harvesting machines were able to conquer the European market, this does not mean that there was no high-level competitive agricultural machine industry in Europe. On the

---

23 *Hutchinson* 1930, 380. Hutchinson tells, how an English visitor in America, Hon. Thos. Tollemache, introduced the Hussey machine to a friend, who later let the firm of Garret & Sons of Leiston make drawings of it. By 1851 the firm had constructed two reapers and exhibited one at Crystal Palace. Ibid. 384.

24 *Hutchinson* 1930, 392-393, 399. In 1851 Hussey made an agreement with Dray & Co. to manufacture for the British market. Garret & Son and William Crosskill also manufactured Hussey's machines for the next harvest. Crosskill also made Bell's reapers. McCormick, on the other hand, arranged with Burgess & Key of London to represent his reapers. Burgess & Key, in turn, engaged Samuelson of Banbury to manufacture the McCormick reaper. By 1855 also Garret & Son of Saxmundham and Ransome & Sims of Ipswich were manufacturing for Burgess & Key; *Saul* 1968, 211. Saul states that still in 1860 Hussey had three or four factories in Britain making his machines.

25 *Hutchinson* 1930, 399 note 67, 404-405.

contrary. English inventors had by 1848 completed a successful threshing machine where a winnowing machine was attached to the thresher. A step further was taken at about the same time when steam power was used to drive the machine.[26] British steam-powered threshing machines won a high reputation all over Europe, and dominated markets to the end of the century.[27] Collins even states that if there had existed demand in the 1850s for reapers, British manufacturers, who had been able to produce such a complicated device as a thresher, could have made just as good reapers too.[28]

According to S.B. SAUL, the history of the reaper in England has been oversimplified. There were intelligent reaper manufacturers who could compete succesfully with the American firms, even at Crystal Palace, where five British makers were given Medals of Honour. In fact, in reapers made largely of metal the British firms led their American counterparts by some twenty years. It was only in the 1870s that Americans began to drive out the British products in Europe.[29] The findings of the present study confirm Saul's results. Although the American companies had sold their machines for years in Europe, the big wave hit in the 1870s, when for example the Walter A. Wood Company began to export thousands of reapers and mowers to Europe. Saul's other statement is true too, in the light of the current study: American companies did not substitute iron frames for wooden ones until the 1880s.

By 1900, the British makers had clearly lost the game. By 1914 the English manufacturers numbered only four reaper, binder and mower makers, catering almost entirely for the home market.[30] Their total output was less than ten percent of that of McCormick alone in America. To clarify his point Saul cites an article in The Times

---

26 *Beaumont-Higgs* 1958, 10; *Fussel* 1952, 221.

27 The predominance of the English threshing machines becomes perhaps best evident by their distribution also to the peripherical agricultural areas in northern Europe. Probably the first English threshing machines reached Finland in 1850 when the owner of the Orisberg estate, Captain Björkenheim, ordered one directly from England. It was followed by the Dregsby estate a year later. The first steam thresher had come to Finland by 1862, when it was acquired from Clayton & Shuttleworth for the Finnish agricultural college at Mustiala. *Liakka* 1920, 130; *Suomen Huoneenhallitusseuran Sanomia* 1851, number 21, 346; *Nya Pressens Landtbruksafdelining* 1890, number 14, 54-55.
In the seventh general meeting of Finnish agriculturalists in 1876, four English threshing machine companies were represented: Clayton & Shuttleworth, Davey & Parman, Richard Hornsby and Robey & Co. *Kertomus seitsemännestä yleisestä Suomen Maanviljelyskokouksesta* 1877, 105.

28 *Collins* 1969, 93.

29 *Saul* 1968, 211; *Aldcroft* 1968, 30.

30 *Jewell* 1976, 128.

describing the Maidstone Agricultural Exhibition in 1899: it "gave the impression that some of our leading firms were becoming implement agents rather than implement makers". Saul points out that not only British but also German and French companies were not quick enough to adapt their machines to local requirements. In Britain, falls in grain prices and in arable acreage, coupled with low labor costs and mixed farming, also made the home market particularly weak. British firms were not able to follow the Americans into mass production methods.[31]

Ever since James Small started to manufacture his swing plows, English plow makers had been active on the European market. As a relatively simple device, the English plow was soon copied all over the world. In Scandinavia, Swedish companies like Överum, Norrahammar and Näveqvarn found significant markets for their implements.[32] English plows were sold also in Germany, where they were copied and adapted to local conditions. In the same manner as in Sweden, local manufacturers soon began to hammer their own plows according to the orginal models, and some of them were lucky enough to expand their production on a prominent scale. Heinrich Eckert began his career by copying American plows initially, but by the 1850s he already had twenty different models in production and expanded his activities to mower production in the 1860s. Rudolf Sack had a very similar history, as had Heinrich Lanz, both of whom began manufacturing in the 1850s, a crucial decade for the German agricultural machine industry, and soon they expanded their operations abroad, especially to Russia. During the last part of the nineteenth century German firms grew into large-scale enterprises. By 1883, Rudolf Sack had made 100 000 plows, and twenty years later one million.[33]

As has been shown above, only the English factories were able to compete with the American manufacturers. German agricultural machine makers were still developing their plants. In other countries on the Continent in the 1870s and early 1880s, such as Sweden, manufacturers were small and most of them were able to make only simple implements like plows and harrows. Of course there were exceptions, such as Th. Munktells of Eskilstuna, who made threshers

---

31 *Saul* 1968, 211.

32 *Eskeröd* 1973, 87; *Kuuse* 1974, 55-57, 60, 64-70; *Lantmannen* 1878 number 24, 372.

33 *Blaich* 1984, 70; *Haushofer* 1972, 112-114, 142-143, 225. The first order, for 120 plows for Russia, Rudolf Sack was forced to give to the English makers due to lack of adequate capacity. English plow making knowhow was also needed in the erection of his first plow factory. *Haushofer* 1972, 142.

and portable steam engines, or J. Thermaenius of Torshälla, who manufactured threshers and horse-powers too.[34]

This situation left the doors open to the English and American factories, as becomes evident in contemporary descriptions of the distribution of the harvesting machines. The first reapers were acquired both in Denmark and Sweden in the same year, 1852. The Danish machine was a McCormick, and the Swedish one a Hussey manufactured by Garret in England. Burgess & Key's McCormick reaper found its way to Sweden in 1857.[35] The American harvester companies soon found their way to Europe too, although many of them licensed English manufacturers to make their machines. Many of these manufacturers began later on to make harvesting machines in their own names: Burgess & Key, Samuelson and Hornsbys to mention the most popular of them. Although the American machines outnumbered their English competitors, there was, however, a feeling that in demanding conditions the English machines were better, due to their stronger construction. From the beginning of the 1870s, however, an expansion of the American harvesters can be seen from one agricultural show to another; new makers and models were exhibited, while the European makers, with a few exceptions, remained the same.[36]

---

34 *Juhlin Dannfelt* I 1913, 304; According to the consular reports of the U.S. consuls, still in 1881 "plows, spades, rakes, hoes, pitchforks, shovels, and hand tools generally, are wretchedly made in France." *Reports from the consuls of the United States No.3, 1881*, 100.

35 *Juhlin Dannfelt* I 1913, 303; *Illustrerad Landtbrukstidnings Årsbok* för år 1876, 64.

36 *Illustrerad Landtbrukstidnings Årsbok* för år 1876, 64-81; *Reports from the consuls of the United States No.20, 1882*, 234. The managing director of a Russian iron foundry also preferred British products. "As to portable engines and thrashing-machines, I prefer the English make - Clayton and Shuttleworth. They are far dearer than the American ones, but also stronger and more solid"; The U.S. vice-consul Geo. W. Sillcox confirms this comment by stating that for "machines constructed of iron, such as engines and thrashing machines, chaff-cutters, plows, cultivators, harrows, etc., preference seems to be given to the English and continental makers." *Reports from the consuls of the United States No. 48., 1884*, 482.

IV

■ Cyrus Hall McCormick Turns Abroad, 1851

4.1. First laurels at Crystal Palace

The fame of the new American harvesting machines began to spread around Europe and from 1849 onward, there are records of McCormick's reaper in Austria; but his greatest victory and lifelong reputation McCormick achieved in 1851 in the Exhibition of the Works of Industry of All Nations, the first World Exhibition, held at Crystal Palace in London. His ridiculed machine struck the audience by surprise with its performance. It reaped laurels, and The Times declared that it was worth the whole cost of the exhibition. Besides the Grand Prize of the World Fair, McCormick gained publicity which lasted for decades not only among the farming community but even among scholars.[1]

But what were the forces that drove Cyrus McCormick to begin foreign business? Alfred D. Chandler has explained that enterprises had the prerequisites to turn abroad when they had taken over the distribution of their products, obtained units for producing raw and semifinished materials and had begun investing in development. In the first phase, companies invested abroad in marketing. Foreign commercial agents were soon replaced by companies' own sales force.[2] Chandler, however, fails to describe the reason for foreign expansion.

Neither has Mira Wilkins found any general reasons for the ante-bellum foreign business of American companies: she merely notices that "motives were diverse, and often highly personal". On another occasion, she explains the post-bellum foreign activities in terms of meeting foreign demand, disposing of surplus output, and obtaining economies of scale, but also of reaching markets or obtaining sources of supply. However, she has carefully described the evolution of the foreign enterprise stage by stage, from independent agents to salaried export managers in the next stage, a branch house or a distribution subsidiary in the third stage, and

---

1 *Hutchinson* 1930, 392, 399-401, 406-407.

2 *Chandler* 1988, 31, 34, 44.

finally a finishing, assembly or manufacturing plant in the final stage.[3]

McCormick knew from previous experience the importance of tests, trials, and exhibitions in the marketing of products. He had noticed that people were most receptive to new ideas when they had tried or seen a machine themselves. This knowledge he immediately put to good use in England, where he sold production rights to several firms, of which Burgess & Key became his permanent business partner.[4]

In the New World, McCormick had abandoned or was abandoning the old organization, based on manufacturing rights. In Europe, however, he found local manufacturers the best way to promote his reapers. Numerous researchers such as Fred CARSTENSEN have explained his decision by the conservatism of European farmers and their preference for homemade products.[5] David A. HOUNSHELL has probably found a more likely reason for the selling of production rights to European manufacturers: as has been previously noted, production technology was in its infancy and factories were mere blacksmith shops. As a consequence, agents made frequent complaints about the workmanship in McCormick's reapers. Furthermore, the McCormick brothers were not unanimous on the company's strategy.[6]

McCormick's were simply not able to produce enough machines, even to meet American demand. It was, therefore, understandable that if Cyrus wanted to expand his trade outside America, he had to find manufacturers near the new markets. This explanation becomes even more plausible in the light of William McCormick's complaint to Cyrus in 1863 of the stress that the changes of models for European conditions put on the factory. He suggested that Cyrus had better find a European manufacturer for his European business. The reasons for William's criticism lie in the production technology, which was not flexible enough to meet new requirements.[7]

---

3 *Wilkins* 1970, 19, 29, 36, 45-46; *Wilkins* 1988, 22.

4 *Hutchinson* 1930, 415, 424; *Carstensen* 1984, 109; *Wilkins* 1976, 29. Burgess and Key made subcontracts with D.L. Laurent and Francois Bella in France, and with a couple of firms in Germany and in Poland.

5 *Carstensen* 1984, 109.

6 From the early days of the business, Cyrus McCormick's younger brother William kept his brother's accounts, and later was in charge of management, whereas his other brother Leander was the Superintendent of the factory. Both of the brothers preferred a very conservative and cautious business strategy. Cyrus, on the other hand, wanted to enlarge production to its limits. *Hounshell* 1987, 155, 157, 159-160.

7 *Leander McCormick to Cyrus H. McCormick* 15.1.1864 Mss 1a, box 53; *William McCormick to Cyrus H. McCormick* 24.1., 7.2., 21.2., 21.12.1864. Mss 1a, box 53; *Hounshell* 1987, 167; *Heikkonen* 1989, 158.

Another possible explanation for the selected strategy could be customs duties. However, among the original material in the McCormick Collection there is no mention of customs at all at this phase.

The selling of production rights freed McCormick from setting up a sales organization of his own for the European trade. His only responsibility was to arrange the collection of royalties. However, McCormick's relations with his English licensee, Burgess & Key, began to cool in the beginning of the 1860s, and he decided to appoint a European agent to look after his business interests. James T. Griffin began actively to canvass continental Europe to find new energetic agents. He also tried to find new manufacturers for the McCormick machines.[8]

Burgess & Key was McCormick's main agent and manufacturer of his machines for England. Relations between the parties were, however, strained, and Griffin tried to find a new partner, but in vain: for example, the respected house of Richard Garret & Son declined to make McCormick reapers under license if they had to compete with other makers. If McCormick wanted to continue his European business, he was forced, under the pressure of his brothers, to continue to rely on Burgess & Key.[9]

The main marketing area for McCormick's reapers was England, but they had found their way also to continental Europe. In the mid-1850s, reapers were being sold in German states and in Warsaw. Russia was from the very beginning a lucrative possibility, with its huge acreage. Some machines were sold there, too and the Russian Consul General, J. de Nottbeck, made a request for two models for

---

8 Burgess & Key almost stopped making McCormick's reapers in the beginning of the 1860s, and consequently machines had to be shipped from America. McCormick's decision in 1862 to form a manufacturing contract with the French house of Albaret et Cie. reflects the tensions between the parties. McCormick still continued to transport his machines in 1862, when he intended to send 50 machines to London and 50 to Hamburg. The machines normally arrived in England through Liverpool, and the recipient had to pay all the charges all the way from Chicago. *James T. Griffin to C.H. McCormick* 28.5.1863. Mss 1a, box 52; *Carstensen* 1984, 110; *Heikkonen* 1989, 145; *T.B. Bunting & Co. to C.H. McCormick* 29.5.1862. Mss 2a, box 23.

9 Burgess & Key tried to benefit from McCormick's name and reputation, but to evade the license fees, by attaching additions of their own to machines; they also developed their own mower, and began to promote its sales by using McCormick's name, while producing only a few reapers. This of course led to a lengthy quarrel. *J.T. Griffin to C.H. McCormick* 3.12.1864. Mss 1a, box 54; *Hutchinson* 1930, 439; Burgess & Key's mower was a good and reliable machine and constructed for European conditions. That is why Griffin asked for permission from McCormick to sell it on continental Europe to compete with Walter A. Wood. The Company's reaper, on the contrary, was not recognized as equivalent to the original McCormick reaper. *J.T. Griffin to C.H. McCormick* 21.1.1865. Mss 1a, box 54; *J.T. Griffin to C.H. McCormick* 9.9.1865 and *Moritz & Joseph Friedlander (Breslau) to J.T. Griffin* 21.1.1867. Mss 1a, box 55.

the new Imperial Agricultural Museum at St Petersburg.[10]

Grain production in England, which was at first the main marketing area for the harvesting machines, began to diminish after the repeal of the Corn Laws in 1846. On the other hand, the U.S. Civil War, which interrupted American exports of grain, and simultaneous good harvests in the 1860s in Europe, gave a stimulus for grain growers, and the area under cultivation expanded until the general depression of the 1870s. The situation deteriorated as a result of a series of wet summers at the end of the decade. Consequently, the situation opened good opportunities for the harvester machine companies, especially in England, where farm labor was for the first time becoming scarce. Besides, the clientele was now expanding from the ranks of the estate owners to include tenant farmers. Furthermore, machine prices had fallen and their workmanship improved.[11]

When McCormick's European agent, James T. Griffin, arrived in Europe, he began energetically to appoint new agents and even had discussions in Prussia about the manufacture of machines. Prospects for larger sales in German states were not promising, however. Griffin had great difficulties in finding willing and competent agents. Dealers and jobbers had tried reapers without any success and spent too much money in vain. In that area the biggest obstacle to the spread of the reaping machines was cheap and abundant labor. In spite of the difficulties, Griffin continued his canvassing from Berlin to Vienna and Budapest, from there to Turin in Italy, and planned a trip to St Petersburg and Moscow. Finally he had appointed agents in German states, Russia, England, France, Austria and Romania. Central Europe and England were, however, the main marketing areas. Griffin had high expectations from Spain, but claimed that Italy was a hopeless country for reapers. In Hungary he saw good prospects for sales in the future.[12]

---

10 *Hutchinson* 1930, 406-407; *Gebrüder Butenose to Cyrus H. McCormick* 20.7.1858. Mss. 1a, box 49; *J. de Nottbeck to Cyrus H. McCormick* 11.12.1860. Mss 2a, box 20; *C.H. McCormick Esqr. Burgess & Key to C.H. McCormick. Royalty Account for* 1863. Mss 2a, box 23. Burgess & Key's royalty account of 1862 confirms sales of 52 reapers, two of which were sold in Hamburg, and one each in Budapest, Wiesbaden and Florence. McCormick got four pounds on each machine as a royalty. In 1863 that made 208 pounds together.

11 *Tracy* 1964, 43-46.

12 After long discussions, agreement was finally reached, and in 1864 Mr. Pintus began to make machines for the Prussian markets. *James T. Griffin to Cyrus H. McCormick* 28.2.1863. Mss 1a, box 53; Griffin assured that in time large sales would be achieved in Prussia. He expressed the importance of wellknown exhibitions such as that at Hamburg and furthermore of mailing circulars directly to would-be customers. Ibid.

Griffin's experiences correlate well with the information on the state of agriculture in various parts of Europe. England and parts of central Europe excluded, the market was not yet ripe in Europe for a large scale reaper trade. Only the wealthy estate owners were in a position to mechanize agriculture on their holdings. In Germany as late as in 1882, the use of the harvesting machines was limited to holdings larger than 180 hectares[13]. For the most part cheap labor connected to the high price of the machine and its inadequate workmanship, however, made the efforts of harvester agents useless.

As a contemporary living amid the events, James T. Griffin either was not able to interpret the market correctly or then he really foresaw a niche for reapers. Anyway, Cyrus McCormick decided to strengthen his presence in the German-speaking area in 1864 by change to the jobbing-house system which was in use in America too. The appointed jobber, James R. McDonald & Co. of Hamburg, spent considerable sums on advertisements and posters, and was hopeful about future prospects.[14]

Griffin's optimism soon changed to disappointment. In Central Europe there were a bunch of agent candidates but sales had not realized: Burgess & Key had sold only 42 machines on the Continent. Besides, Burgess & Key were concentrating more and more on the promotion of their own products in England. For the Continent they

---

James T. Griffin to Cyrus H. McCormick 28.5. 1863. Mss 1a, box 53. Griffin estimates production costs in Prussia and calculates that timber was cheaper in Chicago but labor and iron about a third cheaper in Prussia and coal even half the price that it was in Chicago. Ibid. *James T. Griffin to Cyrus H. McCormick* 14.7.1864. Mss 1a, box 54. Griffin had in 1866 36 agents in Europe, of whom ten were in Germany. In Russia he had agents in Moscow, in St. Petersburg and in Odessa. *Statement of Reaping Machines sold by Mr. Griffin of London.* 1a, box 55.
Hungary was so promising that Griffin even suggested after a dispute with Burgess & Key that production could have been moved to Budapest. *James T. Griffin to C.H. McCormick* 17.3.1866 and 19.1.1867. Mss 1a, box 55.

13 *Blaich* 1984, 72.

14 Griffin had a very ambitious marketing program too. In 1863 he published sales catalogs in his own name both in English and in French. The cover showed the Company's latest reaper, with an automatic sheaf delivery attachment, and the inner pages proclaimed its best abilities in work. The catalog also reveals all the trials where McCormick's reaper was exhibited during 1862, and carefully described all the changes in construction and their benefits. *McCormick's reaping and mowing machine. Catalog 1863.* Mss 5x, box 1. Besides, McCormick's reaper was exhibited at Turin and Cuneo in Italy, Budapest and Vienna, Grignon in France, Gembloux in Belgium, Moscow and St. Petersburg in Russia, Stanford-le-Hope, Hemel Hempsted, and Preston in England and Phantassie, Sterling, Berwickshire and Dunkeld in Scotland. Ibid.
McDonald & Co. suggested that they could only be importers and that selling efforts had to be left to the agents. This way, the Company would take charge of the machine stock and then supply all the agents. *James R. McDonald & Co. to James T. Griffin.* 26.10.1864. Mss 2a, box 25.

did not develop any organized system.[15]

But there were still other problems to come. For years, agents had praised the quality of McCormick's reaper, but it was actually too heavy: a severe drawback in competition with Wood's and Samuelsson's lighter machines. McCormick's success in the agricultural shows began to slow down, too. At the Plymouth Agricultural Fair, Walter A. Wood, Samuelson and Hornsby totally beat McCormick. 1865 also proved how closely interrelated were good weather and large sales of reapers. Griffin reported in August that the harvest in Germany had failed and his agents had suffered great losses; and in September he had to announce the total failure of the European harvest.[16]

McCormick's confrontations with Burgess & Key were continual. Burgess & Key had developed their own mower and reaper models, and would have preferred their own production. Nevertheless, they were bound by agreement to make only McCormick's reapers. All these plans collapsed the next year. In spite of all their promises, Burgess & Key failed to push their business and sales effectively enough, and finally the company went into bankruptcy. The receivers for the bankrupt estate agreed on the continuation of business, but were reluctant to continue production of McCormick's reapers. Burgess & Key's mower was selling well in England, while demand for the McCormick reaper had sunk to 40 machines.[17] The company's bankruptcy was a minor catastrophe for McCormick. James T. Griffin made a contract for the unsold machines of Burgess & Key without McCormick's concession. Griffin and McCormick had also discussed Griffin's salary, which was tied to the volume of sales;

---

15 On these machines Burgess & Key earned four pounds royalty per machine. The machines had no mowing attachment. Griffin also suggested changes in his own terms: he proposed that his salary be raised to 300 pounds a year, plus one pound for each machine sold. The licensee would have paid his travelling expenses if he was not working directly for McCormick. *James T. Griffin to C.H. McCormick.* Mss 1a, box 55. In 1865 Griffin appointed new agents in Leipzig, Oschusleben, Braunschweig and Hannover; in France, on the other hand, he had made no active efforts, and Russia offered also a severe setback. Twenty-eight machines in Moscow had to be returned unsold to London. McCormick's reapers were too expensive and heavy. Besides, there were no resources for such investments in Russia. Griffin's situation darkened further, when during 1865 two of his reliable Central European agents died. Griffin estimated his sales to reach at most 200 machines. Ibid.

16 *J.T. Griffin to C.H. McCormick* 21.1.1865; 11.3.1865; 30.5.1865; 29.6.1865; 8.7.1865; 21.7.1865; 9.8.1865; 9.9.1865. Mss 1a, box 55.

17 In 1865, Griffin made a proposal for a new license, in which McCormick would have withdrawn from the trade and Burgess & Key would have exported all machines to the continent excluding France. The Receivers for Burgess & Key's bankruptcy demanded that McCormick pay in cash for all the machines that the company had made for the next year. Griffin asked McCormick to send 1000 pounds for the payment of about fifty machines. *J.T. Griffin to C.H. McCormick* 6.4.1867. Mss 1a, box 55.

as a result McCormick fired Griffin, but finally the parties agreed that before his final departure Griffin would put McCormick's business in order in Europe.[18]

News of Burgess & Key's bankruptcy and of Griffin's firing aroused serious concern among McCormick's agents. They wanted to know if McCormick intended to continue business, and expressed their preference to buy original American McCormicks rather than Burgess & Key's copies. Consequently, McCormick's sales faded out in Central Europe. In England there was still demand, but because of the ongoing dispute with Burgess & Key's bankrupt estate, the company was not able to supply enough machines.[19]

Disputes with Burgess & Key were not McCormick's only problems. In spite of his reputation Cyrus had to work hard for every single contract. He was not the only reaper manufacturer in Europe, not even the only American reaper manufacturer. Obed Hussey had also made an appearance, though unhappily, at the Crystal Palace World Fair. Although he had little success at the Fair, Hussey managed to make contracts with Garret, William Dray and William Crosskill, and by 1852 claimed to have sold 1500 machines. William Crosskill also made Patrick Bell's reapers, however, and competition grew in 1853 when Manny and Atkins from the U.S. brought their machines to Europe as well. In 1858 Walter A. Wood, who at about that time began production in America, sent his agent to Europe with fifty reapers; the next year the volume increased to two hundred and fifty, and from then on Wood sold yearly over one thousand machines. It is, however, difficult to obtain exact sales figures for any other manufacturers. Moreover, the English companies of Samuelson and Hornsby were powerful competitors not only in their home fields.

It has been difficult to draw an overall picture of the competition in Europe during the first two decades. From 1873, however, we know that at least Warder, Mitchell & Co., Johnston, the Austrian Hoffer's, the American Kirby's, Burdick, Buckeye, The Little Champion of Wisconsin and The Little Champion of Germany and Bradley's were being sold in Europe in addition to those mentioned above.[20] Hutchison estimates that by 1859 Burgess & Key had sold up to about 2000 reapers and states that in 1861 it was still the

---

18 *J.T. Griffin to C.H. McCormick* 24.10.1866, 29.10.1866, 19.1.1867, ?.2.1868. Mss 1a, box 55; *Hutchinson* 1935, 436-438.

19 *Moritz & Joseph Friedlander to C.H. McCormick* 21.1.1867. Mss 1a, box 55; *J.T. Griffin to C.H. McCormick* 5.8.1867 and 20.8.1867. Mss 1a, box 55.

20 *Warder, Mitchell & Co. Catalog* 1873. Mss 4z, box 24; *Bradley Mfg. Co.* Catalog 1873. Mss 4z, box 2; *Adriance, Platt & Co.* Catalog 1875. Mss 4z, box 1.

most popular machine in Europe.[21] If Hutchinson's information is reliable, it is very unlikely that Hussey or even Wood had been really as successful as they had claimed.

American reaper manufacturers continued their rivalry on every possible level in Europe, with all the significant companies taking part. The patent war was also moved to Europe. American companies had to protect their machines both against their American competitors and against illegal European copying. In 1855, J.H. Manny had already taken out patents in Europe. The patent office of Robertson, Brooman & Co. took care of Cyrus McCormick's patent claims in most European countries. In 1868 McCormick held patents at least in England, in France, in Italy, in Austria, in Belgium and in some German states, such as Hamburg.[22] The real value of the patents proved, nevertheless, insignificant. Cyrus McCormick's wife, Nettie Fowler McCormick, noted nearly twenty years later that over more than twenty-five years the Company had spent considerable sums on patents, but these had not protected it from patent infringements. Besides, the Company's European competitors were so irrelevant that she recommended abandoning the remaining patents.[23]

4.2. The European trade fades away

Burgess & Key's bankruptcy and the firing of his salaried European agent James T. Griffin in 1867 were severe setbacks for McCormick's business in Europe. In the aftermath of the Burgess & Key fiasco, Griffin tried to arrange the production of reapers with its successor, Burgess & Co., and with some German manufacturers and one in Hungary.[24] The old models, nevertheless, he was obliged to sell at

---

21 *Hutchinson* 1930, 392, 399-401, 412-413; *Heikkonen* 1989, 144-145; *McCormick* 1931, 55-57; The Eight Census. *Manufacturers of the U.S* 1860, 1865, ccxii; *The Farm Implement News* 21.3.1892, vol. XIII no.3, 19.

22 Patent laws varied from country to country. The German states were especially problematic for foreign companies; for example in 1868 the Patent Office of Prussia stated that they were unable to continue McCormick's patents because the patented device was already in common use. In France annual patent costs were ten pounds and in England renewal of a patent cost 100 pounds. *Robertson, Brooman & Co. to J.T. Griffin* 18.1.1868, 29.9.1868, 2.11.1868, 12.11.1868. Mss 1a, box 56; *Berlin Patent Agent to J.T. Griffin* 17.1.1868. Mss 1a, box 56.

23 *Nettie McCormick to C.H. McCormick Jr.* 13.10.1883. Mss 1b, box 20.

24 Burgess Co. offered machines at a price not to exceed 16 pounds but Griffin had serious doubts whether they would make McCormicks even without the royalty and certainly not with the royalty. He was likewise skeptical about the next summer's prospects. Griffin estimated that the trade should be at least 500 pounds plus expenses

extremely low prices. McCormick's reapers, which normally sold for ca. 30 to 40 pounds sterling, were disposed of for little over five pounds.[25]

Cyrus McCormick did not want to give up, despite all the difficulties that had filed up before him in Europe. He personally took part in the 1867 World Fair in Paris and once again achieved the highest honors. Together with Walter A. Wood, he won the gold medal of the Fair and was appointed a Chevalier in the Legion of Honor. McCormick crowned his triumph with a contract for two reapers with the Emperor of France, Napoleon III. News from the Paris World Fair spread rapidly in Europe. Information was spread at least in France by advertisements in newspapers and agricultural magazines. Newspapers were, however, not always sources of neutral and impartial knowledge. A. Albaret, who was McCormick's manufacturer for the French market, promised to arrange articles by two journalists "in their papers in praise of your (McCormick's) machine".[26]

All these marvelous achievements, nevertheless, failed to save McCormick's situation in Europe. He continued to manufacture most of the machines sold in Europe at Burgess & Co., and Albaret & Cie. manufactured reapers for the French trade. In Germany he appointed James R. McDonald as his main agent in 1868.[27]

---

but the continental sales would not be enough alone. *James T. Griffin to C.H. McCormick* 16.1.1868. Mss 1a, box 56. By April, Griffin, fired by McCormick, was back in America again and with some bitterness wrote to the superintendent of the McCormick Co., Mr. Spring, who probably was his old friend, expressing his wish to see Leander McCormick. Furthermore he predicted that after his departure "all that he (Cyrus McCormick) has abroad will go to the dogs". *J.T. Griffin to Mr. Spring* 6.4.1868. Mss 1a, box 56. Griffin mentioned in his letter that he had a good offer to return to England; this warning was realized when Griffin moved to Walter A. Wood's camp and became Wood's main agent in Europe. *J.R. McDonald to C.H. McCormick* 5.4.1872. Mss 1a, box 56.

25 *J.T. Griffin to C.H. McCormick* 1.2.1868. Mss 1a, box 56; *Sales account of J.T. Griffin*, 1864. 1a, box 55.

26 *Bericht über die Welt-Austellung* zur Paris im Jahre 1867. Herausegaben durch das K.K. Österreichische Central-Comite. Mss 6x, box 1; *Le Moniteur universel*. 6 Janvier 1868. Mss 6x, box 1; *A.Albaret to C.H. McCormick* 20.4.1868. Mss 1a, box 56; *Heikkonen* 1989, 158.

27 McDonald had already been McCormick's agent since 1864 and knew the systems, and was one of McCormick's most reliable representatives who could announce to Chicago the sale of 63 reapers. This was, however, an insignificant amount overall and besides, could hardly guarantee any profits. He had to buy machines in cash from Burgess & Co. and pay a three pound royalty for each machine to McCormick. This left McDonald only two pounds commission, from which he also had to cover advertising expenses. *James R. McDonald to C.H. McCormick* 21.8.1868. Mss 1a, box 56; McDonald paid 20 pounds for each reaper to Burgess & Co. In addition he had to pay three pounds royalty plus two pounds for freights, making 25 pounds for each machine. McDonald considered it too risky a business; especially if the trade expanded,

Burgess & Key continued to press McCormick. They demanded that McDonald should buy at least 500 reapers for cash, declined to finance advertising or show expenses, and declared themselves free from liability for possible breakages in their machines. Cyrus McCormick had apparently begun to have second thoughts about the European business, but McDonald encouraged him to continue in spite of the modest benefits. Reapers had now been introduced to European farmers, and had become a regognized necessity on the Continent.[28]

McDonald's anticipation of a large trade was not realized. In 1871 he complained about poor sales. In 1869 he had bought 200 reapers, 40 of which had remained unsold. Next year McDonald took only 112 machines, and still the number of unsold reapers increased. He noted the possible effects of the Franco-Prussian War, but the biggest obstacle was the machine itself. The McCormick reaper was too heavy, and for this reason unsuitable for German conditions; besides, it was too expensive. These factors gave McCormick's competitors an advantage, and McDonald had great difficulties to find new agents.[29]

It is difficult to obtain accurate figures for McCormick's foreign trade from the first years of business, since this was totally Cyrus

---

it would bind up too much cash money. *J.R. McDonald to C.H. McCormick* 2.10.1868. Mss 1a, box 56.
According to Hutchinson James R. McDonald was the U.S. consul in Hamburg. *Hutchinson* 1935, 423.

28 J.R. McDonald to C.H. McCormick 29.1.1869. Mss 1a, box 56.

29 *J.R. McDonald to C.H. McCormick* 1.4.1871. Mss 1a, box 56. McDonald & Co. worked hard to increase sales, but had to admit weaknesses in the construction of the McCormick reaper. Besides, Burgess & Key could not produce a machine suitable for Germany, and were more interested in their English trade, where they had their own machine on sale. McDonald & Co. was McCormick's main agent for Germany and Austria-Hungary, but was also selling Walter A. Wood's machines. Sales of Wood's machines increased every year. Wood shipped reapers and mowers directly from America and, most importantly, introduced machines that were in demand. It was therefore no wonder that finally, after many warnings, McDonald declined to do further business with machines made by Burgess & Key. *Burgess & Key to James R. McDonald & Co.* 25.4.1872. Mss 1a, box 56; *James R. McDonald & Co. to C.H. McCormick* 5.4.1872, 2.12.1873, 3.2.1874, 22.4.1874, Mss 1a, box 56; *James R. McDonald to C.H. McCormick* 27.7.1875. Mss 1a, box 58; *McCormick's Getraide-Mähmaschine. Catalog of James R. McDonald & Co.* 1870. Mss 5x, box 2. To increase the fame of his machines Cyrus McCormick decided to take part in the Vienna International Exposition of 1873, and despite many difficulties his "Advance" reaper obtained a gold medal of merit. McCormick also received a bunch of applications from new agents for the representation of his machines. *Vienna Awards. C.H. & L.J. McCormick.* Mss 1a, box 56; *Fourteen Highest Prize Medals Awarded the Champion in Europe for 1873.* Warder, Michell & Co. Mss 4z, bobx 24; *Bradley Mfg. Co.* Catalog 1873. Mss 4z, box 2; G.E. *Illingsworth to C.H. McCormick* 18.6.1870 Mss 1a, box 56; *Gülich & Koeppel to C.H. McCormick* 9.11.1872. Ibid; *Henri Lion to C.H. McCormick* 3.6.1876.

McCormick's own private affair, separate from the company business. Furthermore, a large proportion of the machines were made in Europe, but simultaneously reapers were also shipped from America. Although James T. Griffin was McCormick's general agent, it is not sure that all the machines were in his balances. Normally very reliable, William T. Hutchinson estimated that up to 1859 about 2000 reapers had been sold in Europe and that in 1861 McCormick was still the market leader.[30]

Nevertheless, McCormick lost his market position rapidly during the 1860s. In 1866, 267 reapers were sent for sale and 209 of them were sold. The total value of reaper sales was about 8115 pounds, from which we have to subtract agent charges and commissions.[31] James R. McDonald claimed in 1866 to have sold only 40 reapers, and in 1868 his balance shows sales of 63 reapers of which he remitted three pounds royalty per reaper to McCormick. It is of course possible that some machines were imported directly from America, especially when sales the next year increased to 160 reapers. According to Hutchinson, McDonald sold in 1870-1875 altogether 238 reapers. Because information is very fragmentary and partly difficult to interpret, the researcher has to be satisfied with an approximation for total sales in 1868-1875, which seems to be about 600 reapers.[32] The Statistical Abstract of the U.S. reports the total export of mowers and reapers in 1870 as only 537 machines, and for the next year 3342.[33] McCormick's sales in Europe at the end of the 1860s were so minimal that they offered hardly anything else than some mariginal fame for the inventor.

An additional reason for this dramatic drop can be found in the competitive companies. Cyrus McCormick and Obed Hussey were among the first to open doors for American manufacturers in Europe.[34] During the 1870s, the reaper had made a breakthrough

---

30 *Hutchinson* 1935, 412-413; *Heikkonen* 1989, 151.

31 *Statement of Reaping Machines sold by Mr. Griffin of London 1866.* Mss 1a, box 55. Agents' commission was approximately ten percent, which made 525 pounds for 209 reapers. Agents' charges were 2020 pounds. Griffin's statement also reveals the number and distribution of agents. McCormick had 37 agents, of whom ten were in Germany. The number of agents in Russia is astonishing: eight altogether, in St. Petersburg, Odessa, Moscow and Riga.

32 *James R. McDonald to C.H. McCormick* 31.12.1868 and 31.12.1869. Mss 1a, box 56; *Hutchinson* 1935, 440-441; *Heikkonen* 1989, 153.

33 *Statistical Abstract of the U.S. 1878.* p. 91.

34 Other American harvesting machine companies followed their example, and Walter A. Wood succeeded in conquering the leading position. The 1860 Census maintains that in 1858 Wood sent fifty reapers to Europe, and the next year already one thousand. The latter, however, is evidently a huge overestimate. Nevertheless,

among wealthy farmers in Europe. It had an established and growing, although still relatively small, demand.[35] Increasing demand attracted new manufacturers in the field. At least Johnston's reaper had been exhibited in 1871 in Hungary and, in 1873, at the Vienna International Exposition all the leading American and European reapers fought for the victory[36]. The contest spread rapidly to almost every corner in Europe where grain was grown. Adriance, Platt & Co. competed during 1873 at least in Prussia, Holland, Hungary, Norway, Sweden, Posen and Hannover against Samuelson, Hornsby, Wood, Brigham, Bramlette, Burdick, Kirby, Bickerton, Champion, Howard, Hubbard and Excelsior.[37] Three years later, in 1876, in the seventh general meeting of the Finnish farmers, nine machines were exhibited. What was interesting in this show was the two Swedish companies who exhibited their mowers. Westerås had its Buckeye and Palmcranz its one horse mower. Palmcranz had possibly manufactured its own model but Westerås made its mower very likely under license to the Buckeye line.[38]

------

although this information has to be approached with caution, the fact is that Wood expanded his sales, won important trials and gained reputation and reliability. In 1864 Wood seems to have sold at least 300 machines in Moscow and the following year he sold 250 machines in England to Chuttleworth & Co. Since information on Wood's sales comes from Griffin's letters and from other secondary sources, it is obvious that his trade may have been much larger. By 1868 at the latest, Wood had two branch offices in Europe, in London and Madrid, and two years later a third one, in Paris. Although there were also other American manufacturers competing for the slowly growing European trade, the real contest was between Wood, the English firms Samuelson and Hornsby, and in the 1860s during Griffin's period, also McCormick. *James T. Griffin to C.H. McCormick* 10.12.1864 and 9.9.1865. Mss 1a, box 55. Wood too had problems with his business and a bad harvest put obstacles in every manufacturer's way. Because of the failed harvest, according to Griffin, Shuttleworth was able to sell only 6 reapers. Ibid. *Walter A. Wood. Circular for the Year 1868 and 1870.* Mss 4z, box 24. See for instance *J.T. Griffin to C.H. McCormick* 8.7.1865 and 21.7.1865. Mss 1a, box 55. Once again Griffin complains of the weight of the McCormick reapers and claims that Samuelson had sold over 300 reapers during the season.

35 In Germany Gülich & Koeppels alone announced having arranged sales of 1500 machines. *Gülich & Koeppel to Messiers McCormick Brothers* 9.11.1872. Mss 1a, box 56.

36 *Bradley Mfg. Co. Catalogue 1873.* Mss 4z, box 2. Among the companies were Aultman, Miller & Co., Adriance, Platt & Co., Johnston Harvester Co., Osborne & Co., Sprague Mowing Machine Co., Warder, Mitchell & Co., Walter A. Wood and McCormick from America, Hornsby and Samuelsson from England, and some smaller Austrian and German manufacturers. *Fourteen Highest Prize Medals awarded to the Champion in Europe for 1873.* Warder, Mitchell & Co. 1873. Mss 4z, box 24; *Vienna awards 1873.* C.H. & L.J. McCormick. Mss 1a, box 56.

37 *The Premium Harvester of the World. The Buckeye Mower & Reaper.* No date. Mss 4z, box 1.

38 *Kertomus seitsemännestä yleisestä Suomen Maanviljelys-kokouksesta...1876,* 103-105. The companies were Adriance, Platt & Co, Champion, Hornsby, Palmcranz,

The history of McCormick's European business begins with an energetic push into the markets. Why did he not, in spite of all the warnings and advice from Europe, change the construction of his machines, start shipping machines from Chicago and get rid of the annoying partnership with Burgess & Key? In a word, why did he not fight for and defend his position in Europe with the same vigor as in America?

Fred Carstensen stresses that prestige and boosting of domestic trade were the main reasons for McCormick's foreign enterprise.[39] This explanation seems to be accurate. McCormick very aggressively used his success in European Fairs on the domestic American markets.[40] McCormick took part in almost every major international fair after his triumph in London[41]. Triumphs on foreign fields were utilized on the home markets, but so were his competitors' triumphs.[42] Victories in trials and fairs were used to expand foreign

----

Walter A. Wood, Westerås, Aultman, Miller & Co., William Anson Wood and Johnston. Ibid. *Heikkonen* 1983, 84.
In Germany especially Johnston had managed to make large sales and also Champion was gaining new ground. *James R. McDonald & Co. to C.H. McCormick* 5.4. 1872. Mss 1a, box 56.
But, in spite of the increasing competition, Walter A. Wood was able to hold its position as the leading reaper manufacturer in Europe. It had employed James T. Griffin as its main agent and used hard and agressive sales methods. When, in 1872, other manufacturers raised their prices by ten percent, Wood held his old prices to outdo his competitors. The next year Griffin had his son and two travellers in Europe and intended to sell 1000 machines in Germany and Austria alone. For 1874 he had plans for the shipment of 3000-4000 machines to Hamburg, 800 to Bremen and 800 to Rotterdam. James R. McDonald anticipated a great rush of American companies in Europe and was ready to introduce McCormick's new reaper but not the one made by Burgess & Key. Besides, the machine should be light. *James R. McDonald to C.H. McCormick* 4.6.1873 and 2.12.1873. Mss 1a, box 56. In April 1874 McDonald confirmed this information by reporting that Walter A. Wood had agents who were contracting for up to 1000 machines. Furthermore, he himself had a customer in a small province, who had already ordered 124 implements and expected to go up to 200. Ibid. 3.2.1874 and 22.4.1874. Mss 1a, box 56.

39 *Carstensen* 1984, 109.

40 Atkins, for example, attacked McCormick in his 1856 catalog, and declared McCormick's success in the 1855 Paris exhibition was inflated. Furthermore, Atkins claimed that McCormick had to go abroad to get premiums and continued that "your success... was owing less to the work in the field... than to the ability of your agents in humbugging the committee." *Atkins' Automaton* 1856. Mss 4z, box 1.

41 Besides the large international exhibitions machines were also exhibited in numerous agricultural societies' trials and shows. These local fairs were in many cases more important for concrete sales than those for the large audience. Cyrus McCormick won for example the 1853 and 1862 silver medals of the Royal North Lancashire Agricultural Society and the 1863 gold medal awarded by the French Minister of Agriculture at the agricultural fair in Lille. *McCormick Company. Catalog 1882.* Mss 5x, box 1.

42 John H. Manny advertised in 1855 that his machine had also been patented in Europe, and in 1857-1858 Talcott, Emerson & Co. announced that Manny's reaper

trade, too. For example, the news of McCormick's victory at the Paris World Fair in 1855, where he won the Grand Medal of Honor, was imediately publicized to a larger audience through the press. European farmers were as conservative and as reluctant customers as their American counterparts; they wanted to see the machines in practise, and the machines which worked most effectively reaped the sales.[43]

When Cyrus McCormick had disputes with his English lisencee Burgess & Key, he had to send his machines from Chicago. This put a heavy burden on the factory that still relied on crude and simple production methods. It could not adjust its patterns according to the requirements of the European conditions. The European models were seen as a hindrance to the domestic business. During 1851-1876 McCormick sold about 4 000 machines in Europe while they sold more than 4 000 machines in the state of Iowa, alone, in 1875.[44]

McDonald's complaints about the reaper were illustrative of McCormick's reaper business during the early years in Europe, where he tried to sell the same basic American reaper to European customers without taking local conditions into account. A partial explanation for this stubborn attitude can be found from the expanding domestic markets. Coverage was larger and easier to obtain there than in Europe. Especially if McCormick's share of the U.S. trade was only 5 percent as Olmstead and Rhode have calculated[45], there were good incentives to push on at the home front.

During the civil war, American reaper manufacturers enjoyed prosperous times. In spite of the numerous patent cases nearly all the companies could increase their business. Also McCormick & Brothers increased their output. Cyrus himself found it wiser to move abroad where he stayed until 1864. This way he was able to stay neutral in the war and also to push his European trade. When he returned to the U.S. in 1864, he met the combined resistance of his brothers for his ideas of an even larger turnout than before. Although the company was not able to satisfy the demand, Leander

had won the silver medal at the Paris World Fair in 1855 and the gold medal, also at Paris, the next year. *J.H. Manny*. Catalog 1855. Mss 4z, box 14; *Talcott, Emerson & Co*. Catalog 1857-1858. Mss 4z, box 22.

43 *McCormick* 1931, 56; *James T. Griffin to C.H. McCormick* 3.6.1863, 4.8.1864, 21.7.1865. Mss 1a, box 55.

44 *Kuuse 1974*, 272.

45 *Olmstead-Rhode* 1995, 28.

opposed expansion. Leander's and Cyrus' business ideologies were contradictory and their personal relations deteriorated year after year when Cyrus began to take an active part in the business. After William's death in 1865, Charles Spring followed him as General Manager. Spring began to make the company's sales organization more effective and also advocated a more agressive marketing strategy. Not until the great fire of Chicago destroyed McCormick factory were there any possibilities for changes. As Hounshell has shown, the company, and especially Leander as the Superintendent, were not able to utilize the possibilities that the situation offered, when the new and larger factory was built. Production technology and machinery stayed practically the same as before and did not change until the firing of Leander.[46]

When this background is taken into account, also the decline in the European business and Cyrus's obvious reluctance to spend more energy becomes more understandable. He could not provide all the machines even to meet the American demand and had to rely in Europe on Burgess & Key. The McCormick Company was not able or willing to produce reapers for Europe and so they did not listen to McDonalds' requests for modifications in the models and as a consequence lost the battle in Europe.

Why did Cyrus not then simply retreat from Europe? He could still use his achievements in Europe to promote his sales on the domestic field. Besides, European business did not stress Cyrus economically too heavily. McDonald bought his machines in cash f.o.b. in New York or from Burgess & Key and he earned a royalty for each machine sold. Thus, business was in a way self-sufficient, did not strain too much economically and McCormick's name did not vanish totally from Europe.

Alfred Chandler's remark that although the first great businessmen were brilliant strategists, "their moves were personal responses to new needs and opportunities. They did not plan systematically for the continuing growth of the enterprise".[47] This explanation may resolve some of the questions around the early stages of McCormick's foreign enterprise.

---

46 *Hounshell 1987*, 167-175. Fred Carstensen maintains that after his return in 1865 (should be 1864) Cyrus moved to New York and gave little attention to the reaper business "domestic or foreign". Carstensen 1984, 111. However, he had at that time major patent cases pending and fought for the business in that sector. Besides, as emerges from Hounshell's research, the brothers had continuous disputes over production.

47 *Chandler* 1977, 414.

■ The New Start in Europe, 1878-1894

5.1. Slow recovery with the self-binder

In spite of their strained personal relations, Cyrus and Leander McCormick expanded their business and added new models to their production lines. One of the most important and crucial inventions was the wire-binder. McCormick's made their first experimental machines in 1875 and in 1876, and began large-scale production in 1877.[1] The wire-binder also made it possible for Cyrus McCormick to return to the European field.

In 1875, at the time when his European busines was dying off, Cyrus McCormick appointed a Dane, A.A. Westengaard, as his agent in Denmark. Westengaard came to Denmark with high expectations and began energetically to recruit agents and advertise McCormick's "Advance" reaper and mower. In spite of his efforts, Westengaard in 1877 finally had to admit that establishing a market in bad harvest years and against acute competion was too laborious.[2] A.A. Westen-

---

1 *Miller* 1902, 36-37; *Ardrey* 1894, 74-77; *Rogin* 1931, 110-112.

2 *A.A. Westengaard to C.H. McCormick* 1.2.1876, 18.2.1876, 1.9.1876. Mss 1a, box 66. 17.2.1877. Mss 1a, box 70. 6.4.1877 Mss 2x, box 180; *Agreement of James T. Mason and Cyrus H. McCormick* 16.11.1876. Mss 2a, box 56. In the agreement Rush T. Mason agreed to sell on commission the unsold machines of A.A. Westengaard. Westengaard landed in a highly competitive market in Denmark. In the machine test arranged in 1875 in Tastumsö, there were 9 reapers and 8 combined reapers and mowers tested. Unfortunately the manufacturers or brandnames are not mentioned in the report of the test. Jörgensen 1902, 12-21.
Although the Danish enterprise did not find enough wind under its wings, it gives a vivid picture of the complex conditions of the grass-root business. Westengaard had to arrange the shipment of the machines, to assemble and try them out, and of course take care of advertising. Westengaard started in Denmark an American style marketing program. He published advertisements in the newspapers and agricultural magazines, spread posters in hotels and tried to recruit agents. Furthermore, machines were exhibited in numerous field tests and expositions, in Sweden as well. For the most part, the potential customers came from the ranks of the estate owners, and possibly the larger peasant farmers who bought their machines in partnership. Already at that time it was considered essential for further sales to make customers satisfied. A satisfied buyer was the best advertisement. By experience, Westengaard learned to know how fundamental an element weather was in the harvester business. Equally important was to have the machines shipped as early in the spring as possible in order to assemble them and distribute them to the agents, because most sales were made just before harvest time. *A.A. Westengaard to C.H. McCormick* 1.2.1876, 18.2.1876, 1.3.1876, 1.8.1876, 1.9.1876, 1.11.1876. Mss 1a, box 66; 17.2.1877. Mss 1a, box 70.

gaard was a victim of bad luck and probably of bad management. On the other hand, time also worked against him. He arrived in Denmark during great changes in the agricultural production structure. Grain growing was giving way to dairy production, and as a result demand for reapers diminished while demand for mowers increased.

Simultanously with the Danish enterprise, Cyrus McCormick intended to nominate a new main European agent to promote his business. In 1876 he made an agreement with an Englishman, Rush T. Mason, to represent his machines; he had also received offers from numerous applicants for agencies in Europe.[3] These propositions reveal the fact that in spite of his now reduced stance, customers had not forgotten McCormick's name and reputation, and were eagerly waiting for his new harvester.

In his agreement with McCormick, Rush T. Mason undertook to exhibit the new McCormick "automatic binder" and to take for sale at least ten binders, one hundred and fifty "Advance" reapers and four hundred mowing machines. The terms were those customary in McCormick's foreign trade: machines were to be delivered free on board a steamer at an Atlantic port at sixty days' notice. The contract was made for one year only and during its duration Mason had exclusive sales rights in Europe. Furthermore, Mason agreed to procure patents for the new binder in Britain, Germany and possibly in Hungary, too.[4]

Cyrus McCormick had been in the business for so many years that he knew the meaning of publicity for trade, and sent his new binder to the Liverpool Agricultural Show. His son, Cyrus Jr., and an expert mechanic, F.C. Newell, followed the machines to England. On the fair ground, McCormick's foremost competitors were Walter A. Wood and D.M. Osborne. Preparations for the trial showed that the binder needed to be adjusted for the heavy European grains; furthermore, European farmers, like their American counterparts,

---

3 *Agreement between Rush T. Mason and Cyrus H. McCormick* 16.11.1876. Mss 2a, box 56; *J.W. Snedeker to C.H. and L.J. McCormick* 14, 1876 and *Julius Damus to C.H. McCormick* October 1876. Mss 2x, box 196.

4 *Agreement between Rush T. Mason and Cyrus H. McCormick* 16.11.1876. Mss 2a, box 56. The agent's price for the binder was thirty-six pounds, twenty-one for the "Advance" and thirteen for the mower.
When Cyrus McCormick had made the decision to return to the foreign field, he commenced with great vigor. He sent two of his binders at short notice to his old agent in Hamburg, James R. McDonald, who very cordially declined to do any more business with the McCormicks. He had still sixty-one unsold old reapers in his possession and had decided to sell them by public auction. Walter A. Wood had nominated McDonald as his main agent in Germany, and besides, McDonald anticipated major problems in organizing a totally new network of agencies all over Europe. *James R. McDonald to C.H. McCormick* 6.2.1877 and 7.7.1877. Mss 1a, box 69.

also had objections to wire as a binding material. A sale could also depend on minor details, like the nippers to cut the wire before threshing. Moreover, the Liverpool show confirmed how unreliable the results of these tests were. One of the judges was inclined to McCormick's side, the second to Osborne, and the third's opinions were not known before the trial. It was likewise important to make a good expression on the public. For this reason, F.C. Newell had acquired a pair of iron-grey horses, on which the animals' perspiration would be less visible.[5]

The promising start in Britain was followed by a severe setback when Rush T. Mason suddenly died, just before the Liverpool Show. Cyrus decided to continue preparations for the trial and began to seek a new agent. At the same time his machines made an invasion in New Zealand, when Newell sold fifty machines to Morrow, Basset & Co. and made an equal contract with another dealer from the same country.[6]

Newell, however, met obstacles that he could not overcome. Walter A. Wood was too strong and influential a competitor. He had established his name over many years, and knew the tricks to handle the jurors. Moreover, McCormick's binder needed adjustments for the local conditions. Consequently, the Liverpool Show was a defeat. McCormick had made an agreement with Rush T. Mason, but he also had another contract, with the house of Brown, Shiply & Co. Newell complained that these two firms treated each other jealously, were only interested in the commissions, and would sell only for cash. This had sabotaged the sales and neither firm had made any efforts to promote and make the binder known.[7]

Newell also proposed a scenario of the manner in which future business should be conducted. McCormick should avoid commission agents and sell directly to responsible dealers. Dealers would need exact terms of trade, since they had no time to consult with the Chicago factory. Newell also underlined the importance of personal influence, which was lost with the prevailing system. Without a complete system of agencies like Walter A. Wood and Osborne had, business in Europe would only gather expenses.[8]

---

5 *C.H. McCormick to Messrs. Brown, Shipley & Co.* 22.6.1877. Mss 1a, box 68; *F.C. Newell to C.H. McCormick* 11.7.1877. Mss 1a, box 69; *F.C. Newell to C.H. McCormick* 11.7.1877. Mss 2a, box 56; *Cyrus McCormick Jr. to C.H. McCormick* 13.7.1877. Mss 2b, box 15; *F.C. Newell to C.H. McCormick* 15.7.1877. Mss 1a, box 69.

6 *F.C. Newell to C.H. McCormick* 15.7. and 5.8.1877. Mss 1a, box 69.

7 *F.C. Newell to C.H. McCormick* 20.7.1877. Mss 1x, box 181; 20.7.1877, 5.8.1877, 6.8.1877, 27.8.1877 and 12.9.1877. Mss 2a, box 56.

8 *F.C. Newell to C.H. McCormick* 18.7.1877 and 6.8.1877. Mss 2a, box 56.

Cyrus McCormick probably adopted or at least listened to F.C. Newell's advice, for in the fall he appointed O.S. Gage from Chicago as his agent for Britain, Germany and Austria. Gage agreed to exhibit the binder and other machines at the important Smithfield Cattle Show in December 1877, and to establish agencies in the countries mentioned. In return, McCormick guaranteed the services of an expert to assist him at trials and in setting up the machines. O.S. Gage introduced Waite, Burnell & Co. to Cyrus and recommended making an agreement with them for business in Britain for the harvest of 1878.[9]

In spite of these new agents, the McCormick factory was not in a state to provide Europe with adequate machines. The Superintendent of the factory, F. H. Mathews, informed Cyrus that they were able to make only minor changes in the structure of the machines for Europe; otherwise they would be forced to make totally new models, which would have been too expensive at that stage. Another problem was Leander McCormick, who declined to have

---

One of McCormick's machines was sold at 59 pounds and the other at 60 pounds. McCormick asked James R. McDonald to sell machines in Germany too at 60 pounds. McDonald sold Wood's machines at the same price and received a discount of 27.5 to 30 percent on each machine. Newell knew that Wood had closed a deal at 45 pounds on a lot of fifty machines for Australia and that Osborne was selling at 40 pounds in lots from fifteen to fifty also to Australia. *F.C. Newell to C.H. McCormick* 18.7.1877, 6.8.1877 and 12.9.1877. Mss 2a, box 56; *James R. McDonald to C.H. McCormick* 27.7.1877. Mss 1a, box 69.

9 In the agreement McCormick undertook to furnish machines to the ports of entry in Europe, from which points transportation expenses and commission were to be added to the port of entry price and thus collected from the purchasers. Gage's commission was confirmed at ten percent. In addition, McCormick agreed to pay in advance $3000 as commission and a loan of $2000 for Gage to begin the business. *Agreement between Cyrus H. McCormick and O.S. Gage* 1.12.1877. Mss 1a, box 68. McCormick had already sent an expert to set up machines in summer 1877 and now in his agreement with Gage promised at his own expense to provide such advice. This was the first step towards completion of the Company's foreign organization, although as Fred Carstensen has noted this happened only just before the founding of the International Harvester Co. in 1902. *Carstensen* 1984, 117. *O.S. Gage to C.H. McCormick* 2.2.1878. Mss 1a, box 71; *Waite, Burnell & Co. to Richard McCormick* 11.8.1877. Mss 1a, box 70. Waite, Burnell & Co. had previously sold Walter A. Wood's machine and Buckeyes'. The Company asked in its application for sales rights for France, Italy and Algeria. Ibid. *Points for a contract between Cyrus H. McCormick and Waite, Burnell & Co.* No date. In the draft agreement Waite, Burnell & Co. agreed to buy two hundred binders at a price not exceeding 65 pounds for the customers. Ibid. James R. McDonald, to whom McCormick had sent two of his binders, also expressed his willingness to sell again in larger amounts. This was probably because of the outcome of the previous year's business with the Woods. McDonald also recommended the exhibition the next summer in Hamburg, where people would gather from all over Germany and Scandinavia. *James R. McDonald to C.H. McCormick* 16.1.1878. Mss 1a, box 72. McDonald finally had to withdraw his offer when Walter A. Wood refused permission. *J.R. McDonald to C.H. McCormick* 4.4.1878. Mss 1a, box 72.

anything to do with the European business.[10]

Nevertheless, Cyrus Sr. decided once again to send, at his own expense, an expert to help Gage. Louis Frank was in Britain by May. He informed Cyrus of the severe competition in the United Kingdom and also of the inadequate packing of the machines.[11] Cyrus Jr. followed him in June and was instructed by his father to look after Gage and his undertakings. Cyrus Sr. was not yet quite sure of the European business and its possibilities, for he asked his son to "take into account the low wages of agricultural laborers, the value of the grain saved by the use of the machine, the expense of introduction, and all points in connection with the subject". This comment shows how aware Cyrus was of European market conditions and of the factors affecting them.[12]

When Cyrus Sr. gave instructions to his son, he simultaneously had to fight Leander, who had communicated with other manufacturers proposing to them the use of some of the Company's patents. Leander had also raised the question of the originality of Cyrus's patent and opened a decades-long debate over the origin and invention of the reaper. Leander stated that their father was the real inventor of the reaper.[13] The brothers' antagonism had been obvious for decades and had merely worsened over the years; its impact was also felt in the daily functions of the factory, as well as in the foreign trade, which Leader staunchly opposed.

1878 was crucial for McCormick's future foreign business. Cyrus Jr's travel to and preparations for the Royal Agricultural Society's show in Bristol reveal the state of McCormick's establishment in Europe. The nippers which had caused difficulties the previous year were still missing, nobody had sent invoices to the new agent Waite, Burnell, Huggins & Co., nor to McDonald, there were no price lists for spare parts and no price for binding wire. Furthermore, many

---

10 *F.H. Mathews to C.H. McCormick* 1.12.1877. Mss 3a, box 5.
The role of Nettie McCormick begins to become more and more visible in foreign business during 1878. In her letter to Cyrus Jr., who at the time was in college at Princeton, she expressed her good opinion of Gage. Nettie regretted that they (Nettie and Cyrus Sr.) had been able to send only 15 machines adapted for European conditions, since "the others" (Leander and his son Hall) had already stopped making them. *Nettie McCormick to Cyrus Jr.* 23.5.1878. Mss 1b, box 25 vol. 1.

11 *Louis Frank to C.H. McCormick* 1.6.1878. Mss 2a, box 56.

12 *Cyrus H. McCormick Sr. to Cyrus H. McCormick Jr.* 17.6.1878. Mss 1a, box 72. Cyrus Sr. was eager to know what Gage had done so far to promote business in England and what exactly were the terms of the contract between Gage and Waite, Burnell & Co. There were already questions over the use of the money that McCormick had sent (the aforementioned $3000) and the remaining $2000 that had been promised Gage to cover his travelling expenses.

13 *Cyrus H. McCormick to Leander J. McCormick* 17.6.1878. 1a, box 72.

parts of the machines were missing. This shows lack of adequate control in the factory. It was thus no wonder that Cyrus Jr. recommended his father to appoint a secretary for his foreign trade, and at the end of his letter finally suggested that "there can be no doubt but that this English & foreign trade must be made a regular department, and attended to as regards points mentioned, like any other part of the business."[14] This was the first time that the importance of foreign business was openly expressed. It had cost considerable sums when Cyrus Sr. had not been able to take care of all the nessessary correspondence.

Cyrus Jr.'s letter shows the central position that his mother had acquired in the Company business.[15] Nettie McCormick's firm hand and opinion were decisive when Cyrus McCormick got an offer for the manufacture of his binder from the English company of Samuelson. McCormick's binder had won the gold medal and the Grand Prize at the Bristol Show, and also at the Paris International World Fair. In Paris, Cyrus McCormick was also decorated with the Cross of an Officer of the Legion of Honor.[16] This provoked great interest both among farmers and manufacturers. Nettie McCormick, however, very strongly advised her husband to stay away from all British companies, reminding Cyrus of all the problems that Burgess & Key had caused. Nettie's letter also explains some reasons why the American machines had defeated their European and especially English rivals. According to Nettie, the English companies would merely copy the patterns, change them, make experiments, wait a season or two and lose important time at the moment when the fame of the victories in trials should have been utilized in sales. The English makers were simply too cautious and not ready to take risks.[17]

---

14 *Cyrus Jr. to Cyrus McCormick* 7.8.1878. Mss 1a, box 72.

15 *Cyrus H. McCormick Jr. to C.H. McCormick Sr.* 8.7.1878. Mss 1a, box 72.

16 *Cyrus H. McCormick Jr. to Nettie F. McCormick* 8.8.1878. Mss 2b, box 16. According to Cyrus Jr. McCormick, the binder was the only machine to give a good performance or to cut all the test areas. In The Times there was a large article on the competition for "the gold medal of the Royal Agricultural Society, a distinguished honour seldom awarded". *The Times* 9.8.1878. Mss 2c, box 128. Although judges awarded the gold medal to McCormick, they suggested all manufacturers should replace wire by some other binding material: wire could cause harm to animals and to workers in the threshing and stacking of the sheaves. *Haddingtonshire courier* 6.9.1878. Mss 2c, box 128. *The French Minister of Foreign Affairs to C.H. McCormick* 21.10.1878. Mss 1a, box 72.

17 *Nettie F. McCormick to Cyrus H. McCormick* 14.8.1878, 17.8.1878 and 27.8.1878. Mss 1a, box 72.
In his estimates on the English implement manufacturers, Jewell comes to the same conclusions as Nettie McCormick. They were too slow to react to changes and for example plowmakers produced too many models to make the production profitable and efficient. *Jewell* 1976, 130-131, 135.

The volume of the foreign trade had increased and grown beyond the limits of Cyrus and Nettie McCormick. Nettie therefore also suggested, just as Cyrus Jr. had a month before, to turn over the foreign business to the McCormick company and to give Leander his share of it. Nettie's comments make it clear that McCormick's foreign trade had thus far been Cyrus's own private business, which he and Nettie had shared with the assistance of some company officers such as F.A. Matthews. This was the case especially from 1876 to 1878. Nevertheless, it had caused problems, and required too much of their time. Nettie was sure that the Company would be capable of making the machines for Europe and also of taking care of the administrative side too; otherwise she would be greatly disappointed, especially since they already had patterns made for the European binders. Leander insisted on a definite order either from Cyrus or Waite, Burnell, Huggins & Co., and at the same price as charged in New Zealand. In this way, he added pressure for the salvation of the foreign trade problem. Although Nettie was confident of the factory's ability to manage the European business also, Matthews was stressed over the new machines and demanded exact production numbers well in advance. However, for the first time Company officals began to express their belief in the European trade and its possibilities.[18]

1878 began in Europe very favorably for Cyrus McCormick. He had reaped great victories and fame, and his name was once again in the minds of customers. His sales organization was, nonctheless, undeveloped, and he needed the support of a prominent jobbing house both in Britain and on the Continent. Such agents were rare indeed and he had to be satisfied with a contract with O.S. Gage and Waite, Burnell, Huggins & Co.[19] After the agreement, though

---

18 *Nettie F. McCormick to C.H. McCormick* 27.8.1878. 1a, box 72; *F. H. Matthews to C.H. McCormick* 26.8.1878 and 21.12.1878. Mss 1a, box 72.

19 Cyrus made a contract with O.S. Gage for selling his machines in Europe; Gage for his part had made his own agreement in the name of McCormick with Waite, Burnell, Huggins & Co. This agreement McCormick would not accept, and had long discussions with Gage, who had exceeded his authority by settling the cash price of the binders at forty-five pounds instead of fifty. Finally Cyrus Sr. and Waite, Burnell, Huggins & Co. made an agreement in which the latter agreed to buy both for 1879 and 1880 three hundred binders, with a provision for at least two hundred more. The price of the machines was settled at forty-five pounds in Liverpool, and Waite, Burnell, Huggins & Co. had the right to set the price for customers as high as they saw fit. Their firm had exclusive selling rights for Britain, Ireland, France, Spain, Italy and Algeria for two years. Waite, Burnell, Huggins & Co. had to pay for the machines within four months from the arrival of the ship in harbor. Furthermore Cyrus had to promise that the new agreement would not interfere in any way with the agreement made between Gage and Waite, Burnell, Huggins & Co. *C.H. McCormick to Waite, Burnell, Huggins & Co.* 21.9.1878. Mss. 1a, box 72; *Notes for an agreement* between C.H. McCormick and Waite, Burnell, Huggins & Co. No date. Mss 1a, box 72; *F. A. McCormick to Mr. Samuelson.* No date. Mss 1a, box 72; *C.H. McCormick to Waite,*

sick and tired, Cyrus McCormick was satisfied with the state of affairs in Europe. Future business looked bright, he already had a contract for a large number of machines, and prospects for even larger sales were good. Although everything was developing promisingly, the greatest obstacle was at home. Nettie was still afraid that Leander would not allow the machines to be sent to Europe.[20]

In spite of all the good prospects and work done in Europe, once again everything was fading away. Gage and Waite, Burnell, Huggins & Co. could not settle their dispute over Gage's commission. Besides, Matthews had sent 110 machines to England and 70 to France on terms of four months' payment, but in July Waite & Co. announced their inability to fulfil their responsibilities. Furthermore they wanted to revise articles of the contract to carry on the business in Europe.[21]

Burnell, Huggins & Co. 4.10.1878 and 15.10.1878. Mss 1a, box 72; *Agreement between C.H. McCormick and Waite, Burnell, Huggins & Co.* 30.11.1878. Mss 1a, box 56.
Waite, Burnell, Huggins & Co. had printed for 1878 a flyer announcing them to be McCormick's agents for Great Britain, Ireland and the Colonies. The price of the machine was set at sixty-five pounds and the firm promised expert help from the manufacturer free of cost to put up and start machines. *Waite, Burnell, Huggins & Co.* 1878. Mss 4z, box 22.

20 *Nettie F. McCormick to Cyrus* Jr. 1.10.1878. Mss 1b, box 25.

21 *Nettie F. McCormick to Cyrus Jr.* Jan. 1879. Mss 1b, box 25, vol.1; *F. H. Matthews to C.H. McCormick* 30.5.1879. Mss 1a, box 74; *Waite, Burnell, Huggins & Co. to C.H. McCormick* 22.7.1879 and 2.10.1879. Mss 2a, box 57.
Shipments of machines were made either with the Anchor Line or Inman Line, both of which also had agents in Chicago for foreign freights. There was some variation in freight costs. Through Inman Line the ten first machines went to Liverpool at 17 shillings and 6 pence per ton or 40 cubic feet. London shipments cost 22 shillings and 6 pence per ton and then the freight to Liverpool rose to 20 shillings per ton. Matthews calculated that on average freight per machine was about $12 from New York to England, five percent primage being added to all freights mentioned above. Freights from Chicago to New York were $5 per machine or $50 for a car load. *F.H. Matthews to C.H. McCormick* 30.5.1879. Mss 1a, box 74.
As a reason for their unsatisfactory achievement Waite & Co. pointed to the general economic depression both in Britain and in France and the rainy season that had spoiled harvests. Their information is probably not quite sincere, however, for private information reveals some misappropriations on the side of Waite & Co. The firm had for instance charged ten dollars for packing cases which McCormick had sent free, for the shears they charged one dollar, and five pounds for the extra parts that accompanied every machine. In addition, Davies complained of the incompetency of the firm's travellers. Furthermore Waite & Co. had used McCormick's old stock in Europe as a source for extra parts. All this had raised bad feelings among farmers, many of whom had canceled their orders after having heard of the state of affairs and of the business manners of Waite, Burnell, Huggins & Co. *James Davies to C.H. McCormick* 11.10.1879. Mss 2a, box 57.
When it became evident that Waite, Burnell, Huggins & Co. could not fulfil their liabilities, Cyrus McCormick at first limited Waite & Co.'s trading area to cover only Britain, Ireland, France, Italy, Spain and Algeria; the next year, after the firm had gone into liquidation, he had to find new agents and resolve the problem of presentation in Europe. *C.H. McCormick to Waite, Burnell, Huggins & Co.* 12.11.1879. Mss 2a, box 57. Before the bankruptcy McCormick and Waite, Burnell, Huggins & Co. made an agreement on the Gage case by admitting to Gage 30 percent

McCormick's sales methods through an independent jobbing house were not effective enough. His competitors had their own representatives and offices in Europe in direct and close touch with customers and their wishes. McCormick's, on the contrary, continued its cautious policy. It was not ready to expand in Europe, which was at the moment only a minor market compared with America.

The growing importance of overseas business, on the other hand, is shown by problems caused to the Company in that field. F.A. Matthews wrote a warning letter to Cyrus McCormick in which he informed the inventor that the Company had needed to borrow much more money than he had anticipated because of the falling-off of foreign trade.[22]

In 1877, Cyrus tried to persuade Leander to take a positive attitude toward the foreign business, by offering him a share of the profits. Furthermore, a separate office was established to administer foreign trade. Before anything permanent could be settled, Cyrus had to resolve his controversy with Leander. Leander McCormick had been connected with the manufacturing of machines since 1835 and been in charge of the manufacturing operations since 1859. His experience was, however, from the time of blacksmith shops where simple machines were hammered together. Cyrus and Leander had had continued disputes over the volume of turnout, which Leander did not want to increase. Leander McCormick was a blacksmith who was interested in the machines, and did not want to take any risks. He could not understand his brother's more managerial approach and demands for ever-increasing production. In the great fire of Chicago, which also destroyed the McCormick works, Leander lost the chance to build a technically and organizationally modern factory.[23]

Cyrus wanted to stabilize the situation, and suggested the formation of a joint-stock company. Finally, after long discussions, Cyrus, Leander and his son Hall agreed to close down the old firm

---

of the sales price exceeding forty-five pounds. Furthermore McCormick agreed to supply his machines to Waite & Co. up to July 1st 1880. *Draft for an agreement between C.H. McCormick and Waite, Burnell, Huggins & Co.* 1879. Mss 1a, box 74.

22 In the older research this side of the role of the foreign trade has been omitted. See for example Kuuse 1974, footnote 170, page 272 and table 60, page 273 and Carstensen 1984, page 114. The Company had borrowed so far $30 000 but older loans from Connecticut Mutual were falling due for $100 000, $150 000 and $100 000. *F.A. Matthews to C.H. McCormick* 30.5.1879. Mss 1a, box 74.

23 Charles Colahan, a patent expert whom Cyrus had hired in 1877, reported several defects in the factory organization and in the quality of the products. *Officials of the McCormick Harvesting Machine Company.* Mss Special Reports File, box 14; *Hounshell* 1987, 172-176.

of C.H. & L.J. McCormick on August 1st 1879, and on August 11th 1879 agreed on forming a joint-stock company, which was ratified next month at a meeting of the Board of Directors. The new company was called the McCormick Harvesting Machine Company.[24]

The formation of the new Company did not resolve any of the old problems. Cyrus and Leander could not agree on the terms of the foreign trade. Leander now wanted to have his share of the business, but Cyrus on the other hand resisted this, as he saw it to be totally his own private business. At the height of the controversy Leander dictated in the company records a clause in which he objected to the establishing of foreign agencies and the selling of machines abroad. He stated that such activities were beyond the activities of the Company.[25] Leander did not have any interest in the company either and hardly spent any time at the factory. Company officials began in increasing measure to show their discontent with the state of affairs, and even demanded the dismissal of Leander and Hall. Finally, in February 1880, Cyrus left a statement to the Board of Directors about the poor management of the factory, a statement which led to the resignation of both Leander and Hall.[26]

Organizationally and technically one of the most important events in the history of the McCormick Company was the hiring of Lewis Wilkinson as the new Superintendent. He was an experienced

---

24 Three-fourths of the property was to be owned by Cyrus and one fourth was in the hands of Leander and Hall. Its capital stock was two million five hundred thousand dollars. The Board of Directors consisted of six persons, of whom Leander and Hall could apppoint two. Leander McCormick was appointed as Superintendent of the manufacturing department, and as Vice President of the Company. His son Hall was appointed as an assistant superintendent in the same department as his father. Although Hall was nominally appointed in a key position in the Company, Cyrus Jr. had also long been involved in the business, and company officials regarded him as the successor of Cyrus Sr.. This idea can also be found in the correspondence of Nettie and Cyrus McCormick. *Agreement between Cyrus H. McCormick, Leander J. McCormick and R. Hall McCormick 1.8.1789.* Mss M/I, box 2; *By-laws of the McCormick Harvesting Machine Company.* Mss 1a, box 74; *Meeting of board of the directors 9.9.1879.* Mss M/I box 3. In this meeting the purchase of property from C.H. & L.J. McCormick by the McCormick Harvesting Machine Company was approved. *Nettie McCormick to C.H. McCormick 22.8.1878.* Mss 1a, box 72; *W.J. Hanna to Cyrus Jr. 8.4.1879.* Mss 1a, box 1879.

25 *Minutes of the board of directors 13.2.1880.* Mss M/I box 3.

26 As an example of the state of affairs between Leander and Hall McCormick on the one hand, and factory officials on the other, can be mentioned a quarrel with Charles Colahan over his journey to New York on patent affairs. Matthews, who was in charge of correspondence and was left by Cyrus in charge of the business, complained that Hall is "the most outrageous fool" and wanted him to be displaced from the business "to which he is only a drag & a hindrance". Leander and Hall had threatened to establish a factory of their own. Furthermore, Matthews had to write to Cyrus in the evening from home because Leander and Hall wanted to know everything about the correspondence between Cyrus and the factory. *F.H. Matthews to Cyrus H. McCormick 14.3.1879.* Mss 1a, box 74. *Directors meeting 6.4.1880.* Mss M/I, box 17; *Hounshell 1987, 178.*

engineer who had worked at the Colt Armory, at the Connecticut Firearms Company and at the Wilson Sewing Machine Company. He brought with him the new system of manufacturing developed and used in the New England factories. His arrival meant also an end to the old blacksmith system. Production was for the first time organized in a practical fashion, new special-purpose machines were bought and the factory began to use nightshifts.[27]

5.2. From private to corporate enterprise

Cyrus McCormick had to admit once again how difficult it was to win a place in the sun. On the home field he had fought a continuous battle against Leander to increase the Company's production. The McCormick Company had established a working sales organization of its own throughout America, but in Europe it was using jobbing houses. For some reason Cyrus had kept foreign trade as his own private business, to which he unfortunately could not devote enough time.

During the crisis of Waite, Burnell, Huggins & Co., two of its older employees, George Glough and Percy Lankester, suggested important changes to the way the business should be performed. They noticed that although the new harvester had now been introduced to farmers, there were still crucial drawbacks. Firstly, there was no control and acquaintance of the business, and secondly, there was no contact between the manufacturer and the consumer. Too many hands between factory and farmer increased the price of the machine. As a consequence, Glough recommended that McCormick open in London his own business house, which could also serve as a central depot. The manager should have a fixed salary. Furthermore, although the wire-binder had attracted wide publicity, few farmers could afford to buy one; they needed a light, strong mower or a combined machine.[28]

Waite, Burnell, Huggin & Co. signed a contract with an agent for the representation of McCormick's machines in Russia. He, however, soon found it impossible to make profitable trade when he had to pay in cash sixty-five pounds for a binder, plus freight and packing.

---

27 *Extract from the minutes of Board of Directors.* 4.5.1880. Wilkinson's salary was set at $3000 p.a.; *Hounshell* 1987, 178-180. Wilkinson's salary has to be regarded as very low when his position and importance to the company are kept in mind. It can be compared with the $4000 per year at which sum the Moline Plow Company made a contract with its first branch manager in 1878. *Thomas* 1976, 41.

28 *George Glough to C.H.McCormick* 9.1.1880 and *Percy Lankester to C.H. McCormick* 9.1.1880. Mss 1a, box 76.

These charges raised the retail price up to eighty pounds, which was unthinkable in Russia, even for estate holders. Perrin could not meet the clauses of his agreement and began negotiations for a direct agency with McCormick.[29]

The appointment of Perrin as a salaried agent also meant a clear shift from McCormick's earlier conduct of business. Now, instead of a jobber who thought only of his own benefit, he had an agent of his own who should take care only of the Company's interests. A salaried agent was actually a form of direct investment in the sense defined by Mira Wilkins, and could be seen as a first step on the way to representation abroad.

Russia was a large area to cover for an inexperienced man, and so the Company decided to send George A. Freudenreich, a practised agent from the Red River Valley, to take charge of the Russian business. Perrin was to work in the future under his control. Freudenreich's territory was also extended to include Rumania, where W. Staadecker had already represented McCormick for a year. The McCormick Co. assured Staadecker that Freudenreich had full power of attorney and negotiations concluded with him would have the same effect as if made directly with the Company. Behind Freudenreich's nomination, Cyrus Jr. was clearly taking more and more charge of affairs in the Company.[30]

---

29 *Waite, Burnell, Huggins & Co. to A.V. Perrin* 10.6.1879. Mss 2a, box 57; *A.V. Perrin to C.H. McCormick* ? 1879. Mss 2a, box 57. Perrin's contract with Waite & Co. reveals that Waite & Co. took twenty pounds profit per machine and furthermore charged agents for freights and even for packing the machines although McCormick normally took care of packing in Chicago. Machines and extras had to be paid in advance in London. Perrin also informed McCormick of the big profits that the other harvester companies were reaping in Russia: at least Johnston, Wood, Shuttleworth, Samuelson and Osborne were represented there by their own men.

30 *McCormick Co. to Emil Liphart & Co.* 7.1.1880, *McCormick Co. to W. Staadecker* 8.1.1880 and 17.1.1880. Mss 1x, LPCB 455.
Before McCormick Co. approved Staadecker as its agent it examined his background very carefully. Company officals sent letters to the American consul in Bucarest, Simon J. Sina and M. Hofher in Vienna and to R. Hornsby & Sons in England. 7.11.1879. Mss 1x, LPCB 455. This practise was also continued on numerous other occasions and displays the Company's stricter control over affairs. Foreign trade was no longer private business, but part of regulated corporate management. *A.V. Perrin to Cyrus Jr.* Mss 1a, box 77.
The McCormick Company simultaneously continued negotiations over representation in Britain with Waite, Burnell, Huggins & Co., who were seriously indebted to the Company. After a long correspondence, the McCormick Company finally decided to terminate the contract. The junior partner in Waite & Co., George Glough, had visited Cyrus McCormick in New York and later offered his own services to the old harvester king; when McCormick's contract with Waite & Co. expired, Waite's old employee Percy Lankester took their place.*Percy Lankester probably to C.H. McCormick* 9.1.1880 and *George Glough probably to C.H. McCormick* 9.1.1880. Mss 1a, box 76; *C.H. McCormick by Cyrus Jr. to Clarkes, Rawlins & Clarke* 1.2.1880. Mss 1x, LPCB 455 and 10.2.1880 Mss 1a, box 76; C.H. *McCormick by Cyrus Jr. to Albaret & Cie* 29.6.1880. Mss 1x, LPCB 455. From all of these letters there can be seen the constantly increasing touch of Cyrus Jr.'s fingers in the business. Cyrus Sr.'s sickness began to

At this point, the McCormick Company planned a total change in its European marketing structure. The concept was to concentrate the European business in the hands of George Freudenreich, in Odessa in Russia. Although McCormick now had two salaried men in Russia, the base structure of the business remained unchanged. Machines were delivered only to reliable dealers f.o.b in New York; the Company did not want to bind its hands with "unsatisfactory credits in a country so far away".[31]

The McCormick Company's business in Europe at the end of 1880 once again looked brighter after new arrangements were made both at home and in Europe. The Company did not want to enlarge its activities too fast. It had to reject suggestions for representation of its machines from numerous candidates.[32] Behind the Company's somewhat conservative policy can be detected Nettie McCormick's influence. On the other, she understood the vast possibilities which Europe offered and complained about the miserable state of business.[33]

---

worsen and although he keenly followed business matters, the daily conduct of it was transferred to company officials and to his son Cyrus Jr. In his letter to Albaret, Cyrus McCormick offered his reaper for tests and implicitly also for sale by Albaret, who had previously also made McCormick reapers; *George Glough and Percy Lankester probably to C.H. McCormick 22.7.1880. Mss 1a, box 76.* The McCormick wire-binder had met resistance from other makers and also from customers, who wanted to have the more comfortable twine-binder. In summer 1880 there were already about half a dozen twine-binders competing for the European trade. *Ibid*; *Percy Lankester & Co. to C.H. McCormick 16.12.1880. Mss 1a, box 76.*

31 Two salaried agents in the same country did not work well. Although Freudenreich was the manager of the business, Perrin, as a native Russian, presented himself to the subagents as the main agent. Freudenreich complained over Perrin's extravagancy, but nevertheless suggested renewal of his contract. Freudenreich was not too confident over the future of the trade in Russia. The country was so undeveloped that he could not expect any large trade with binders in the near future and so the McCormick company had to rely on trade with its other machines. *McCormick Co. to Geo. A. Freudenreich 31.5.1880 and 1.7.1880. Mss 1x, LPCB 455; Geo. A. Freudenreich to McCormick Co. 1.8.1880 and 22.9.1880. Mss 1a, box 76.* Freudenreich wrote the Company to send for the next year 525 machines of which four hundred were mowers and only seventy-five were various kinds of binders. *Geo. A. Freudenreich to McCormick Co. 22.9.1880. Mss 1a, box 76.* Perrin saw the situation in a totally different light. He anticipated large sales of binders, but sales needed hard work and it would take a couple of years before large trade would materialize. *A.V. Perrin to C.H. McCormick 8.10.1880. Mss 1a, box 77.* Perrin was appointed as George Freudenreich's general traveling agent for two years at the same salary as before. His powers to make contracts and handle money were, howerever, limited. *C.H. McCormick to A.V. Perrin 24.12.1880. Mss 1x, LPCB 455; A concept for agreement between Geo. A. Freudenreich and A.V. Perrin. Mss 1a, box 77.*

32 *See for example Cyrus Jr. to H. Pizzolli in Italy. 14.5.1880. Mss 1x, LPCB 455.* All what Cyrus Jr. was ready to offer was cash trade f.o.b. in New York.

33 She could not approve of the plan to give the general agency for Europe to George Freudenreich; she considered that it was too large an area for one man and certainly too large for Freudenreich. Nettie had serious doubts about his talents, commenting

A harvester with a wire-binding mechanism was a big step forward, and generated new sales in America and opened a new start in Europe for McCormick. Wire as binding material had, however, defects which did not make it so popular as had been expected. Farmers were afraid that their animals might get pieces of wire in their stomachs and millers expressed their dislike as well. Already in 1879 some experimental twine-binder models had been exhibited at agricultural shows in Europe. In America the next year was a period of intensive experimentation and development of various binding mechanisms: the continuation of business was dependent on it. The McCormick Company, too, spent much time and energy to develop the best possible knotter. Cyrus Jr. was actively engaged in the process and regularly sent reports to his father.[34]

The McCormick Co. had solved the mechanical problems of the twine-binder, and its markets were promising in America. In Europe, the Company's situation at the beginning of 1881 was still unsettled. After the bankruptcy of its agent, Percy Lankester had marketed the machines in Britain; nevertheless, Lankester was not McCormick's agent: the Company denied any connection with him and tried to find a new agent for Britain. In the same manner, the McCormick Company tried to find agents for France, too.[35] In Russia a natural solution to the problems in Russia was found when the Russian

---

that "F. is not a great man" *Nettie F. McCormick to Cyrus Jr.* Monday 1880 (no other date available). Mss 1b, box 25, vol. 1. Nettie's letter was confidential and she asked her son to destroy it. The letter also reflects both how closely Nettie followed business and also her close relations to her son.

34 In its circulars to agents McCormick Co. in spring 1880 affirmed that the twine-binder was only something which firms that had not succeeded with wire-binders had began to propagate. After these excuses it gradually had to admit that it was experimenting with twine, too and then asked every agent to keep his eyes open and carefully examine every twine-binder they could find and report to the factory how they were constructed. In June, Cyrus Jr. examined the work of Wood and Marsh harvesters and in August was anxiously experimenting with three different binding mechanisms. In August 1880, Cyrus Jr. was already ready to exhibit their first twine-binder in public but was not yet ready to send it to the agents, who for 1881 should still push the sales of wire-binders "to the fullest extent" and twine-binders only in critical cases. The explanation for this procedure was simple: the machine was not yet ready for the market but the pressure of the competitors forced McCormick to put it on sale. *Circular to agents* 20.3., 22.3, 31.5., 18.12.1880. Mss 1x, box 472; *Cyrus Jr. to C.H. McCormick* 16.6., 9.8, 12.8., 16.8.,18.8., 23.8.1880 Mss 1a, box 77

35 *Cyrus Jr. to Louis Franck* 8.1. and 21.4.1881. Mss 1x, LPCB 455. Franck was told to travel to France to see Albaret & Cie. to test their willingness and conditions for an agency and to find out what kinds of mowers and other implements they produced. Franck should also try to find out if Albaret used McCormick's machines only to promote sales of his own mowers. Franck was also asked to contact Burnell & Co. and examine its status and the likelihood of it becoming agents for France.

governement tried to arrest Perrin, who was a Russian naturalized in America. Perrin had to flee the country, and at this moment Cyrus Jr. dissolved the contract with him. For a while, it was left open whether it was nessessary at all to keep a permanent salaried agent in Russia;[36] but Russia was such an important and growing area that the McCormick Co. decided to push on.[37]

The McCormick Co. had hammered together a working model of a twine-binder and decided to exhibit it also in England. The natural venue to do so was once again the Royal Agricultural Society's show, this time arranged at Derby.[38] McCormick's victory in Derby was a severe blow to other companies, but especially to Walter A. Wood, who had invested money and energy in the European trade.[39] Cyrus Jr. understood straight away the full meaning and advertizing value of the achievement. He ordered leading officers of the company to visit editors of the Chicago newspapers, to discuss proper publication of the news. Furthermore, news of the victory was also cabled to the New York papers, and special attention was paid to

---

36 *Cyrus Jr. to Geo. A. Freudenreich* 3.8.1881. Mss 1a, box 80; *Cyrus Jr. to Cyrus H. McCormick* 4.8.1881. Mss 2a, box 57. Cyrus Jr. was not satisfied with Perrin after having heard complaints of his behavior among dealers and how he was constantly exceeding his powers. *Ibid; A.V. Perrin to Cyrus H. McCormick* 6.8., 13.8., 17.8., 20.8. and 26.10.1881. Mss 2a, box 57. Perrin blamed the situation on "jealous individuals" who wanted to destroy his efforts to sell McCormick's machines. Walter A. Wood's agent shared the same fate with him. Both Freudenreich and Company officials handled Perrin very carefully, wanting to avoid open conflict with him, but at the same time made it clear to other parties that Perrin had no longer anything to do with the McCormick Company, even if he might have letters of recommendation from the Company. *Geo. A. Freudenreich to A.V. Perrin* 23.8.1881. Mss 1a, box 79; *Cyrus Jr. to Percy Lankester* 13.12.1881. 1x, LPCB 456.

37 At the end of 1881 McCormick Company had eight agents in Russia: in Moscow and Charkoff (same agent), Simperopol ?, Taganrok, Ekatermoslaff, Rostoff, Poltava, Elizabethgrad and Odessa, plus one agent in Bucarest in Rumania. *McCormick Co. to Cyrus H. McCormick* 15.12.1881. Mss 2a, box 57.

38 The Company decided to send Britain three representatives, including their leading expert, Louis Frank, who had already won a handful of gold medals in Europe. Unfortunately the steamer which was carrying the reapers was shipwrecked, and the machines intended for the show sank along with the ship. When the ship was raised, the binders were rusty and ugly-looking. This situation, nevertheless, McCormick's turned to benefit. The machines were left unpainted and rusty, except that all the bearings were polished and fitted to run as smoothly as possible. Furthermore, Cyrus Jr. ordered Louis Franck to acquire small horses, to make the contrast with other machines as large as possible. As a consequence, the McCormick binder reaped the gold medal. *Cyrus Jr. to W.H. Town* 6.7.1881. Mss 1x, LPCB 455; *C.H. McCormick to R.C. Ransom* 12.7.1881. Mss 1a, box 80; *An Incident Pertaining to the McCormick Reaper by Dr. Gray*. Mss 1a, box 115.

39 *Louis Franck to C.H. McCormick* 11.10.1881. Mss 1a, box 79. For instance, in Russia, the other companies already in 1880 accused of the McCormick Co. of coming to a ready-set table. They had advertised and made the reaper known in Russia and now McCormick intended to take the whole steak. *A.V. Perrin* 26.5.1880. Mss 1a, box 77.

it in the fair circulars.[40]

The triumph in Derby was not simply a victory for the McCormick Company. It was also a great day for Cyrus H. McCormick, whose private business the European trade still remained. For this reason, his name was listed in the advertisements before the Company name.[41] Cyrus McCormick, on his wife's and son's recommendation, had turned the correspondence and practical matters of the foreign business over to the Company; however, the business itself had remained under his control. This probably also explains in some measure the lack of continuity, which appears for instance in the uncertainty of the future of business in Europe. Even in August 1881, Cyrus Jr. still stated that "our desire is not to push the European business any further than Russia and Roumania at present..."[42] The triumph at the Derby Show apparently changed McCormick's mind, for two months later Cyrus made an agreement with Percy Lankester on the exclusive agency for Britain and Ireland;[43] he gave also a positive answer to Waite, Burnell & Co.'s inquiry for an agency in France in the event that the firm could provide proper guarantees.[44]

The fact that the European business was Cyrus's own private affair caused one large and final confrontation between Leander and Cyrus. Leander McCormick began to ask for compensation for the machines sold in Europe in 1878 and 1879. Futhermore, he demanded that the price should be the same as to any outside wholesaler. After month-long discussions, Cyrus agreed to pay for 200 machines which had already been manufactured, but declined to pay for 300 which had only been ordered. Eventually, the dispute was settled at 200 machines, for which Cyrus paid to the Company; Leander's share of the sum was $14,800.[45]

---

40 *Cyrus Jr. to McCormick Co.* 11.8.1881 and to *E.K. Butler* 12.8.1881. Mss 8c, box 11.

41 *Cyrus Jr. to McCormick Co.* 11.8.1881. Mss 8C, box 11.

42 *Cyrus Jr. to Geo.A. Freudenreich* 3.8.1881. Mss 1x, LPCB 456.

43 *Cyrus H. McCormick to Lankester & Co.* 12.10.1881. Mss 1x, LPCB 456. The contract was made for one season only; it comprised 12 iron mowers, 28 harvesters and binders and 26 twinebinder attachments. The machines were to be paid free on board in New York and prices varied for binders from $315 for a five-foot machine to $330 for a six-foot binder. The price of an iron mower was $85. From these prices Cyrus promised a discount of 35 % and from repairs and extras a discount of 33 %

44 *C.H. McCormick by Cyrus Jr. to Waite, Burnell & Co.* 12.12.1881. Mss 1x, LPCB 456. Conditions and prices were the same as for Lankester & Co.

45 *Diary of Cyrus H. McCormick Jr.* 26.1., 10.2., 22.3., 26.4. and 23.11.1882. Mss 4c, box 2; *Conversation between C.H. McCormick, W.J. Hanna and C.H.McCormick Jr.* 14.3.1882. Mss 1a, box 85; *A Statement by C.H. McCormick 20.3.1882* and *A Statement of C.H. McCormick. No date.* Mss. 1a, box 85. According to these State-

In spite of the conflict, the European business was continued. Waite, Burnell & Co. took care of the French sales, but notwithstanding all the guarantees, the French part of the firm failed and the McCormick Company had to turn to Albaret & Cie. and Roche Papillon to take its machines for sale. Percy Lankester, who had been the agent for Britain only, was at first asked to take care of the negotiations in the French question and in 1883 the supervisory power in France was turned over to him. Thereafter all the French transactions went through Lankester, who also had the power to decide which orders should be filled.[46]

The first years in Europe were a long learning process for the McCormick Company. It wasted energy in trying to find agents and in finding the proper ways to manage the trade. In France, for instance, Albaret & Cie., who had previously manufactured

---

ments Cyrus had bought the machines at manufacturing cost and paid all expenses from his own pocket. Expenses came from patent fees for Britain, Germany, New Zealand and Australia. Also advertising, trials, exhibitions and freights had caused considerable costs.

Leander's attack was his last effort against the foreign trade, and was aimed more to gain money and to cause harm to Cyrus than to stop the sales. Although the affair itself was not very serious, the statements and memoranda made reveal important information about the true value and meaning of the European business to the Company. In the use of this documentary material it has to be remembered that it is not unbiased, because of its origin and purpose. According to Cyrus McCormick, the entire European business had achieved economically nothing but expenses, which had been greater than the value of all the machines sold in Europe since 1864. The real reason for such an unprofitable business was the publicity and advertising value which was won through the great victories in Europe. These spread all over the world through the press, and the name of McCormick was made familiar to everybody in the agricultural community. The benefits for the home trade had been enormous. Conversation between...14.3.1882, *A Statement by C.H.McCormick* 20.3.1882 and *A Statement by C.H. McCormick* (no date). Mss 1a, box 85. Cyrus was extremely disappointed with Leander, since both Leander and William McCormick had declined to take any part in Cyrus's European sales. When Cyrus had succeeded in the shows, however, Leander wanted to have the victories in the name of the Company without doing anything for it.

Albeit Cyrus certainly was right in his outburst, it was only one side of the story. Leander had undeniably as good reasons for his behavior. He had managed the production plant for decades, made his own inventions and improvements without any compensation, and had been stressed by the new demands of the foreign trade. The two brothers simply had a totally different kind of approach to the business. Leander was satisfied with the existing state of affairs, while Cyrus saw no limits for his business.

46 Lankester was, nevertheless, informed of the proper ways of trade. Machines should be ordered as early in the season as possible and preferably in car loads. This amount varied by machines but was normally about ten. *C.H. McCormick by Cyrus Jr. to Waite, Burnell & Co.* 7.1.1882. Mss 1x, LPCB 456; *C.H. McCormick by Cyrus Jr. to Albaret & Co.* 13.1.1882. Mss 1x, LPCB 456; *C.H. McCormick by Cyrus Jr. to Lankester & Co.* 14.1.1882. Mss 1x, LPCB 456; *McCormick Co. to Drexel, Harjes & Co.* 8.8.1882. 1x, LPCB 456 and 11.8.1882. Mss M/I, box 3; *McCormick Co. to J.W. Sully* 1.10.1882. Mss 1x, LPCB 456; *Cyrus Jr. to Albaret Co.* 27.11.1882. Mss 1x, LPCB 456; *C.H. McCormick by Cyrus Jr. to Lankester & Co.* 5.1. and 26.2.1883. Mss 1x, LPCB 456; *Cyrus Jr. to Lankester & Co.* 2.8.1882. Mss 1x, LPCB 456.

McCormick's machines under license, was known to be a copier of others' implements. That did not prevent the McCormick Co. from entering into a contract with the firm; but from the very beginning, it turned out to be a complicated partnership. Albaret exhibited McCormick's binder as his own machine and had even replaced McCormick's name with his own.[47]

Victories in the famous shows stimulated sales, not only in America but also in Europe. After the Derby triumph there was a surge of potential agents, none of whom the McCormick Company rejected outright. The conditions were the same as they had been for years; the Company did not want to take any risks, and agents had to buy their machines free on board in New York. For their services, agents received a commission which during the first years was normally around thirty-five percent.[48]

Both for the McCormick Co., and for other companies, too, the trade in Europe caused totally new kinds of problems than they had met in America. Machines had to be fitted for European conditions; but that was not all. The names of the machine parts had to be translated into various European languages, which was difficult when both the machines and parts were new and there were no existing words for them.[49]

When the McCormick Company appointed Percy Lankester as its

---

47 *Cyrus H. McCormick by Cyrus Jr. to Lankester & Co.* 26.12.1882. Mss 1x, LPCB 456. 21.7. and 17.8.1883. Mss 1x, LPCB 457.

48 See for example *McCormick Co. to H. Hendrickx & Sons in Belgium* 26.1.1882. Mss 1x, LPCB 456; *McCormick Co. to Richard Weibull in Sweden* 20.11.1882. Mss 1x, LPCB 456; *McCormick Co. to Antonio de Sarmento in Portugal* 31.1.1883. Mss 1x, LPCB 456; *McCormick Co. to J. Börsum in Norway* 29.6.1883. Mss 1x, LPCB 457; *McCormick Co. to Mateo Funon de Lara in Spain* 19.9.1883. Mss 1x, LPCB 457. During 1883 McCormick Co. made a contract with A. Cosimini & Sons in Italy and the firm was awarded the sole agency for the whole country. *Cyrus Jr. to Cosimini & Sons* 21.12.1883. Mss 1x, LPCB 457. Another contract was made with a Canadian firm, "The North American Agricultural Implement and General Manufacturing Company of London, Canada". Draft for an agreement 9.11.1883. Mss 1a, box 90. An interesting feature in the answers to the applications is the variation in prices. For Hendirickx in Belgium in 1882, the price for a 5-ft twinebinder was $315, for a 6-ft machine $330 and for an iron mower $85. The next year the price for Antonio de Sarmento for the same machines was $198, $208 respectively. For Börsum in Norway the machines cost: 6-ft twinebinder $210, iron mower only $55. In Sweden prices were $250 for a 5-ft binder and, what was amazing, prices for the same machines in Canada were $230 for a 6-ft twinebinder and $65 for mowers. Ibid. Prices for Percy Lankester in Britain were extremely low and he was asked to keep this strictly confidential. For a 6-ft binder he paid $175 and for a mower $47.50. *E.K. Butler to Lankester & Co.* 10.11.1883. Mss 1x, LPCB 457.
These variations cannot be explained by differences in distance and freights. If the price of a harvester in Canada was higher than in Europe, the question might be of competition. If these prices are compared to prices in America, harvesters in Canada seem to be the same as McCormick charged in the U.S., but the price of mowers in 1883 was $75. Other manufacturers charged in the U.S. from $80 to 85 for mowers and for binders from $260 to $275. See Table 6. These are nevertheless the list prices, and do not tell about the actual situation.

agent in Britain and then expanded his responsibilities to include France and most of Western Europe, for the first time since the 1860s its management of the business on the Continent became somewhat concentrated and organized. In 1883 Lankester made a journey to the countries under his supervision, and reported on developments.[50]

In Russia, the McCormick Co. had its own salaried agent and expected good progress.[51] In 1882 the McCormick Co. began to send machines directly to Odessa in large shipments, the first of them comprising 700 machines. The greatest obstacles for trade arose from Freudenreich's apparently too lavish management. The Company pressed him to make only written contracts, and to confirm the existing ones in writing. The McCormick Co. had to pay dearly for its experiences. It wanted to get rid of the old stock of machines in Russia and in Rumania at any price Freudenreich was able to obtain in cash. For future business, the Company took much more careful steps, reducing shipments of the new reaper model to very moderate numbers to avert further growth in the stock of unsold machines.[52]

---

49 *McCormick Co. to Freudenreich* 13.12.1882. Mss 1x, LPCB 456.

50 In France, Lankester had made an agreement with a new agent, Paul Francey; in Italy he had acquired a new agent, and the Company expressed its satisfaction, since it stopped other manufacturers from getting good agents. Spain proved to be an unprofitable area, but in Portugal Lankester appointed an agent. Sales were also extended to Algiers, which was, however, under Freudenreich's supervision. *Cyrus Jr. to Lankester & Co.* 27.7.1883 and *E.K. Butler to Lankester & Co.* 10.11.1883. Mss 1x, LPCB 457.

51 Nettie McCormick's assessment of Freudenreich began to be verfied, however. He could not stand the competition and began even himself to believe that Johnston's machines were better than those he represented (which could be true but was not allowed for an agent to acknowledge). The weather also caused problems for the harvest and for sales. Nevertheles, the Company decided to continue his contract. *Directors Minutes* 7.4.1882. Mss M/I box 24, vol.35; *E.K. Butler to Freudenreich* 6.10.1882. Mss 1x, LPCB 456. Freudenreich did not get along well in Russia and wanted to visit America which was, however, refused in the current market situation. Ibid.

52 Freudenreich asked the Company to loan one of his major agents for a year the balance due, of $4000; this the Company could not understand, especially since it had already sent machines to Mr. Mazewski on consignment so that he paid that fall only for the machines that had been sold and the company carried over the unsold machines. *McCormick Co. to Emil Liphardt* 16.2.1882. Mss 1x, LPCB 456; *E.K. Butler to Freudenreich* 21.2.1882. Mss 1x, LPCB 456; *Cyrus Jr. to Freudenreich* 21.10.1882. Mss 1x, LPCB 456.
The McCormick Company's dissatisfaction with Freudenreich accumulated during the next winter. In Rumania, where Wm. Staadecker had an agency under Freudenreich's supervision, well over thirty machines were left on hand from the season of 1882. Still more anger arose from the shipping of the machines that were left in McCormick's charge. "Your report of our business at Bucharest, Roumania, under date of Dec. 9th. makes us sick of shipping machines all over Russia and Roumania on our own account." *E.K. Butler to Geo. Freudenreich* 1.3.1883. Mss 1x, LPCB 456.

Nettie McCormick's stern hand had been visible at the rudder of the McCormick Company for several years. When the old inventor's health began to deteriorate at the end of the 1870s, her position became central in decision making. Company officers reported directly to her on daily activities. These reports could be very detailed, concerning new production methods, improvements in the machines' construction, or the financial state of the Company.[53] Cyrus Jr., who had taken charge of the business after the firing of Leander and Hall McCormick, was also almost daily in contact or in correspondence with his mother.[54] In 1883 it was already clear that no key decisions were made without Nettie McCormick's consent. She approved the new agressive management philosophy of the newly appointed General Manager, E.K. Butler. Decisions had to be made fast. "Delay is loss - strike while the iron is hot." On the other hand she stood for centralized management. "Where five or six have the management - like the Rockford parties no one can decide and so nothing is done."[55]

Nettie McCormick also had the final word in the European affairs. In spite of the good progress in Europe, she was unwilling to renew the European patents. She stated that sales in Europe were so small that it would not pay back the heavy costs of the patents. Furthermore, in Europe there were only a few manufacturers, and European patents would not give protection against the American competitors.[56] Nettie's approach in the business seemed to be more practical than her husband's. There was no room for emotions. If some operations did not bring profit, they were cut out.

Although Nettie McCormick was a key figure in the Company, Cyrus Jr. had taken the leading role in daily activities in the business. He had been engaged with the business since his college days at the end of the 1870s and had gained practical experience under Lewis Wilkinson. All this was training for the final transition of power after Cyrus Sr., who finally died on May 13th, 1884. Cyrus Jr. was elected as the new President of the Company, and E.K. Butler was nominated also as a Director of the Company. In this way, some of the powers

---

The Feelings of E.K. Butler become clear from the following note where he urged Freudenreich for immediate actions. "For gracious sake. If he is good, sell them to him: we will name as a price $150 for harvester and binder. We want them closed out." Ibid.

53 *See for example E.K. Butler to Nettie McCormick* 10.13.1183. Mss 3b, box 2.

54 *See for example Cyrus Jr. to Nettie McCormick* 11.1.1884. Mss 3b, box 3.

55 *Nettie McCormick to Cyrus Jr.* 4.21.1883. Mss 1b, box 25 vol. 1.

56 *Nettie McCormick to Cyrus Jr.* 13.10.1883. Mss 1b, box 25, vol.1.

of the old inventor and owner of the McCormick Harvesting Machine Company were handed over to a professional manager.[57]

Figure 2. Organization of the McCormick Harvesting Machine Company, 1886.

Organization of the McCormick Harvesting Machine Company, 1886

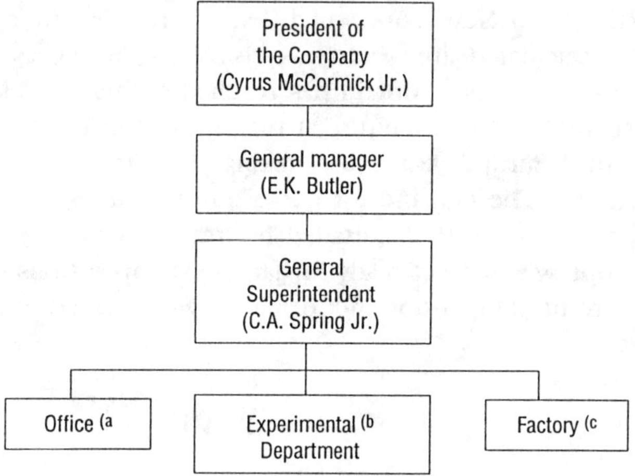

Source: E.V. Crouse, list of the known members of the McCormick Reaper Works organization year 1886. Mss Special Reports File, box 14.

[a] In charge of the office was the superintendent assisted by the assistant superintendent. In addition, there was an office manager and officals for painting, packing and shipping, timekeeper, shipping clerk and a clerk.

[b] The head designer directed the experimental department with the help of a draftsman, a utility man and a pattern foreman.

[c] In the factory worked a bunch of various foremen: a master mechanic, tool foreman, drafsman, grey iron foundry foreman, foundry timekeeper, forge shop foreman, press room foreman, binder room foreman, mower room foreman, packing department, wood shop, assembly department, machine & roller department, receiving office, lumber inspector & yard, yard & stable, grey iron shipping, millwright, cupola, car loading, repairs and fire & watch.

---

57 *Special meeting of the Board.* Thursday May 22nd, 1884. Directors' Minutes. Page 70-72. Mss M/I Box 24, vol.35.

Butler's position had already been central before his official nomination as General Manager. For example, in 1883 Nettie McCormick stated her wish that Butler take over the patent pool negotiations: "Mr. Butler I tell you that you are required in this matter, and I am right in my position, and I want you to take hold of it." *Nettie McCormick to E.K. Butler.* 4.11.1883. Mss 1b, box 24.

The McCormick Co. had re-established itself in the European market through new innovations in harvesting machinery. Cyrus Sr. had great difficulties to find reliable, prominent agents in Europe, and his efforts ended in failures. He succeeded, nevertheless, to organize the foreign trade as part of company business and appointed in Russia the Company's first salaried representative. The American homemarket clearly surpassed the overseas market in importance. Although foreign trade affected to both Cyrus and Nettie, it had to give way for domestic requirements. Foreign sales were not allowed to cause uncalculated expenses; machines were delivered f.o.b. in New York which left the risk on the agent. The same very calculated approach was visible in Europe as it was in the U.S. In Europe only Britain has reached a considerable level in the mechanization of agriculture at the end of the 1870s, while the rest of the Europe was just taking its first steps towards mechanization. The demand for harvesting machines was therefore limited and it was further curtailed by imports of cheap American grain. Europe was not yet ready for large-scale operations but under the pressure of competition McCormick's were forced to show the flag there too.

VI

■ New Fields to Conquer

6.1. Opening up in Australia and New Zealand

Information about the new inventions spread rapidly through the farming community. The McCormick Company had hardly experimented with its wire-binder in 1876 when news of it had spread all the way to New Zealand. The machine had been presented at the Continental Exhibition in Philadelphia, and the Government of the colony of New Zealand decided to purchase one as a sample.[1]

The introduction of the new machine, and the great future prospects in the opening lands of the Pacific colonies, tempted Cyrus McCormick to enter that scene. He had patented his binder in several European countries, and followed the same strategy in the Pacific. Patents were used to restrict competitors' operations and to reserve time for McCormick's own actions.[2]

The Company had sent one of its most able experts, J.C. Newell, to the Liverpool show. The show was a disappointment, but opened the door to the Colonies. Newell arranged a contract for fifty binders with Morrow, Bassct & Co. from New Zealand, and a similar deal with another party from the same colony.[3] Besides New Zealand, Australia was also growing in importance. Cyrus planned to send to the new territory machines that had been left unsold in America. Australian commission agents, for their part, felt it necessary to have new machines and were ready even to pay half of the costs of an expert to help them to set up and start them. These plans were realized and a master mechanic, E.C. Beardsley, was sent already during the fall of 1877 to supervise activities in both countries.[4]

---

1 A.M. Greenfeld to C. & L. McCormick 11.11.1876. Mss 2x. box 196.

2 *C.H. McCormick to F.H. Matthews* 7.11.1877. Mss 1a, box 68. According to McCormick, 450 binders had been sold in colonies during 1877. *Haseltine Lake & Co. to Morgan & Co.* 1.11.1878. Mss 1a, box 71.
Reino Kero in his study on foreign enterprise at the turn of the century in Finland has found that even in such a small and remote market area the big companies secured there interest by patents, either to hinder other companies to reach the market or to restrict the domestic production. *Kero* 1987, 160, 165, 174-175.

3 F.C. Newell to C.H. McCormick 7.15 and 8.5.1877. Mss 2a, box 56.

4 C.H. McCormick to F.H. Matthews 7.21.1877. Mss 1a, box 68. F.H. Matthews to Cyrus H. McCormick 10.8.1877. Mss 1a, box 68.

E.C. Beardsley's reports from New Zealand give a vivid insight into the many difficulties which awaited in a new environment. Machines had to be modified to local conditions, which turned out to be difficult. On many occasions information did not reach the factory or the agent in time, or was worthless due to unintelligible telegrams.[5] Beardsley had to be ready to travel through vast areas at all times of the day. Competitiors had also discovered the colonies and the fierce fight was transferred from America and Europe to the Pacific. The familar maneuvers were repeated on the new fields. Tests and shows were the most important occasions to display the benefits of one's own implements and to blacken competitors' devices.[6]

Beardsley started the business energetically, but was replaced by the more experienced F.C. Newell,[7] which says much about the growing interest of the McCormick Company. After his arrival in Australia, Newell estimated he could sell as many as 1500 binders in Australia alone.[8] The realized orders, only 500 harvesters and binders for 1879, were far less than he had anticipated. Factory officials expected, however, to get further orders for at least 500 more. Since on the other hand, European orders were only for 300 machines, the colonies had already superseded Europe as the main foreign selling area during the two first years of trade.[9]

The Pacific trade casts new light on the question of the role of McCormick's foreign enterprise. Even in the present study, during the first years, the advertising value of overseas business has been found to be of most interest for McCormicks. F.H. Matthews, superintendent of the McCormick factory who, during the absence of Cyrus McCormick, was in charge of the Company operations, maintains in his report to Cyrus in March 1879 that the Company had in the bank about $150 000.00, and by the time that amount is

---

5 The McCormick Co. began to use special codewords to shorten its telegrams to make them cheaper. For example, "McCormick, Chicago: Danger, Ryland" meant "ship us three five foot wide Harvesters with String Binder attachment: the Harvester to have Iron wheels". *A.H. Town to McCormick Co.* 8.9.1880. Mss 1a, box 78.
Codes were used throughout the trade: in 1886 the words "Hamilton charming carrot" meant: "our price for the 100 machines ordered is $155 each net in New York". *McCormick Co. to McLean Bros. & Rigg* 4.27.1886. Mss 1x, LPCB 458.

6 *E.C. Beardsley to C.H. & L.J. McCormick* 1.31.1878. Mss 2a, box 56. Beardsley had to cast some parts himself because the original ones were of inferior quality. The victory of the Osborne in the New Zealand trial was a severe setback to McCormick. It diminished, according to Beardsley, the next year's orders by as much as one hundred machines.

7 *C.A. Spring to C.H. McCormick* 9.26.1879. Mss 2a, box 52.

8 F.H. Matthews to C.H. McCormick 12.21.1878. Mss 1a, box 72.

9 F.H. Matthews to C.H. McCormick 4.14.1879. Mss 1a, box 74.

gone "we will be getting money out of the New Zealand & Australian sales & perhaps some from Waite & Burnell H & Co." The McCormick Co. charged $205.00 cash in New York for ordinary harvesters. If the Company was able to sell nearly 1100 machines, as it calculated, this would mean over $225 000 in revenue. During 1879 the McCormick factory produced about 13 600 harvesters and binders which means that Oceanian trade was 8 percent of the total sales. In this light, foreign sales were not a mere mariginal affair for the Company, and had an important role in the Company balances.[10] A similar event can be found in the European trade too (see page 117).

This information also makes Leander McCormick's demands for a share in the foreign business more understandable. It could therefore be suspected that for Cyrus it would have been lucrative to understate the foreign field, and keep it separate from Company business.

The Company's serious approach becomes clear from its persistent attitude in Australia. It had to lower somewhat the prices of binders and binding wire, under the pressure of competition. Furthermore, the factory decided to send a permanent representative to the area and made preparations in good time beforehand for the coming trials in Melbourne. Sales were increasing in the Pacific. Osborne had cleaned out all the machines they had shipped over and could have sold more. McCormick's agents, on the contrary, had to carry over 90 machines. Wood had met even greater problems and had to sell its harvesters by auction.[11]

As had been stated before, machines intended for American markets did not necessarily fit foreign conditions. Even small alterations meant extra costs, and no company would have made them without hope of compensation.[12] The factory responded to agents' demands with frequent changes in models, made specially fitted machines and machine parts, but was not ready to produce a totally new machine for the Pacific trade.[13]

---

10 *F.H. Matthews to C.H. McCormick* 4.14.1879. Mss 1a, box 74; *McCormick Machines Built since* 1841. Mss M/I box 18.

11 *E.K. Butler to McCormick Co.* 9.30.1879. Mss 2a, box 52; *McCormick Co. to McLean Bros. & Rigg* 10.17.1879. Mss 1x, LPCB 455.

12 The average production cost per harvester and binder at the McCormick works in 1883 was $42.58. This consisted of material costs of $27.55, labor costs of $12.97 and running expenses of $2.06. According to the company officials, changing the patterns alone raised production costs by about $8 a piece compared with the previous year's model. *C.A. Spring to Nettie McCormick* 9.20.1883. Mss 3b, box 2; the Company returned to this question again in 1884, when it reminded McLean Bros. & Rigg that "changes or improvements of any character are not done without a money-cost".*McCormick Co. to McLean Bros. & Rigg* 1.13.1884. Mss 1x, LPCB 457.

13 See for example *McCormick Co. to Edward Ackerman* 4.1.1883. Mss 1x, LPCB

Foreign fields were also important in one more respect. When the factory developed a new machine, as in 1879, it was possible to send a sample for preliminary tests to Australia, to gain practical experience, and then to make the necessary alterations in construction for the coming season at home.[14]

In the McCormick Company's Pacific trade the same trends were reflected as in its domestic and its European business. Farmers demanded twine-binders and agents pressed the factory to supply these or they would take Deering's instead. Company officers repeatedly had to give assurances that the factory was experimenting with one and simply wanted to have the utmost confidence in the machine before taking any action.[15]

By 1880, when McCormick's were assuring their agents of the coming success, there were already at least five companies[16] in full swing selling twine-binders. Some of McCormick's agents had to give up under pressure from farmers. Rival companies pushed hard to gain a foothold and at least Wood, Deering and Osborne had each sent two men to the colonies to promote their business. McCormick had only one man there, who had great doubts if he alone could even earn enough to cover all the expenses he had caused. The

---

456; *McCormick Co. to McLean Bros. & Rigg* 4.27.1886. Mss 1x, LPCB 458 and 4.5.1887 Mss 2c, box 112.
In the correspondence between the McCormick Company and its representatives and its agents the question of adapting machines to local conditions rose to the surface time after time. One of the frequent complaints was that grain and hay were too heavy for the machines or straw was too long. Farmers therefore wanted, for example, binders with a deeper platform. Edward Ackerman, the Company's representative in the area, had numerous suggestions for future changes and was an important part of the business also on that front. He could, for example, write how "The divider is worse on this year's machines than last; the grain wheel leaves a large track every swath... This point as well as lowness of cut are the two important points for Australia, and unless the Steel machines have these improvements, it will not become a favorite among the Australian farmers." *Edward Ackerman to McCormick Co.* 11.30.1884. Mss 3b, box 3.
For the statement on a special machine for the Colonies see *McCormick Co. to Edward Ackerman* 1.3.1891. Mss 1x, LPCB 460.

14 *C.A. Spring to C.H. McCormick* 9.26.1879. Mss 2a, box 52; the same habit continued over the years and for example in 1886 the Company put its new knotting device to the test first in Australia. *McCormick Co. to Edward Ackerman* 10.13.1886. Mss 1x, LPCB 458.

15 *McCormick Co. to McLean, Brothers & Rigg* 4.2. and 6.25.1880. Mss 1x, LPCB 455.

16 The first five were Wood, Deering, Osborne, Johnston and one unknown colonial company. *A.H. Town to McCormick Co.* 10.4.1880. Mss 1a, box 78. Agents were in such desperate need of the new machines that some of them asked if the Company could send them at least the necessary parts to change wire-binders to twine machines. *Morrow, Basset & Co.* 10.9.1880. Mss 1a, box 77.

same comment was repeated six years later by E.K. Butler.[17] These notes once again raise suspicions of the statements concerning the value and reasons for the Pacific trade. On the other hand, in spite of Butler's pessimistic opinion, the McCormick Co. tried to find a better competitive position by expanding its trade in auxiliary articles.[18]

The contest continued in the Pacific harder than ever. Field tests and agricultural shows followed each other, and the results were reported to the home office for use in America and Europe. Positive news from the other continents was also printed in the colonial papers, even though local farmers preferred what they saw with their own eyes and did not trust foreign trials. For all the leading harvester companies, by the beginning of the 1880s the whole globe had become one single market place. Consequently, competition expanded in full force to every part of that market. Smaller factories, as in the colonies, tried to hold on by copying the larger ones' machines. In the field tests all the companies bribed the jurors, and made special machines for these trials. Besides, there were numerous special ways to cause harm to the other competitors.[19]

---

17 *A.H. Town to McCormick Co.* 8.9. and 10.4.1880. Mss. 1a, box 78; *E.K. Butler to McLean Bros. & Rigg* 4.27.1886. Mss 1x, LPCB 458.

18 McCormick Co. made an inquiry about the possibility to add the Aultman & Taylor Co's threshing machines to its line for sale in the Pacific. *McCormick Co.to Aultman & Taylor Co.* 4.22.1888. Mss 3a, box 5.

19 *McCormick to Edward Ackerman* 10.10.1883. Mss 1x, LPCB 457.
The English firm Hornsby's local agents had acquired some of the first models of McCormick's harvesters and then masked them as new machines. Furthermore, they had removed some of the key original parts and replaced them with Hornsby's and other makers' components. The mixtures were then put on sale as original McCormicks. It caused a lot of damage and extra work for McCormick's local representatives to explain to the angry farmers why the machine did not function properly. *Edward Ackerman to McCormick Co.* 9.6.1885. Mss 3b, box 4.

Table 8. Volume of foreign sales of the McCormick Company, by machines and importing countries, 1884-1898.

Country	Harvesters	Mowers	Reapers	Total (A)	Percentage of A
Africa	775	423	1 059	2 257	2.6
Australia	4 619	645	131	5 395	6.1
Denmark	93	2 244	2 704	5 041	5.7
Egypt	2	–	5	7	0.0
England	4 408	6 488	540	11 436	13.1
Finland	–	900	–	900	1.0
France	3 302	8 848	2 558	14 708	16.1
Germany	623	4 834	2 713	8 170	9.3
Mexico	43	21	225	289	0.3
New Zealand	3 741	305	90	4 136	4.7
Norway	37	1 692	311	2 040	2.3
Russia	3 565	4 591	8 986	17 122	19.4
South America	8 916	2 689	699	12 204	13.8
Sweden	100	3 476	916	4 492	5.1
Sum	30 224	37 036	20 937	88 197	100

Source: McCormick Estate Papers. Mss M/I box 18.

From the above it has become evident how the harvester business expanded rapidly to the colonies. It is also apparent that foreign trade began to have a distinct value for all the competing companies. Its material value, as in the case of the McCormick Co., was not merely nominal if compared with the domestic. The total number of reapers, mowers and harvesters sold on foreign markets in 1884-1898 was 8.1 percent of total sales. At the same time the Pacific trade was 10.8 percent of total foreign sales as can be seen in Table 8. That amount certainly meant a considerable sum of money even to such a prominent factory as the McCormick Co.

Table 9. Unit prices of machines sold by the McCormick Company in the Colonies, 1882-1886 (U.S. dollars).

Year	Harvester and binder	Reaper	Mower
1882	305-340	170	85
1883	305-340	170	85
1884	305-315	170	85
1886	145-155	–	–

Source: McCormick Co. to McLean Bros. & Rigg and Morrow, Basset & Co. 12.8.1881. Mss 1x, LPCB 456; McCormick Co. to Edward Ackerman 4.1.1883. Mss 1x, LPCB 456; E.K. Butler to McLean Bros. & Rigg 4.27.1886. Mss 1x, LPCB 458.

Although the sales figures already confirm the growing value of the export for the McCormick Co., further evidence can be found by comparing machine prices in America and in the Pacific. The intention of the comparison is to find out if there were any differences in the Company's coverage. Prices also reveal information about the terms of trade of the various firms.[20]

In America, Cyrus McCormick was among the first to introduce the hire purchase system. It was not transfered to the Company's Pacific trade, but machines were, with minor exceptions, sold f.o.b. in New York or in some other port as was the case in the European trade. The other companies followed the same lines, at least in the Pacific.[21] This meant that the factory had to bear the transportation costs from Chicago to the port, but from there on agents had to pay

---

20 In 1880 Champion dealers sold combine reapers and mowers at $100, single reapers at $85 and mowers at $52. *N.H. Town to McCormick Co.* 8.9.1880. Mss 1a, box 78.
A notice survives from 1884 that Buckeye and Hornsby were selling their machines at $45, and Howard and Deering at $50 and $55 respectively. *Morrow, Basset & Co. to McCormick Co.* 12.6.1884. Mss 3b, box 3. Altough it is unknown which machines these prices refer to, it is probable that the quotations were for mowers. These prices were about $30 cheaper than what McCormick could offer.

21 *McCormick Co. to Morrow, Basset & Co. and to McLean, Bros. & Rigg.* 12.8.1881. Mss 1x, LPCB 456.
For 1882 McCormick offered its agents in Oceania a discount of 35 percent from the prices in Table 8. On spare parts, the discount was 33 percent. Two years later the terms were more liberal: the discount on machines remained the same but for parts it was raised to 37.5 percent. Cash sales were also changed in 1883 to a more liberal 'time sale' in which payments were to be made in three months from the date of Bill of Lading. *McCormick Co. to McLean Bros. & Rigg.* Mss 1x, LPCB 457;
Competition forced the English firm Hornsby, at least, to send its goods on consignment to their agents and even to pay all expenses for advertising and exhibiting the machines. *Edward Ackerman to McCormick Co.* 9.6.1885. Mss 3b, box 4.

for freight and insurance.[22] Domestic and export prices are therefore comparable, but the comparison itself, due to the lack of information, is not totally reliable. However, it appears that at first prices were higher than in America but then, for some years, were considerably lower. If this interpretation is correct, it leaves for the manufacturer a larger marginal in export trade than in domestic business.

Although factories set fixed prices for their agents, it depended on the agent what price he charged his customers. The McCormick Co. had to warn one of its agents not to raise the prices of the spare parts to such a high level as he had done. The Company's own business ideology had been the contrary. The cost of the extras had to be on such a level that customers were satisfied.[23] When the competition became more acute, the McCormick Co. had plans to fight back with prices, but did not put the cuts into effect in the fear that agents would not pass them on to customers.[24] Equally surprising was the Company's comment on how its representative

22 Although the agents had to pay for freights, normally the factory made the arrangements and also negotiated the rates. Freights were calculated in tons or cubic feet. In 1880 the rates were $7.60 + 5 percent primage per ton or 40 cubic feet. *McCormick Co. to McLean Bross. & Rigg* 4.2.1880. Mss 1x, LPCB. Three years later the rates were $7.00 per ton or 40 cubic feet. *McCormick Co. to McLean Bross & Rigg* 2.12.1883. Mss 1x, LPCB 456. Freights were an important part of overall business competitiveness. If one company could get lower rates than its rivals, this was a significant advantage which in turn lowered pressure on agents and expanded their profits. The importance of freights becomes clear when in 1888 the McCormick Co. threatened to charter a vessel to deliver its machines, in order to break the shipbrokers' common prices. This advantage was not intended for those agents who wanted to take care of their own freights. *McCormick Co. to Edward Ackerman* 5.26.1888. Mss 1x, LPCB 459.
At first the machines were sent to the Pacific through Boston or New York and from there via London or Liverpool, but by 1883 some machines went through San Francisco. *McCormick Co. to McLean Bross. & Rigg* 4.2.1880. Mss 1x, LPCB; *McCormick Co. to Edward Ackerman* 1.3.1883. Mss 1x, LPCB 456; *McCormick Co. to McLean Bross & Rigg* 2.12.1883. Mss 1x, LPCB 456.
The question of insurance is more complicated. McLean Bross. & Rigg did not, however, at first want to insure the machines it had ordered. Another agent, Morrow, Basset & Co., on the other hand complained that the McCormick Co. had not insured its machines as requested, at 10 % on invoice value, and if there had been an accident, it would have meant considerable loss for them. *Morrow, Basset & Co. to McCormick Co.* 10.9.1880. Mss 1a, box 77. Morrow, Basset & Co. continued to insure its shipments. *McCormick Co. to New Zealand Insurance Co.* 4.27.1888. Mss 1x, LPCB 459.

23 Furthermore, the company actually ordered its agent to lower the prices to a similar level as the factory itself had done. This action was justified by the large discounts which the factory had passed on to its agents, the idea being that they could trasmit part of that reduction to their own prices. *McCormick Co. to McLean Bross. & Rigg.* 1.12.1884. Mss 1x, LPCB 457.

24 *Edward Ackerman to McCormick Co.* 10.30.1884. Mss 3B, box 3; *McCormick Co. to Edward Ackerman* 2.7.1885. Mss 1x, LPCB 457.
E.K. Butler expressed the Company's beliefs by stating that "we would not, of course, expect to give them a reduction and have them put it in their pockets, and not increase the sale of the machines by making use of the reduction". *E.K. Butler to Edward Ackerman* 9.17.1890. Mss 1x, LPCB 460.

had been laboring with old machines long enough against competitors' newer models. Local agents seemed to be, for some reason, very conservative, for E.K. Butler hoped "that our Melbourne friends will at last consent to allow us to ship them our latest, with which you need have no fears of any of our competitors".[25]

Cyrus McCormick had patented his machines in the colonies, but patents did not prevent copying. Even the large companies kept a keen eye on each other's improvements and made similar modifications in their own models. In this sense it is no surprise how calm Edward Ackermann was about local copies of the McCormick machines in the colonies: a more dangerous threat came from the pirate part makers, who sold their castings as original McCormick spare parts.[26]

In contrast to what it did in Europe, the McCormick Co. in the colonies appointed several agencies but gave none of them sole rights. In Australia, Lassater & Co. had New South Wales and Queensland, while McLean, Bros & Rigg managed Victoria, South Australia and Tasmania.[27]

To strengthen agents' morale and efforts, officers told them confidentially how the factory was running at full capacity and could not fill all the orders even in America. In the next sentence agents were asked to put in their orders as early as possible so that they could be filled.[28] Although the McCormick Company's orderbooks

---

25 *E.K. Butler to Edward Ackerman* 10.13.1896. Mss 1x, LPCB 458.

26 *Edward Ackerman to McCormick Co.* 9.6.1885. Mss 3D, box 4. The Company accused of copying was Reid & Gray of Dunedin.
The situation in Oceania was a special case, since the countries in that region were British colonies and the Company held patents in England. The English factories could not copy McCormick's machines in England and ship them over to the colonies, and manufacturers in the colonies were too insignificant to take seriously. *E.K. Butler to Edward Ackerman* 10.13.1886. Mss 1x, LPCB 458.

27 The two agents could not resist the temptation and began to sell in each other's areas. Competition went on with price reductions, but at this stage headquarters in Chicago took a firm stand in the matter. The Company declared strictly that it wanted its agents to keep prices as high as possible, as has been the case in America. The McCormick Company justified its argument with the good quality of its products. McCormicks could be sold even at higher prices. This argument perhaps explains why the McCormick Company, like the others, took tests and trials so seriously. To keep its word the Company had to defeat its competitors over and over again. *McCormick Co. to Lassater & Co. Limited* 12.13.1879 and 1.12.1880. Mss 1x, LPCB 455. When Lassater & Co. repeated its trick some years later, its rights to the McCormick agency were withdrawn. *McCormick Co. to Lassater & Co.* 9.11.1882 and *McCormick Co. to Edward Ackerman* 4.1.1883. Mss 1x, LPCB 456.
To get rid of its unsold stock of outdated wire-binders, McCormick's former European representative, Waite, Burnell & Co., sent them to New Zealand at prices which "materially interfere with their (other agents') trade". The McCormick Co. took a strict stand on the question. It sent stern letters to Waite, Burnell & Co. and also to its new European agent, Percy Lankester, in which it prohibited such actions in the future. *McCormick Co. to Waite, Burnell & Co and to Percy Lankester & Co.* Mss 1x, LPCB 456.

28 *McCormick Co. to Lassater & Co. Limited.* Mss 1x, LPCB 455.

were full, and there was a growing demand for harvesting machines, there remains a doubt as to its real intentions: was this only a trick to encourage agents to make further orders?

Irrespective of all the efforts of the agents and the factory, the realized sales depended to a considerable extent on the weather and the farmers' prospects from the harvest.[29] Any competing harvester company could win the tests and shows and agents could expect large sales, but if the weather was too wet or too dry, farmers were unwilling to invest in expensive machinery. This was one of the key factors to consider when agents made decisions about the next year's orders.[30] In one respect, McCormick's agents were in a disadvantageous position, since they claimed that competitors charged their representatives only after machines were sold; if the year turned out bad, it was the factory that carried the risk. The f.o.b. system that McCormick used forced its agents to avoid unnecessary risks, which in some conditions also decreased sales. Agents simply could not trust even the "brightest prospects".[31]

6.2. The Pacific falls into oblivion

The big expectations which the McCormick Company had of the Pacific trade were not realized. It had to fight the same companies as on the home front and in Europe. The McCormick records leave an impression that in the Pacific it was a market leader and other companies tried to usurp its place. Not all years were, however, equally good, and a marked fluctuation in sales is visible, especially during the 1880s.

After a brisk start in 1878, McCormick's sales began to drop. Machines remained unsold on agents' hands. At the beginning of

---

29 Bad weather could also have a positive influence on sales, as happened in America: in 1878 the weather had been very rainy throughout the whole Northwest and consequently farmers had hurried up their harvests by ten days. For the McCormick Co. it meant an increased demand for binding wire which the company had to collect in a hurry from all possible sources to prevent farmers from returning their newly bought wire-binders. *F.C. Matthews to C.H. McCormick* 8.3.1878. Mss 1a, box 72.

30 See for example *Edward Ackerman to McCormick Co.* 11.30.1884. Mss 3b, box 3 and 9.6.1885 Mss 3b, box 4. Since after the long drought, crops were very light, farmers bought 'strippers' instead of heavier harvesters, since these were the only machines by which such crops could be reaped. Strippers were domestic Australian machines that began to compete with the exported reapers.
See further: Thompson, Alan: The Origins of a Harvester Revolution: The Development of the Combine Harvester in Australia, Canada and U.S.A. (Tools and Tillage. Ed. by Axel Steensberg, Alexander Fenton and Grith Lerche. Vol. III:2 1977). Copenhagen 1977.

31 *McLean Bros. & Rigg to McCormick Co.* 1.22.1897. Mss 2x, box 298.

1883 hopes for large trade were still bright, and headquarters requested Edward Ackerman to stay in the area.[32] Ackerman was able to defeat his competitors at most of the shows, but during 1884, in spite of his efforts, about half of the machines remained unsold and had to be carried over to the next year. The English Hornsby had outdone its archrival, with its machine carefully built for these markets. For the moment McCormick and Wood had to fight for the second prize.[33] The McCormick Company acknowledged the situation but a letter to Edward Ackerman probably gives an answer to why it continued its foreign trade: "We do not propose to abandon the trade, nor the ground; neither are we disposed to take a back seat in the future and allow the opposition to occupy the field that up to this time has been so succesfully held by our Australian and New Zealand agents."[34]

Business had spread to every part of the globe and it had become impossible to give it up even in the most remote places. No company could anymore afford to allow rumors of a retreat from some markets in a situation where the news of the results of field tests in Europe was used for advertising in Australia and vice versa. Mira Wilkins has noticed that American and German multinationals often invested abroad to sell and to make goods, based on new technology. They expanded over borders either to reach the market or to obtain sources of supply. In the case of the American firms Wilkins stresses these two factors even over the financial returns.[35]

In spite of the doubtful outlooks for the future, within a year McCormick had regained its position and its agents could once again report victories at shows.[36] With its new machine, the Company

---

32 *McCormick Co to Edward Ackerman* 1.3. and 4.1.1883. Mss 1x, LPCB 456.
Exact numbers of sales to Oceania by year are not available before 1891. In 1883 McLean Bros. & Rigg from Melbourne ordered 250 harvesters. Ibid.

33 *McCormick Co.to McLean Bros. & Rigg* 1.12.1884. Mss 1x, LPCB 457; *Edward Ackerman to McCormick Co.* 10.30.1884. Mss 3B, box 3.
During the season of 1884-1885 McCormick was awarded seventeen first prizes in the Colonies. The competitors were Hornsby, Wood, Buckeye, Deering, Johnston, Samuelson, Esterly, Champion and Howard. *McCormick Co. to Lankester & Co.* 5.5.1885. Mss 1x, LPCB 458.
To its European representative McCormick boasted to be "further at the front to-day than we have ever been heretofore both in New Zealand and in Australia"; the English companies could not compete with American manufactures. *E.K. Butler to Lankester & Co.* 1.8.1884. Mss 1x, LPCB 457.

34 *McCormick Co. to Edward Ackerman* 2.7.1885. Mss 1x, LPCB 457.

35 *Wilkins* 1988, 22-24.

36 *McCormick Co. to Edward Ackerman* 4.7.1885. Mss 1x, LPCB 457; *Edward Ackerman to McCormick Co.* 9.6.1885. Mss 3B, Box 4.
Ackerman hurried to post local newspapers which had published results of the tests and shows to headquarters.

declared the next year that "what we have produced is something they will have to work a long while at before they can equal".[37] These proud words were fulfilled, and in 1886 the only relevant competitor was the English Hornsby. Next year the McCormick Company anticipated severe rivalry, but expected to knock out its competitors with its new steelbinder, as it had done on the home front.[38] The positive trend continued to the end of the decade and the Company rejoices over its large orders. Among its competitors the same trends were visible as in other countries. The Deering Company had increased its share of business, but the Canadian Massey Company had become the strongest competitor. Massey proved to be an aggressive rival, that threatened McCormick several times for infringements of its patents.[39]

By 1890, Edward Ackerman, who had been in Oceania since 1881, began to devote some of his time to European problems, arranging exhibitions and trials in France and in Belgium.[40] Ackerman was still in the Colonies in 1892, but thereafter his services were needed in

---

37 *C.H. McCormick Jr. to Edward Ackerman* 7.24.1886. Mss 1x, LPCB 458. The McCormick Company's serious approach to the Oceanian business becomes evident from Cyrus Jr.'s personal interest in its problems.

38 *McCormick Co. to Edward Ackerman* 10.8.1886 and 7.21, 1887. Mss 1x, LPCB 458 and 459.
The McCormick Co. trumpeted that they had sold over 21 000 steel-binders on home markets in 1887 and estimated to have lost the sale of at least 5000 more due to their inability to meet the demand. For their competitors, this success had been a serious blow and between the lines can be read how McCormick's expected new efforts from its rivals.
The New Steel Binder won 16 first prizes in the Colonies during the season of 1886, the English Hornsby being the only significant rival. *E.K. Butler to Lankester & Co.* 12.8.1886. Mss 1x, LPCB 458.

39 *E.K. Butler to Edward Ackerman* 4.29 and 9.17.1890. Mss 1x, LPCB 460.
In 1883 the Massey Company sent its first twenty-five machines to Australia on consignment. Consigment business was soon found to be a total failure and three years later the company sent over its own traveler to oversee sales and set up the machines. *Denison* 1949, 95-96, 98, 100-101.
The McCormick Co. was eager to get all possible details of its competitors' machines. When Deering introduced a new Chain Drive Harvester, McCormick reminded Ackermann not to fail to post all the information available on the working of the machine. *E.K. Butler to Edward Ackerman* 4.29.1890. Mss 1x, LPCB 460.
Even in 1888 E.K. Butler did not foresee Canadian companies as a serious threat to Lankester in the Colonies, since their machines "are not calculated for that work". *E.K. Butler to Edward Ackerman* 11.7.1888. Mss 1x, LPCB 459.
By 1892 the situation had changed totally. The new Massey-Harris Co. and Wood did their utmost to capture the Australian trade and other firms, Deering, Hornsby and Buckeye were on the decline. *E.K. Butler to Edward Ackerman* 7.18.1892. Mss 1x, LPCB 461.
On patent infringements see *E.K. Butler to Edward Ackerman* 7.18. and 10.29.1892. Mss 1x, LPCB 461; *D.M. Osborne & Co. to McLean Bros. & Rigg* 2.3.1896. Mss 2x, box 282; *McLean Bros. & Rigg to McCormick Co.* 10.27.1897. Mss 2x, box 298.

40 *McCormick Co. to Edward Ackerman* 10.9.1881. Mss 1x, LPCB 456; *Edward Ackerman to Stanley McCormick* 8.2.1890. Mss 1a, box 109.

Europe and in South America.[41] The reasons for this changed focus of activity have to be sought in the volume of trade. Relative to the European business, the Oceanian trade had lost significance, and it was understandable that the Company directed one of its most able men to more lucrative markets.[42]

Table 10. Sales of harvesting machines and mowers by the McCormick Company in the newly colonizing areas, 1878-1901.

Year	South America[1]	Australia	Cape Colony
1878	–	596	–
1879	–	255	–
1880	–	101	–
1881	–	270	–
1882	52	–	2
1883	106	1 011	–
1884	157	1 096	17
1885	281	75	17
1886	56	198	36
1888	171	–	–
1889	87	–	–
1890	832	1 159	98
1891[+]	100	170	110
1895[+]	646	274	420
1896[+]	594	1 181	237
1898[*]	1 950	965	317
1899[*]	2 617	898	242
1900[*]	2 140	839	238
1901[+]	1 310	554	169
1902[+]	2 168	1 043	461

1) South America includes here Argentina, Chile, Mexico and Uruguay.

Source: Statement showing machines sold to foreign countries, 1874-1901. Mss 1a, box 72; Machines sold to foreign countries counted by agents. Mss 3x, box 26. Marked with [+]; Statement showing machines sold to foreign countries and profits on foreign machine business, 1898-1900. Mss M/I, box 18. Marked with [*]

---

41 *E.K. Butler to Edward Ackerman* 7.18 and 11.29.1892. Mss 1x, LPCB 461.

42 *Machines sold to foreign countries counted by agents.* Mss 3x, box 26.
According to the Machine Ledgers, which state the number of machines sold by the agents, in 1891 there were only 170 machines sold in the Colonies and 274 in 1895. This information must be approached with caution, since there is no record at all of McLean Bros. & Rigg's orders for 1891, but from other sources it is clear that this firm had three houses in Australia and had made orders and sales both in 1891 and 1892. *McCormick Co. to Edward Ackerman* 1.3.1891. Mss 1x, LPCB 460 and *E.K. Butler to Edward Ackerman* 11.29.1892. Mss 1x, LPCB 461.
In spite of these comments the overall trend is visible. Europe had become the main foreign sales area.

The sales figures in Table 10 do not totally coincide with the information that has been collected from the McCormick Company's correspondence. At the end of the 1880s Ackermann announced the victories in trials and agricultural shows and the large orders from the Company's agents; but in the Company records these orders seem to be considerably smaller. Only in 1884 and in 1890 did the sales exceed 1000 machines, and in most years sales were totally missing or were minimal. It is, however, obvious that the reports of Edward Ackerman are in this case more reliable than the Company records.[43]

Simultaneously with its Pacific trade the McCormick Company also sent out its tentacles to South America and South Africa. A first shipment of twelve wire-binders was sent in 1879 c/o Dumeresq, LeBas & Company of Montevideo in the Oriental Republic (today Uruguay). This firm obtained the rights to the sole agency in South America for one year, extended until 1882. At that time McCormick sent its travelling agent, Lee Borrel, to oversee the business. On his recommendations, the agency was transferred to Buenos Aires, where a permanent agency was established in 1883.[44] The earliest information on the African trade is available from 1882, when the McCormick Company sent its first shipment to Smuts & Koch in Malmesbury, Cape of Good Hope.[45]

In Buenos Aires the McCormick Company was a latecomer that

---

43 Statements showing the volume of the McCormick Company's production, sales and profits are very fragmentary. Ledgers are available only for 1891, 1895, 1896, 1901 and 1902. The various extant statements were produced by company officials for various purposes over a long duration of time. For this reason, not all the available material is totally consistent, and has to be approached with caution. Nevertheless, it was produced for internal Company use and when the purpose of the material is kept in mind, it gives at least outlines of the development of business.
It is, however, obvious that not all sales and transactions are visible in the statements, as becomes clear by comparing sales figures in Table 10 and information collected from the Company's correspondence.
Scholars' use of the material has varied. Jan Kuuse includes machine parts in the Company's production figures. *Kuuse* 1974, 273. In this study this has been avoided, since such a practise can lead to misinterpretations. In some years an agent might order a large amount of extras and only a few machines, but if in the statistics these are calculated together, the result can be misleading.

44 The Buenos Aires agent was J. Mohr, Bell & Company and it continued to act as McCormick's agent until 1887, when the agency was transfered to Agar, Cross & Company. *Memorandum February 1931.* Mss Special Reports File, box 14. This note was probably written for the use of Cyrus McCormick III, grandson of the inventor, who at that time was completing his book "The Century of the Reaper".
This information is partly confirmed by a letter to *Linus R. North care J. Mohr, Bell & Co.* 11.30.1883. Mss 1x, LPCB 457.

45 *McCormick Co. to Smuts & Koch* 8.5.1882. Mss 1x, LPCB 456.
The first shipment met with considerable obstacles. The goods were delayed owing to a strike of railroad employees in New York and had to be sent to South Africa via London. There is no record of the machines that were sent, but the number was apparently only one or two judging from the general invoice of $316.30.

had to fight for its place. In spite of its position McCormick's stuck to its principles in foreign business: machines were sold only on f.o.b. terms in New York.[46] On the other hand, prices were considerably lower for South America than they were in Oceania[47] and even on the home fields, as can be seen from Table 5, whereas McCormick's competitors sold their goods at the same prices as in the U.S.

The McCormick Company regarded South America as an important and promising market. It sent its travelling experts as soon as possible from Europe to assist South American agents. Sales were increasing and by the end of 1884 Linus B. North reported that demand had been larger than the Company was able to fill.[48]

In the beginning of the 1880s, there is a visible expansion in the McCormick Company's foreign business. It had established itself in Europe and tried aggressively to find new markets. By the beginning of the 1880s, it had become a world wide concern, but it was not alone in the field: its competitors had also extended their networks into the same areas, and the contest for farmers' dollars continued in every part of the world.

Jan Kuuse explains the rapid expansion of the harvester companies to the newly colonized areas by emigration. The new lands were smooth and suitable for large-scale farming; moreover, these areas were short of labor and there was therefore a heavy demand for machines. Kuuse states furthermore that these new immigration areas succeeded Europe as the principal customer for agricultural technology, and only after the mass emmigration had sucked up the surplus population from Europe did it once again, at the end of the 1890s, became the largest selling area for American agricultural machines.[49]

Kuuse's interpretation seems at first sight very realistic. There were new fertile lands waiting for European emigrants, who were in need of tools and machines. On the other hand, as Europe discarded its extra hands and mouths, it needed to invest on

---

46 McCormick warned its representative, Linus R. North, to make only cash sales. "You will understand that whatever orders you may be able to get shall come through James E. Ward & Co. or some other equally as responsible house in New York, who will pay for the goods they may want when delivered in New York." *McCormick Co. to Linus R. North* 11.30.1883. Mss 1x, LPCB 457.

47 The price for a regular twine-binder was $190.00, for a combined reaper and mower $110.00 and for an iron mower $52.00. *McCormick Co.to Linus R. North* 11.30.1883. Mss 1x, LPCB 457.

48 *Linus B. North to McCormick Co.* 11.21.1884. Mss 3b, box 3. North also reported that Argentina was too large an area for a single agent to cover it properly.

49 *Kuuse* 1974, 274, 278-279.

machines. Nevertheless, in the case of Australia the emigration statistics and figures for the growth of population do not confirm Kuuse's theories. According to Olavi KOIVUKANGAS, the mean annual growth rate of population through net migration was 1.60 percent in 1881-1890 and only 0.08 percent in 1891-1900. Furthermore, Koivukangas maintains that fertility was the major factor in population growth during the latter part of the 19th century; and the two leading sectors that brought settlers into Australia were mining and railway construction. Koivukangas confirms, however, the impact of climate on the economic situation.[50] On the other hand, consular reports seem to confirm that Australasia as a whole suffered from a shortage of labor, which pressed farmers to invest in labor-saving machinery.[51]

In discussing emigration, we have to remember age and gender distribution. Not all of the emigrants were young men at their best working age. Besides, as has been noticed in the case of American agriculture, farm-making was not a simple task: it required money and years of backbreaking work before the fields were in a shape to allow the use of a harvester.

The situation was similar in South America and South Africa. Argentina was the main grain-growing state in South America, but not even there was immigration a major factor. The migration peak in the 1880s did not reach 150 000 migrants, and immigration passed this figure only after the turn of the century. Argentina tried to induce immigrants during the 1880s by free tickets, but without visible success. Besides, the large estate holders prevented the enactment of a homestead act according to the American model. This of course limited the number of possible farmers but on the other hand wealthy holders were freer to invest in machines.[52] Immigration to South Africa did not rise to the same levels as in the other newly colonizing areas.[53]

This information does not deny the fact that there was a large reclamation process going on in the Colonies, but it explains the difficulties and setbacks which McCormick encountered in Oceania. Sales did not reach amounts that would justify Kuuse's arguments.

The McCormick Company's expansion to the Pacific, and later in

---

50 *Koivukangas* 1986, 49-50. According to Koivukangas "drought and depression combined in 1892-93 and 1898-1900 to occasion higher out-migration than new immigrants arriving". Ibid, 50.

51 *Agricultural machinery in their several districts...* 1885, 751-752.

52 *Lähteenmäki* 1989, 65-66, 275 appendix 1.

53 *Kuparinen* 1991, 368-370 appendix 1 and 2.

South America and South Africa casts new light on the role of foreign trade. Although its financial returns from overseas business were only around 8 percent of the total sales in the 1880s, which in fact was not a minor income, its importance was growing in another way. Foreign sales had in the beginning of the 1880's become an integral part of normal harvesting machine trade and companies could not afford to leave some markets without consideration. The whole World had become one market for the harvester firms. In this light the McCormick Company's interest also in the European trade becomes more understandable. It was not very eager to expand it activities but it had to follow its competitors. Common to all its foreign actions was its aspirations to minimize its risks. It made a direct investment by sending its representative to the Pacific as it did to Russia. But otherwise trade continued on the f.o.b. basis. Furthermore, it can be argued that all the competing companies rushed to these newly colonizing areas with future hopes. Immigration and population in the Pacific, South Africa and South America in the 1880s and 1890s were still too small to allow large-scale trade, but factories had to secure their positions there in hopes of a brighter future.

VII

IV

■ Founding of the Foreign Market

7.1. New winds begin to blow

7.1.1. A jump into the new era

At this phase of the study, it is time to turn again to Mira Wilkins' and Alfred Chandler's theories on the evolution of foreign enterprise. Was there in America in the 1880s already overproduction looking for new markets? Was the McCormick Company in the 1880s a modern enterprise and did it seek economies of scale and broadening of market through foreign business?[1]

In the beginning of the 1880s, the harvester machine industry had to adjust to a new generation of harvesting machines. After the invention of the twine-binder, all the main principles of the modern harvester were combined. Subsequent inventions were more or less merely improvements on the basic design and materials: wood was substituted for iron and steel and the knotting apparatus was simplified. This does not diminish the importance and value of these changes, and it would be an underestimation to call them only marketing tricks.[2]

Competition between the harvester companies intensified at the same time, and the remaining twenty-one companies tried to fight each other through prices, trials, shows and other marketing manoeuvers. Prices of reapers sank from $200 in 1872 to $160 in 1883, and whereas the twine-binder cost $350 in 1882, by 1888 only from $160 to $150 was charged (Table 5). Although all the competitors understood how injurious the situation was for all of them, they were unable to reach an agreement over production quotas. During the 1890s, competition expanded to every continent and was transformed to what was called a reaper war.

The McCormick Harvester Machine Company was prepared to meet the competition. It had rationalized its production systems after

---

1 *Chandler* 1988, 31, 34, 44; *Wilkins* 1970, 19, 29, 36, 45-46.

2 *Hounshell* 1987, 185.

the firing of Leander and Hall McCormick, and had changed to a new branch house system. It had also obtained rights for the Appleby binding mechanism for its new twine-binder

Table 11. Main indicators describing the McCormick Company's economic activity, 1880-1902.

Year	Net sales[1]	Net profits	Dividends	Notes	Notes on hand	Capital invested	Net profits on repairs[a]	Net profits on twine[b]	Number of machines sold[c]
	$	$	$	$	$	$	$	$	
1880	2 429 278	1 192 733	..	1 537 913	70 111	..	..	..	26 786
1881	3 010 942	1 254 961	..	1 601 956	94 330	..	..	..	32 353
1882	4 269 653	1 761 226	..	2 044 526	174 664		..	..	44 848
1883	4 425 467	1 486 632	..	2 457 992	236 824	4 213 180	..	..	50 376
1884	4 469 271	1 776 506	500 000	2 735 755	275 546	5 856 163	151 602	154 346	54 922
1885	3 691 110	841 007	400 000	2 472 949	254 438	6 790 542	66 339	116 091	51 439
1886	2 851 926	679 924	250 000	1 740 785	176 928	7 502 121	76 090	121 447	44 103
1887	3 855 643	1 007 767	400 000	2 415258	261 646	8 061 750	73 141	301 944	71 363
1888	4 616 057	1 473 986	200 000	2 853542	385 954	9 655 270	120 170	324 456	91 881
1889	4 687 093	1 803 319	..	2 860 771	615 020	11 291 084	112670	508 828	104 114
1890	5 051 291	1 543 037	625 000	3 087 249	1 394 362	12 324 170	115 727	369 620	127 654
1891	6 180 153	1 867 058	625 000	3 406 038	2 785 388	13 634 961	131 536	151 773	170 666
1892	7 356 123	2 550 322	1 000 000	..	..	15 507 670	128 354	340 119	202 350
1893	6 485 193	2 056 481	625 000	..	..	17 209 492	131 935	361 442	179 643
1894	5 183 167	1 502 581	..	..	..	15 815493	230 546	..	143 143
1895	7 449 770	2 397 862	2 500 000	..	..	16 197 297	337 680	195 461	206 488
1896	7 244 627	2 284 814	2 500 000	..	..	16 603 891	292 803	263 042	129 100
1897	8 016 089	3 321 666	2 000 000	..	..	19 083 550	274 930	206 138	151 885
1898	11 094 464	4 695 010	1 000 000	..	..	21 844 689	365 925	378 135	208 346
1899	12 358 905	4 677 733	1 000000	..	..	..	..	..	266 849
1900	14 203 873	3 292 997	..	..	..	..	..	..	312 128
1901	17 085972	..	..	..	..	..	..	..	402 362
1902	17 275 106	..	100 000	..	..	..	..	..	432 100

Source[2]: For net sales; McCormick Harvesting Machine Co. Net sales of machines, repairs, twine and wire for seasons 1880 to 1902 inclusive. Mss M/I box 18. For net profits; Statement showing net profits for the years 1880-1891. Mss 2C, box 29. Ozanne 1962, 366 for the years 1892-1897. Comparative statistics of McCormick and Deering companies for 1898-1900. Mss M/I, box 1. For dividends; List of Stockholders of Record on date of each Dividend. Mss M/I, box 21. For reaper notes and notes on hand; Statement. Reaper Notes to May 1st 1892. Mss 2c, box 29 [a] and [b] are from Net Profits on Repairs sold, 1884-1898. inc. and Net Profits on Twine sold, 1884-1892. inc. Mss M/I, box 18. [c] is from Statement showing number and kind of machines sold during the years 1880 to 1902 inclusive. In this figure are included also attachements and other supplementary implements.

[1] In this case figures include also repairs or extra parts because these were displayed in net values and as such demonstrate more accurately the real situation.

[2] Figures are only instructive. McCormick Company's officials have produced a substantial amount of accounts on machines produced and sold which, however, differ from each other. For example statements *Machines manufactured 1884-1898 inc. and Machines sold 1884-1898* have omitted totally binders. Mss M/I, box 18. On the other hand its information on net profits on twine are in line with other sources (Kuuse 1974, 285) and the same can be expected also on repairs. See also *To the President and Board of Directors of the McCormick Harvester Machine Co.* 4.1.1887. Mss 2c, box 29 which confirms the informations of column number of machines sold. Some new light in the question comes from the letter of G. Freudenreich where he explains how, "you (McCormick Co.) figure machines, counting each H.& B. as two machines." *Geo. A. Freudenreich to McCormick Harvester Machine Co.* 12. 6-18. 1884. Mss 3b, box 3. This explanation makes more understandable remarkable low figures in some tables.

It has to be taken into account also that every year some of the machines were left unsold on hand.

and besides had developed new mower and reaper models.[3] As a result of these exertions, the Company reaped unparalleled net profits and could declare relatively high dividends to its shareholders in the beginning of the 1880s. A noteworthy feature in Table 11 is that over the period about half of the sales were made on credit. Collection of these revenues was burdensome, and as can be seen from the two final years, 1891 and 1892, in the first year only a marginal share of the notes were paid. It was therefore understandable that all the companies preferred cash sales and warned their agents against bad notes. Notes bound a significant allotment of a company's investments for a long time in the future. Another item to be noted in the McCormick Company's revenues concerns its income from twine and repairs. Early in its infant years, the Company had understood the significance of repair services, and after the invention of wire- and twine-binders, took care of the distribution of the essential binding materials too.

McCormick's profits dropped dramatically more than 50 percent from 1884 to 1885 and dropped 19 percent more during 1886 hand in hand with its sales. The same trend is visible also in its dividends. In spite of these setbacks, the Company remained profitable and continued its investments.

A similar process is also visible in its arch-rival's development. The Deering Company's net profit in 1884 was $1 508 649, in 1885 $834 405, in 1886 $627 575, in 1887 $940 180 and in 1888 $1 227 582. These figures show how alike the companies were. Their profits followed the same general trend, but it is interesting that in 1885 the Deering Company blamed the drop in profits on the bad harvest.[4]

The explanation for the rapid decline after the profit peak of 1884 was the business recession, which also hit the McCormick Company. In concert with other employers, the McCormick Co. decided to hit back by cutting all day workers' wages by 10 percent and all piece workers' and machinists' wages by 15 percent. Cyrus McCormick had not anticipated his employees' reactions: after fruitless negotiations the molders went out on strike in March 1885 and in April the entire plant followed them. McCormick's main competitor, the Deering Company, responded to the workers' demands and restored the wages. The McCormick Company tried to keep the factory open with non-union molders but as tensions began to grow and the Company was not able to get police protection, it had to surrender, restore the wages and dismiss the scab molders.[5]

---

3 *McCormick Harvesting Machine Co. catalog* 1880, 1882, 1884. Mss 5x, box 1. Catalog 1887, Mss 5x, box 2.

4 An undated and unnamed memorandum of the Deering Co., 1890. Mss w, box 2.

5 *Ozanne* 1962, 362-363; *Schonberger* 1964, 20-22.

Cyrus McCormick Jr. had no intention of giving up, and he waited for a second round in the game. Remodeling of the factory systems in the McCormick Company had been under way for some years and keeping this in mind it was understandable that Cyrus and E.K. Butler decided to replace molders by pneumatic molding machines.[6] As a result, not a single one of the 91 striking molders was on the payroll for 1886. The Company was able to manage new strikes, but the situation deteriorated and finally developed into the infamous Haymarket incident on May 3. 1886.[7]

Robert OZANNE, who made a pathbreaking study on McCormick's labor policy, tried to show that trade unionism played a major role in the Company's wage policy. According to Ozanne, it was the molders' union that pressed the McCormick Company to increase wages, and consequently owners wanted to replace harmful labor by machines.[8]

David Hounshell opened a new approach to the subject by pointing out the implications of the new production systems. As has been already noted, Leander McCormick had had to give way to Lewis Wilkinson in 1880. During his short time in the Company, Wilkinson brought with him a new management and production ideology which has been called the American Manufacturing System. The factory was turned over to two work shifts and various jigs, fixtures and gadgets were introduced into production. The Company also began to use in increasing measure special-purpose machines, which in many cases it had to develop itself. Moreover, the new manufacturing system offered means for control over labor.[9]

---

6 New Molding machines did not work accurately enough and castings did not fit together. Nettie McCormick expressed her concern over the matter but gave her backing to Butler's handling of the matter. *Nettie McCormick to E.K. Butler* 11.11.1885. Mss 1b, box 24.

7 *Schonberger* 1964, 22-24.

8 *Ozanne* 1962, 362-364, 373-375.
The son of Cyrus McCormick Jr. passes over the whole event with a couple of words in his book and blames the harsh Superintendent, who was dismissed shortly after the incident. *McCormick* 1931, 90. According to the Directors Minutes, the General Superintendent of the McCormick Harvester Machine Company, C.A. Spring Jr., tendered his resignation on June 1st 1889. *Directors Minutes*. June 1st, 1889. Mss M/I, box 24. vol. 35. C.A. Spring might be the person whom McCormick means. In spite of this fact, in the light of new evidence, McCormick's explanation of the strike is oversimplified and partisan, and fails to take into account changes in the factory. There are some small errors in Ozanne's calculations. He has not found labor costs for 1883-1885. For 1884 costs were $11.66 per machine, in 1885 $11.89 and in 1886, according to Company statements, $12.91 instead of $16.16, which Ozanne used as evidence for the aims of the molders' union. *Cost to Manufacture and Sell 1886 Machines*. Mss M/I, box 4. For 1885 *C.A. Spring to Nettie McCormick 9.24. 1885.* Mss 2b, box 31.

9 *Hounshell* 1987, 178-182; *Kobayashi* 1974, 205.

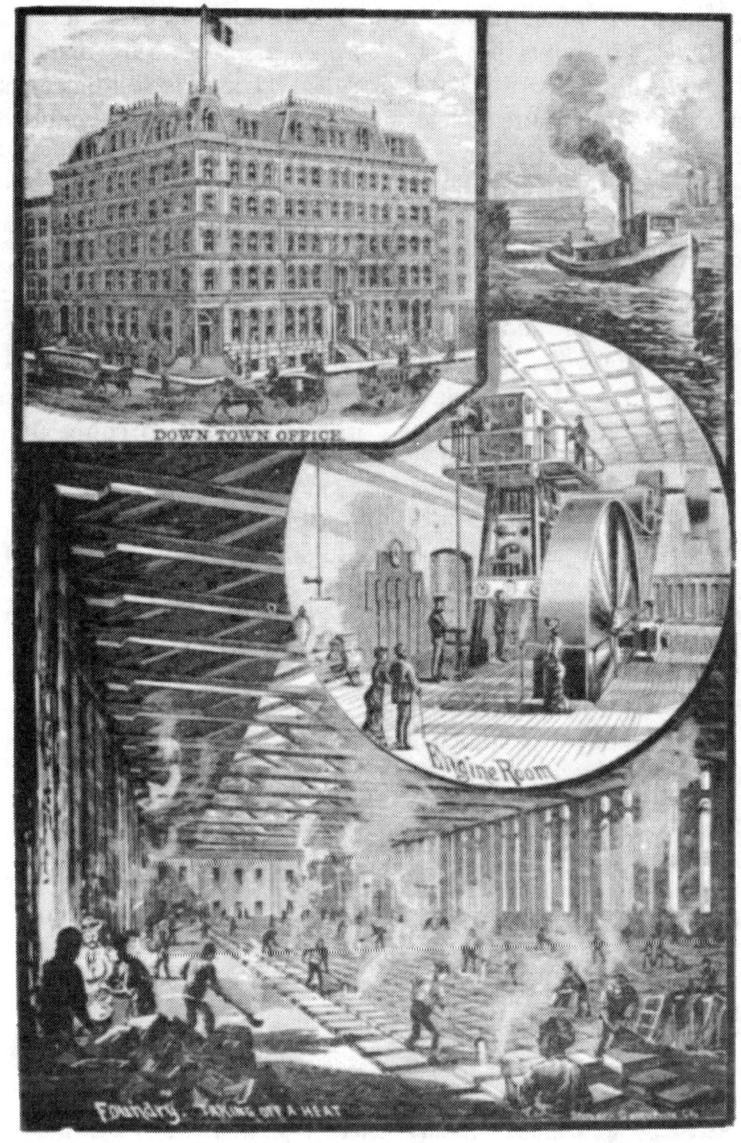

In a careful study, Hal HANSEN has taken Hounshell's reasoning
a step further by showing the introduction of cost accounting in
1881 at the McCormick works. Furthermore, he stresses the
expansion of piecerates and growing managerial supervision. Hansen
also points strongly to the enhanced significance and development

---

Although Hounshell underlines the importance of the two-shift system, Nettie
McCormick stressed her preference for one shift. She wanted to avoid night work,
which raised the cost to manufacture; she would rather see "the need of improving
the time in the *first* part of the season, instead of pushing for dear life in the *last*
part." (spacing Nettie McCormick). Nettie McCormick to E. K. Butler 10.15. 1885.
Mss 1b, box 24.

of the machine shop, which finally became an independent unit.[10]

Table 12 shows how the cost of manufacturing machines fell during the first part of the 1880s. The change is still more visible if it is compared to the $62.11 in 1876. The clear jump in expenditures from 1880 to 1881 may be due to increases in material prices, or, equally possibly, to rising labor costs, as Ozanne has stated.[11] The McCormick Company included under manufacturing costs all material, labor and running expenses. To the overall costs, there must also be added sales expenses, which included the categories agents, special and office. In 1886, sales expenses raised the expenses by $42.25,[12] which means $77 overall costs per machine. Because the McCormick Co. summed all the various kinds of machines in the average cost, it is impossible to count net profit per machine.

Table 12. Average unit cost per machine in the McCormick Company, 1880-1886.

Year	Cost of manufacture ($)
1880	41.21
1881	44.28
1882	41.78
1883	42.15
1884	38.72
1885	40.41
1886	34.75

Source: C.A. Spring Jr. to Nettie McCormick 9.24.1885. Mss 3b, box 5. For 1886, Cost to Manufacture and Sell 1886 Machines. Mss M/I, box 4.

---

10 *Hansen 1989(?)*, 6-7, 15, 18-23.
Hansen's ideas are supported from C.A. Spring's letter to Nettie McCormick, where he calculated the average cost of a machine to be $1.69 higher in 1885 than in 1884. The reason for the increase was not labor costs, but more expensive raw materials: it simply cost more to make the new steel machines. *C.A. Spring Jr. to Nettie McCormick 9.24.1885. Mss 3b, box 5.*

11 *Ozanne* 1962, 369 table 6.

12 *Statement of Cost to Sell 1886 Machines.* Mss 2c, box 29.

It is clear that by 1884 the McCormick Company had entered a new phase in its history. It was on its way to become a modern corporation with a professional management. It had invested major amounts in new labor-saving machinery, and reorganized its production. Although it is evident that at the McCormick plant the supply side of the business was in better shape than it had ever been before, there remains the demand side and various factors affecting it.

Some of the key elements in the agricultural machine trade were the prospects for coming crops and for grain prices. In the late 1870s, rising prices of wheat encouraged farmers to invest in land. Even investors smelled money in the air, and organized the famous Red River Valley bonanza farms. After 1883, however, prices of agricultural products began a downward trend to 1896. Export prices of wheat also followed the overall trend. During the 1880s, acreage sown in wheat stayed approximately constant, but wheat output sank considerably, largely due to bad weather conditions.[13]

The business recession that began in 1884 had severe side-effects on the McCormick Company. Its management tried to cope with the depression by wage cuts, but was hit by strikes that cut into productivity and profits. Since American agriculture was simultaneously facing falling grain prices, there seemed to be good reasons for the expansion of trade abroad. On the other hand, American companies had to encounter the European agricultural depression, caused by the cheap American grain. How did they manage the situation? Did it affect their operation? Or was the agricultural crisis visible at all in the McCormick Company's activities?

7.1.2. Sales through jobbing houses

In 1884, when the first effects of the depression were apparent in America and the McCormick Company decided to resort to wage cuts in its production plant, its foreign trade also began to faulter. In the Colonies it was losing ground to its competitors, especially to the English firm Hornsby. In Europe the situation was not much better. McCormick's representative in Russia, George Freudenreich, with his continuous demands and complaints, pleased neither Cyrus Jr. nor E.K. Butler. In 1884, Freudenreich visited the Chicago headquarters, probably for new ideas and to gather encouragement for his future undertakings.[14]

In England, low prices of wheat had collapsed the demand for harvesters, and was forcing English manufacturers to seek customers

---

13 *Shannon* 1945, 156-160, 292-295, 417; *Hughes* 1987, 277, 280-282.

14 *McCormick Company to J. Maszewski* 2.28.1884. Mss 1x, LPCB 457.

on almost any terms. Percy Lankester reported considerable reductions in prices and accused his English competitors of provoking prejudices against American machines. Accordingly he asked for more liberal terms of trade and lowering of prices.[15]

In the midst of its domestic problems, the McCormick Company decided to fight back. It had developed new machines, the "Daisy" light one-horse reaper. To boost the efficiency of its organization, the McCormick Co. transferred Algiers from Freudenreich's to Lankester's control. The Russian trade was given a new injection by sending a couple of able mechanics to help Freudenreich for as long as they could be of service. Freudenreich, for his part, intensified his efforts by cutting the trading areas of his agents and tightening the stipulations.[16] Furthermore, Percy Lankester was shown the green light for the appointment of an agent in Hungary.[17]

In the beginning of 1885 the McCormick Co. had twelve agents, who had appointed a network of their own subagents in various parts of the world.[18] Yet the Company had not decided on its future strategy in Europe. George Freudenreich had made frequent complaints over bad management in the Old World. In its replies, the McCormick Co. had to admit that it had not done enough on canvassing and appointing agents, as the other companies had. The extension of trade to Romania, Hungary and Syria were due to

---

15 *Percy Lankester to McCormick Co.* 12.15.1884. Mss 3b, box 3.
The list price of the McCormick harvester for Lankester was 60 pounds. Walter A. Wood and Howards fixed their price at 52.10.0. , for Samuelson 50 and Hornsby 55 pounds. Lankester received 20 % off 55 pounds or 15 % off 52, with 2.5 % for cash in October. Now he complained that he could only get 43 pounds for his binders instead of 50, because the binder cost him 40 pounds in New York with shipping charges. Three pounds profit marginal on $15 was not enough for Lankester even to cover his working expenses.
In Paris the English makers delivered their mowers at $50 while Lankester had to pay the same in New York. Consequently Lankester asked for his price to be dropped to $47.50 at least. Ibid.

16 *E.K. Butler to Messrs. Lankester & Co.* 4.15.1884. Mss 1x, LPCB 457; *McCormick Co. to Geo. A. Freudenreich* 5.6.1884 and 6.5.1884. Mss 1x, LPCB 457; *De Franquefort to the Directors of McCormick Harvesting Machine Co.* 7.11.1884. Mss 3b, box 3; *Cyrus Jr. to Geo. A. Freudenreich* 1.31.1885. Mss 1x, LPCB 457.

17 *E.K. Butler to Cyrus Jr.* 1.3.1885. Mss 3b, box 4; *McCormick Co. to Messrs. Lankester & Co.* 1.8.1885. Mss 1x, LPCB 457.

18 *List of foreign agents 1885.* Mss 2b, box 31.
The agents were Bell, J. Mohr & Co. in Buenos Ayres, Geo. A. Freudenreich in Odessa, Lankester & Co. in London, Le Bas, Dumaresq & Co. in Montevideo, Emil Liphardt & Co. in Moscow and Charkow, McLean Bros. & Co. in Melbourne and in Adelaide, Morrow, Basset & Co. in Christchurch, J. Maszewski in Odessa, Wm. Staadecker in Bucharest and Jules Thiollier & Co. in Algiers. Ibid. To the list of agents there should also be added Cosimini & Sons in Italy.
With some of the agents McCormick Co. had direct contacts, but in the other cases, like Emil Liphardt and J. Maszewski, information went through the local representative.

Freudenreich's own activity. Cyrus Jr. also agreed that it would be "profitable and wise to do something more in the way of extending our business in Western Europe". From Cyrus Jr.'s letters to Freudenreich, it becomes evident that the McCormick Co. had managed European affairs in an ad hoc way: it had no overall plans for the future, but on the other hand, Cyrus Jr. denied any prejudice against the European business.[19] Freudenreich, who was seemingly getting tired with the slow movements in Chicago as well as with long journeys in Russia, was ready to move his headquarters from Odessa either to Berlin or Vienna. This Cyrus Jr. was not ready to accept, in spite of low sales figures in Russia.[20] On the contrary, general manager E.K. Butler repeatedly assured Freudenreich of the future possibilities in Russia, albeit the breakthrough might be slow.[21]

As a friendly gesture, the McCormick Company sent its machines and one of its mechanics, Lee Borrell, to the Budapest trial in Hungary where he was able to defeat his rivals. Success in the Show opened doors for sales and attracted several agent applicants in Hungary, including the famous English machine and implement factory of Clayton & Shuttleworth, who, however, represented Walter A. Wood in other countries.[22]

---

19 *Cyrus Jr. to Geo. A. Freudenreich* 5.29.1885. Mss 1x, LPCB 458. Cyrus Jr. was confident over the future possibilities and continuation of business in Russia. On the other hand he was not ready at that point to transfer the American sales area system with a main agent and the Company's own houses to Europe. Ibid.
Cyrus Jr. to Freudenreich 7.24.1885. Mss 1x, LPCB 458. Cyrus assures Freudenreich that "we are as anxious to push forward in bringing our machines before notice of European agriculturists, as is possible".

20 *Cyrus Jr. to Geo. A. Freudenreich* 5.29.1885, 7.24.1885 and 8.6.1885. Mss 1x, LPCB 458.
Freudenreich blaimed cheap labor as his main obstacle to large sales in Russia. According to Cyrus Jr. he had even suggested that the only way "to improve the trade is to kill off a couple of thousand stout and healthy young fellows." *Cyrus Jr. to Freudenreich* 8.6.1885. Mss 1x, LPCB 458.

21 *Butler to Freudenreich* 9.15.1886. Mss 1x, LPCB 458. Butler's opinion was that "there is more money in the business to us in Russia, even though it may seem slow for the present, than any other countries upon the continent or in England." Butler restated his argument some time later in about the same words. *Butler to Freudenreich* 11.10.1886. Mss 1x, LPCB 458.

22 *Cyrus Jr. to Nettie McCormick* 7.9.1885. Mss 2b, box 31; *Cyrus Jr. to Lee Borrel* 7.9.1885. Mss 1x, LPCB 458; *Cyrus Jr. to Messrs. Lankester & Co. and to Geo. Freudenreich* 9.3.1885. Mss 1x, LPCB 458.
George Freudenreich objected to such a move, referring to the basic sales strategy of the McCormick Co. "As it has not been your habit to make arrangements with firms which handle other reaping machines, especially if the Woods". Consequently he closed an agreement with Emil Müller. *Geo. A. Freudenreich to McCormick Co.* 10.4.1885. Mss 3b, box 4. Freudenreich was able to get rid of commission trade in the negotiations with Emil Müller, but otherwise had to give the same terms as in the Russian trade. That meant payments in three years, so that by the third year every machine had to be paid for, whether sold or not.

The strikes that hit the McCormick Company are not visible in the foreign correspondence. Mostly they were felt only in machine shipments, which were delayed, or the Company was not able to supply goods at all.[23] It would, nevertheless, be attractive to contemplate whether the recession that began in 1884 had any influence on the foreign trade. A dramatic drop in the production and sales figures is apparent. Was it due to the strikes or due to the slump? If the recession is to blame, the logical outcome for the McCormick Company should be the expansion of its foreign business. On the other hand, it should be remembered that there was a recession in Europe too.

Nevertheless, during the summer of 1886, E.K. Butler, General Manager of the Company, made the first of his subsequently annual tours to Europe.[24] He wanted to meet Percy Lankester, and settle the question over a Hungarian agency. Butler extended his trip to the main grain-growing areas, to obtain an idea of local conditions. After his journey he was convinced of the future possibilities in Russia, which he considered more promising than any other part of Europe, though he did not expect there any rapid expansion. In Hungary he was not ready to enlarge business beyond the small agency.[25]

Still, Butler did not want to close doors against possible customers from other countries.[26] The next year Cyrus Jr. himself made a trip to Europe, where he attended trials in England and in France.[27]

---

23 *E.K. Butler to Messrs. Lankester & Co.* 4.5.1886. Mss 1x, LPCB 458. Butler declared that the Company had "lost a full month" in the strikes, but at the time of the writing of the letter the works were in operation with between 700 and 800 men at work. The situation was, nevertheless, serious for the Company for Butler noted: "This labor question is becoming a very serious matter, and at present places our business entirely out of our control".

24 *E.K. Butler to Cyrus Jr.* 8.11.1886. Mss 3b, box 6.

25 *E.K. Butler to Cyrus Jr.* 8.24.1886. Mss 3b, box 6; *E.K. Butler to Geo. Freudenreich* 9.15.1886. Mss 1x, LPCB 458.
Freudenreich had build an extensive agency network that covered the central grain-growing regions in Eastern Europe. Freudenreich had fourteen agents under him. In Russia they were: F. Trepke in Poltava, Stoll & Co. in Voronesh, A.G. Riedel in Rostov-on-Don, Rahm & Co. in Kazan, John Maszewski in Odessa, Emil Liphardt in Moscow, Hamm & Schmidt in Armavir, Alfred Grodsky in Warsaw, F.K. Ewert in Saratow, P. Van D's Nachfolger in Riga, R.K. Ehrt in Saratov and Koenitzer & Co. in Samara. In addition, Emil Müller in Budapest, Wm. Staadecker in Bucharest and Joseph Friedlander in Vienna worked under Freudenreich. *McCormick Co. to Geo. Freudenreich* 8.9.1886. Mss 1x, LPCB 458.

26 *E.K. Butler to Salicaths Efterfolgere, Esq.* 4.9.1886. Mss 1x, LPCB 458. Salicaths Efterfolgere was a Danish agent candidate to whom Butler announced the Company's normal terms: f.o.b. in New York in cash. Butler remainded the applicant that the Company did not have encouraging memories from Denmark, but for the time being had no agent in that country.

27 *Cyrus Jr. to E.K. Butler* 8.13.1887. Mss 3b, box 7.

If the McCormick Company really expected new growth for its trade from Europe, it needed to acknowledge many setbacks and defects both in its business contracts and in its organization.[28]

Simultaneously with its Hungarian operations, the McCormick Company bargained with Percy Lankester over his future role in the European trade. So far, Lankester had sold machines in various European countries as a jobber.[29] Cyrus Jr., while visiting Europe again in 1888, proposed in a long letter to E.K. Butler a change in Lankester's status. Lankester should take care both of his own business and of McCormick's interests in Europe. In their correspondence Butler and Cyrus Jr. defined the future strategy of the McCormick Company in Europe. Lankester's own jobbing house should take care of the British Isles, Italy and Spain, but the rest of Europe would thereafter be directly under the Company's own control, with Lankester as its salaried manager.[30]

---

28 Butler had to remind Freudenreich in 1887 of Müller's poor collections in Hungary and of his late payments. *E.K. Butler to Geo. Freudenreich* 6.22.1887. Mss 1x, LPCB 459.

Concern about Hungary began to accumulate. The next summer McCormick's travelling expert, H. Poppe, sent a worried message from Budapest: only a few sales had been made, and Poppe judged that the business was not being run properly and asked either Cyrus Jr. or Butler to come and investigate the matter. *H. Poppe to McCormick Co.* 6.9.1888. Mss 2c, box 112.

The affair was transferred to Lankester, who after long negotiations was finally able to settle it. Freudenreich was released from the Hungarian business and Lankester made a contract with the Prager Maschinenbau Actien Gesellschaft of Prague for a new agency in Budapest. Emil Müller continued his career in Prager's service, as its manager in Budapest, but simultaneously sold McCormick's machines on commission. This arrangement offered the McCormick Co. opportunities to get rid of the unsold stock of machines and at the same time continue business in Hungary.

Lankester to Cyrus Jr. 8.3. and 8.14.1888. Mss 2c, box 112; *Geo. A. Freudenreich to Cyrus Jr.* 8.26.1888. Mss 2c, box 112; *E.K. Butler to Cyrus Jr.* 9.8.1888. Mss 3b, box 9; *Lankester to Cyrus Jr.* 10.19.1888. Mss 2c, box 112; *Lankester to McCormick Co.* 11.3.1888. Mss 2c, box 112.

The poor outcome of the Hungarian business arose partly from the insuffient economic standing of Emil Müller, but also from the growing disagreement between him and George Freudenreich.Ibid.

Calculations by Percy Lankester, who was sent to make an investigation, showed that it would be totally unprofitable for the McCormick Co. to continue business in Hungary under their own management. Even if Freudenreich had been able to sell all the estimated 40 harvesters, 30 Daisy reapers and 10 mowers for 1889 with a profit of $9320, the total expenses of $13 000 would have have meant about $3680 deficit. The total loss caused by Müller to the McCormick Company Lankester was estimated to be $16 000. *Lankester to McCormick Co.* 11.3.1888. Mss 2c, box 112.

29 A jobber bought his machines at a specified price either for cash or on credit. He ordered from the factory only the number of machines he judged it possible to sell. The Company's responsibilities ended when it had delivered the machines aboard a steamer in one of the Atlantic ports. The jobber of course thought mostly of his own advantage, and did not want to increase his stock which might be left unsold. Jobbers favored long contracts, because as trade began to grow, companies tended to cut their sales areas and to contract with new jobbers. *Lankester to Cyrus Jr.* 10.5.1888. Mss 2c, box 112; *Schonberger* 1964, 25-26; *Heikkonen* 1989, 162-163.

30 *Cyrus Jr. to E.K. Butler* 8.24.1888. Mss 2c, box 112; *Butler to Cyrus Jr.* 8.28. and

The new management ideology that the McCormick Company had introduced since the time of Lewis Wilkinson now also became visible in its foreign trade. Organization and decision making were concentrated in the hands of professional managers. When E.K. Butler was appointed as the General Manager of the Company, a clear division of responsibilities must have been made between him and Cyrus Jr., although no formal agreement has been found. In the correspondence of Cyrus and Butler during Cyrus's trip to Europe, it becomes evident how dependent Cyrus was on Butler's opinions. On almost all the key decisions, he asked Butler's views. Finally Butler took foreign trade totally under his own control, which was a logical extention of his travels in Europe.[31] This left Cyrus Jr. more time to concentrate fully on the overall business as the President of McCormick Company, while the everyday functions were left to professional managers.

Hand in hand with the new management structure, the McCormick Company's attitudes towards the European trade also began to change.[32] While Lankester was still discussing the terms of his own contract with the McCormick Company, he was also negotiating with the French house of Mot & Co. on the future business in France. He did not know that his contract was bound to the results of his

---

9.8.1888. Mss 3b, box 9; *Cyrus to Butler* 9.12.1888. Mss 2x, box 201; *Lankester to Cyrus Jr.* 10.21.1888. Mss 2c, box 112; *Cyrus Jr. to Lankester* 1.26.1889. Mss 1x, LPCB 459.

Lankester and Cyrus Jr. found a common base for the contract in the fall when Lankester finally agreed to manage the European countries for $2000. This amount would compensate for the loss of his own trade in France and in Germany. *Cyrus Jr. to Butler* 9.12.1888. Mss 2x, box 201; *Lankester to Cyrus Jr.* 10.21.1888. Mss 2c, box 112; *Cyrus to Lankester* 1.26.1889. Mss 1x, LPCB 459.

In his letters Cyrus Jr. also made estimations on the outlook for business in some countries. In Germany and Austria-Hungary he regarded prospects for large sales in the near future as minimal, though these countries had potential and should therefore be under the Company's own management. See for example *Cyrus Jr. to Butler* 8.24.1888. Mss 2c, box 112.

31 For example on 8.24.1888 Cyrus sent an eight-page letter to Butler where he explained in detail the European situation and his own ideas, and asked Butler to send a telegram in code. *Cyrus to Butler* 8.24.1888. Mss 2c, box 112.

On decision making, cf. the following examples. "There is no real necessity your going to Budapest; I have no doubt Lankester can do as well..." Take the whole Budapest business off Fr (Freudenreich) + put in chg L. (Lankester) and let Fr. return Odessa..." *Butler to Cyrus Jr.* 8.28.1888. Mss 3b, box 9.

32 The Company had for some time felt dissatisfaction with George Freudenreich. Now, when the McCormick Co. was ready to move, Cyrus was inclined without hesitating a second to throw over Freudenreich, who was regarded as incompetent to deal with sharp businessmen. *Cyrus Jr. to Butler* 8.24.1888. Mss 2c, box 112. Cyrus did not feel himself responsible to offer Freudenreich a new place in case he was fired, although Cyrus noted that Freudenreich "will be entirely in the cold if we don't offer him some place".

bargaining in France.[33]

An agency contract was signed, in spite of the fact that Mot & Co. also represented a competing harvester company, the Johnstons Harvester Company. The McCormick Company also broke another of its basic rules: its contract with Mot was for three years, whereas it normally made only one-year agreements. McCormick's action was defensive: it wanted to prevent other companies from possible agreements with Mot;[34] but this policy was soon found to have been a mistake. The Company had bound its hands for three years in France, and when Mot & Co. did not promote its machines as anticipated, could only protest.[35] This episode showed the McCormick Company the defects of jobbing houses and forced it to re-evaluate its marketing strategy.

The McCormick Company's activity also spread to other European countries, but it responded to initiatives rather than actively searched for new markets. Between 1888 and 1890 the Company made agency contracts in Germany, Denmark, Sweden and Norway. The expansion of its operations also revealed the Company's new organization structure. Headquarters in Chicago forwarded incoming applications to the care of Percy Lankester, and informed all parties of his role as the European Manager of the McCormick Company. The daily conduct of business, and bargaining with agents, were left to Lankester, who reported on his undertakings to Chicago. The Company's subagents acted in their respective countries as sole general agents.[36] Now the McCormick Company had constructed a

---

33 *Cyrus Jr. to Butler* 9.12..1888. Mss 2x, box 201; *II.T. Mot & Co. to Cyrus Jr.* 9.16.1888. Mss 2c, box 112.

34 Before a formal agreement was made, the McCormick Co. checked the background of Mot & Co. Since the results of these investigations were positive and Mot & Co. offered greater sales than Paul Francey, E.K. Butler showed the green light for continuation of the discussions with Mot. *E.K. Butler to Cyrus Jr.* 10.17.1888. Mss 3b, box 9; *Copy of a proposition for an agreement with Mot & Co.* 9.27.1888. Mss 2c, box 112; *Lankester to Cyrus Jr.* 10.5.1888, 10.7.1888, 10.10.1888, Mss 2c, box 112; *Draft for an agreement between A.Mot and McCormick Harvester Machine Co.* No date. Mss 2c, box 112.
McCormick's new approach to the business in Europe was also clear in the French case. Paul Francey, the former agent, had not sold enough machines and had to give way to a more energetic and larger firm. *Cyrus Jr. to Lankester* 1.26.1889. Mss 1x, LPCB 459.

35 *E.K. Butler to Percy Lankester* 1.17.1889. Mss 1x, LPCB 459. Hardly had the ink on the contract dried before Mot & Co. bought machines from Osborne "at a ruinously low price, for less than we can manufacture", as Butler stated. Later on Mot also bought mowers from the Buckeye concern. *E.K. Butler to Lankester* 6.24.1889. Mss 1x, LPCB 459. At the end of 1889 the contract with Mot & Co. was a total disappointment. The firm had to carry over unsold machines and could not make large orders. *E.K. Butler to Lankester* 12.6.1889. Mss 1x, LPCB 460.

36 *Butler to Lankester* 7.31.1889. Mss 1x, LPCB 460.

working organization and could effectively control its achievements.[37]

As Manager, Lankester wrestled with one constant problem. Agents bought their machines in cash f.o.b. in New York. As a result, the marketing structure was inflexible. When there was either a surplus or shortage of machines in some areas, it was difficult to move machines from one agent to another. The reason for this situation was that the McCormick Company did not keep a free stock of extra machines or machine parts in Europe.[38]

Since the main lines of the McCormick Company's European trade after the recession of 1884 have now been examined, it is time to return to the effects of the depression on foreign trade. In the newly colonizing areas of the Pacific and of South America there was no visible growth in sales, as can be seen from Table 10, nor did the Company devote any more energy to these regions than before. In Europe we can observe a change in the marketing strategy of the McCormick Company. Although sales were still made through independent jobbing houses, a special European Manager had been

---

37 From Germany an agency application came from P.H. Mayfarth & Co., whose application was directed to Lankester. *E.K. Butler to Cyrus Jr.* 9.26.1888. Mss 3b, box 9 and to *P.H. Mayfarth & Co.* 9.26.1888. Mss 2c, box 112. Here too Lankester asked for information on the applicant's financial standing, this time from Clayton & Shuttleworth. *Lankester to Clayton & Shuttleworth* 10.12.1888. Mss 2c, box 112. Finally Lankester did not contract with Mayfarth, who demanded machines on consignment. *Butler to Lankester* 7.26.1889. Mss 1x, LPCB 460. Nevertheless, at least F. Vogeler in Prussia acted as McCormick's agent in Germany. *McCormick Co. to F. Vogeler.* Mss 1x, LPCB 459.
Appelberg & Co. from Gothenberg in Sweden also expressed its interest in McCormick's machines. *Butler to Lankester* 6.11.1888. Mss 1x, LPCB 459. Koefold & Haugberg from Copenhagen in Denmark applied for the whole of Scandinavia but had to be content with Denmark. *Butler to Lankester* 8.12.1889 and 9.11.1889. And *McCormick Co. to Koefold & Haugbergs Maskinudsalg.* Both Mss 1x, LPCB 460. In 1890 Koefold & Haugberg got a competitor, Geo. S. Bendix also from Copenhagen. *Butler to Lankester* 8.6.1890. Mss 1x, LPCB 460 and *Butler to Geo. S. Bendix* 8.26.1890. Mss 1x, LPCB 460.
Conrad Knudson from Arendal represented McCormick's machines in Norway and Sweden was finally allotted to Andersson & Mattson of Malmö. *Butler to Lankester* 7.26.1889. Mss 1x, LPCB 460; *McCormick Co. to Lankester* 1.28.1890. 1x, LPCB 460.

38 See for example for complaints about the situation *Lankester to Cyrus Jr.* 7.22.1889. Mss 10c, box 10.
McCormick Co. recognized the problem and promised to arrange some extra machines. *Butler to Lankester* 8.12.1889. Mss 1x, LPCB 460.

Table 13. Number of machines sold in the main European agricultural countries by the McCormick Company, 1877-1902.

Year	England	France	Russia	Germany
1877	4	-	-	-
1878	20	-	-	2
1879	100	70	58	-
1880	19	-	504	
1881	-	-	-	-
1882	-	-	-	-
1883	-	8	-	-
1884	200	-	647[g]	-
1885	126	13	233[h]	-
1886	120	26	621	-
1887	-	-	-	-
1888	-	-	-	1
1889	-	-	150[i]	-
1890	325	110	561	77
1891	(407)	(75)	-	(150)
1892	-	-	-	-
1893	-	-	1008[d]	-
1894	-	1025[f]	1131[e]	-
1895	(1126)	(1141)	(2093)	(866)
1896	1443	1587	2171	1211
	(1443)	(1587)	(3672)	(1309)
1897	-	-	2310[c]	-
1898	-	-	-	4500[a]
1899	-	9100[b]	195	10
	-	-	-	5600[a]
1900	-	-	-	-
1901	3287	10428	15865	13125
	(3134)	(11434)	(12191)	(14806)
1902	(3780)	(11562)	(10156)	(12386)

Source: Figures in the paranthesis are from Machines sold in foreign countries counted by agents. Mss 3x, box 26. Other figures are from Statement showing machines sold to foreign countries, 1874-1801. Mss 1a, box 72. For the realiability of the information see chapter 5.2. note 12. [a] is from William Couchman to Cyrus Jr. 2.23.1899. Mss 2x, box 306. [b] is from R. Wallut to McCormick Co. 10.14.1898. Mss 2x, box 304. [c] is an estimate. Tracy to McCormick Co. 4.17.1897. Mss 2x, box 294. [d] is George Freudenreich to McCormick Co. 7.18.1893. Mss 2x, box 235. [e] George Freudenreich to McCormick Co. ? 1894. Mss 2x, box 235. [f] Lankester to McCormick Co. Received 12.16.1894. Mss 2x, box 235. [g] Freudenreich to McCormick Co. 12.6-18.1884. Mss 3b, box 3; Cyrus Jr. to Freudenreich 1.31.1885. Mss 1x, LPCB 457. Figures are only for Emil Liphardt of Moscov and for Maszewski of Odessa. [h] Cyrus Jr. to Freudenreich 1.31.1885. Mss 1x, LPCB 457. Figures are only for Maszewski in Odessa. [i] Freudenreich to Cyrus Jr. 9.21.1888. Mss 2c, box 112. Maszewski's order for 1889.

appointed to oversee the business. In this way, one could state that by intensifying its operations, the McCormick Company was attempting to recoup compensation from abroad for its setbacks in the States. This explanation follows in general the outlines of Mira Wilkins' concepts on the behavior of American enterprises.[39] It also displays several of the characteristics of a modern enterprise as defined by Alfred Chandler:[40] a managerial hierarchy, a new factory system, and transformed production technology; yet it retained many of the older structures, foremost of these, control by the owning family. Besides, there was no surplus to sell abroad.

It has been impossible to cover systematically the McCormick Company's foreign, especially European business, in the 1880s. Figures constructed from the primary material are only indicative and give at best only a suggestion of possible developments. However, it does appear that there was no rapid, visible growth in foreign sales during the recession years of the 1880s. Organization of the European trade did not lead to any drastic changes in the volume of trade. Consequently, earlier explanations have to be modified in this respect.[41] Europe had economic problems of its own, and in spite of falling grain prices, the American farmer still had land to conquer, and could compensate for low prices with increased acreage. As long as there was demand on the domestic market, it was far more important than foreign fields.

Although there was no significant growth in the volume of McCormick's exports, its sales organization was founded on a stable footing, notwithstanding Fred Carstensen's statements.[42] Foreign trade had been made part of Company business directly under the General Manager, who supervised it through two salaried agents in Europe. McCormick's also sent its traveling experts abroad from the very beginning, and the question of a central depot for machines

---

39 *Wilkins* 1970, 45.

40 *Chandler* 1987, 57-61.

41 See for example *Schonberger* 1964, 24. In one case Schonberger states that "the great attraction of Europe was a World's Fair, an exhibition, or a trial." p. 15. On the other hand, he writes on the effects of the depression and drop in McCormick's sales that "one discernable effect was an increased interest in and expansion of the Company's foreign business." p. 24.
Schonberger's statements are contradictory and my own research does not confirm either of them. Fred Carstensen also came to the same conclusion. *Carstensen* 1984, 252, note 13. On the other hand Carstensen supports Schonberger's views on the significance for foreign trade of the depression of the 1880s, *Ibid, 112-113,* and in fact has adopted them from the latter.

42 *Carstensen* 1984, 113-114. Here Carstensen either relies on secondary sources or "survey of correspondence, SHSW/McC. ser. 2x" and "this summary is based on extensive research in McCormick records". Ibid, 252 notes 14-16.

and repairs had been taken under consideration. In this way, the foreign trade was transformed during the 1880s as an integral part of McCormick's business and organization, although it did not have all the same resources as McCormick's domestic organization.

7.2. The fight continues in Europe

7.2.1. Old enemies, new circumstances

In the 1880s, the McCormick Company had made fundamental changes in its organization structure. Professional managers replaced family members in the daily operations of the factory, and its business ideology was becoming even more aggressive, if possible, than before. Were these changes also reflected in its relations towards competing firms, and how successful were they? As has been shown, the McCormick Company relied on jobbers in its foreign trade, and, except in Russia, did not extend long credit to farmers. Was the McCormick Company in this relation a typical representative of American harvester companies abroad, or did it follow its own marketing policy?

Among the American harvester manufacturers, perhaps Walter A. Wood had best established itself in Europe. Even by the 1870s, it already dominated the field: in 1874 Wood had exhibited or taken part in trials and shows in ten countries.[43] By 1882, the Company had either branch offices or general agencies in London, Paris, Buenos Ayres and Valparaiso.[44] In 1876, the Johnston Harvester Company also proclaimed victories in several countries and in 1882 it had founded its branch office in London.[45] In 1876, D.M. Osborne & Co. listed branch offices and depots in Bremen, Liverpool and Paris, and three years later it had expanded to Christ Church, Sydney,

---

43 *Wood's Mowing and Reaping Machines, 1874*. Mss 4z, box 25. The countries were Great Britain, France, Germany, Holland, Belgium, Russia, Sweden, Norway, Denmark and Switzerland.
In 1878 Wood boasted to have beaten, between 1873 and 1877, in Europe alone, Samuelson & Co. at 146 trials, Hornsby & Sons at 104, Osborne & Co. at 133, Johnston at 71 and Buckeye at 58 trials. *Wood's Mowing and Reaping Machines*, 1878. Mss 4z, box 25.
Wood's catalog does not mention the McCormick Co. at all, which reflects the state of the business in Europe.

44 *Fair circular, 1882. Walter A. Wood's Harvesting Machines*. Mss 2z, box 25. Because of the fragmentary material on competing companies, no definite date can be stated. Although catalogs do not mention foreign agencies before 1882, it is obvious that Wood had established them in Europe before that year.

45 *The Johnston Harvester Company, 1876 and 1882*. Mss 4z, box 14.

Melbourne, Adelaide and Montevideo.[46] Similarly, in 1878 William Anson Wood listed in its catalog a branch office in London.[47] In addition, at least Champion and Buckeye had foreign branches.[48] Furthermore, it should be kept in mind that European manufacturers were also struggling for the same markets. In addition to the English firms, Hornsby and Samuelson, there were numerous smaller manufacturers in almost every country that were making copies of the American harvesters.[49]

During the early years of the foreign trade, all the American manufacturers were very similar in their functions. They all established themselves in Europe through jobbing houses that oversaw the entire business in the Old World. Another typical feature in these early foreign undertakings was the central role played by the Oceanian and South American markets. In this respect the McCormick Company was no different from its competitors; but some of the other companies had already extended their business beyond this phase, by sending their own salaried agents to Europe or by establishing their own branch houses or depots. The real role and functions of these branches remain unclear, due to the lack of information. Although details in the Company catalogs have to be approached with caution, they are nonetheless indicative. Percy Lankester confirms, for example, in 1884, that Champion and Osborne had opened agencies in London.[50] In this respect the McCormick Co. maintained its careful strategy. It was not ready to take risks in foreign business and in spite of extension abroad, priority was given to the home market.

------

46 *D.M. Osborne & Co. 1876 and 1879.* Mss 4z, box 17.

47 *William Anson Wood's Sweep Rake Reapers and Improved Eagle Mowers, 1878.* Mss 4z, box 5.

48 *The Champion Harvesting Machines, 1883.* Mss 4z, box 24. Champion's branches were in Bremen and in Valparaiso.
Buckeye Harvesting Machines. C. Aultman & Co. 1884. Mss 4z, box 1. Buckeye registered among its branch offices and principal depots Paris, Matanzas in Cuba, Buenos Ayres and Sydney.
Adriance, Platt & Co. of the Buckeye combination had in 1881 also a general agent of its own in Liverpool. *Adriance, Platt & Co. 1881.* Mss 4z, box 1.

49 *Samuelson & Co. Mowing and Reaping Machines 1878.* Mss 4z, box 21. Unfortunately only one copy of Samuelson's catalogs has been preserved in the McCormick Collection. Up to 1878 Samuelson & Co. had exhibited its machines mostly in France but also in such remote places as Sweden, Norway and even South Africa.
On the imitations of American agricultural machines see for example *United States Consular Reports No. 38, 1884,* p. 555-556.

50 *Butler to Lankester 1.8.1884.* Mss 1x, LPCB 457.

As a latecomer on the European field, the McCormick Company had a major task to accomplish if it wanted to turn the situation to its benefit. As can be seen in Table 15, the McCormick Company was one of the leading harvester companies in America during the 1880s, but not the leader. In its 1884 catalog, the Walter A. Wood Co. published still larger sales figures than the McCormick Co. did. Although it is impossible to inspect the reliability of figures of Wood's sales, they certainly are indicative. In addition, the Deering Company was also rapidly increasing its share of the markets.

Table 14. Total number of harvesting machines and mowers produced in the U.S., 1870-1904.

Year	Total production
1870	163 085
1875	159 410
1880	188 974
1890	324 779
1900	674 199

Source: Census of the U.S. Manufacturers 1880, 1900, 1920.

Table 15. Sales of the harvesting and mowing machines of the leading harvester manufacturers, 1870-1888.

Year	McCormick Co.		Walter A. Wood Co.	Deering Co.	
	Number	$m	Number	Number	$m
1870	9 033[a]	..	15 000	..	..
1875	11 476[a]	..	23 507	..	..
1880	26 786	..	27 903	..	..
1881	32 353	..	..	7 197	..
1882	44 848	..	..	12 197	..
1883	50 376	..	..	17 130	..
1884	54 922	4.5	48 315	22 709	4.3
1885	51 439	3.7	..	25 480	3.0
1886	44 103	2.9	..	26 564	3.6
1887	71 363	3.9	..	31 715	4.2
1888	91 881	4.6	..	41 095	5.1
1889	104 114	4.7	..	..	4.9

Source: For the McCormick Company: McCormick Machines Built since 1841, Statement showing number and kind of machines sold during the years 1880 to 1902 inclusive. Mss M/I, box 18. For Walter A. Wood: Harvesting Machines. Thirty-second Annual Circular, 1885. Mss 4z, box 25. For the Deering Co.: An unlabeled and undated memorial of the Deering Co, 1890. Mss w, box 2; Deering Co. Financial State. Schedule "A" showing the net sales for each year. Mss w, box 2.
[a] machines manufactured. The McCormick Company's figures also include attachments, carriers, grinders and similar machines. In this respect the figure is not entirely comparable with the others.
[b] In reading the Deering figures, it has to be taken into account that the Company has not necessarily counted the harvester and binder as two distinct machines as the McCormick Co. did.

In 1884 Deering's net sales were $4 251 506,[51] as compared with McCormick's sales of $4 469 271.[52] In summary, we can with certainty state that McCormick, Walter A Wood and Deering controled at least half of the American harvester market during the 1880s. Although it is not possible to make far-reaching interpretations, due to the nature of the material, it seems that during the strike years the McCormick Co. lost its leading position to the Deering Co., but was able to regain it at the turn of the century.

---

51 *Deering Co. Financial state.* Schedule "A" showing the net sales for each year. Mss w, box 2.

52 McCormick Harvesting Machine Company. *Net sales of Machines...* Mss M/I, box 18.

Did the European trade also follow the same main lines as in America? How did the smaller companies react to the situation? Did they compensate for their losses on the home field with expansion abroad?

When McCormick appointed Percy Lankester first as its agent and then later on as its salaried European Manager, its position in Europe also began to grow to the same scale as in America. In one of his first letters to Chicago, Lankester noted that Wood did not sell half the binders he did. On the other hand, France was Johnston's territory.[53] In Russia, the competitors were the same; at least Wood, Deering, Osborne, Champion, Johnston and the English makers.[54] The worst rivals for the McCormick Company were the English firm of Hornsby and Wood,[55] but Osborne was losing ground, at least in France, where it had to close its large office and warehouse and withdraw its representative.[56] By 1888, the Canadian Massey Company had arrived in Europe, and seemed to be following the same policies as the U.S. companies.[57]

In Russia, by the end of the 1880s, the foreign harvester companies began to meet local competition. George Freudenreich complained of cheap Russian reapers, that "have taken away a large proportion of his (Maszewski's) customers".[58] Local makers, with agressive price policies, were also becoming an obstacle in France to all American firms: by 1888 there were at least three French binder makers. Besides, copying of machines and machine parts appears to have been common.[59]

---

53 *Lankester to McCormick Co.* 12.15.1884. Mss 3b, box 3.

54 *Freudenreich to McCormick Co.* 12.6-18.1884. Mss 3b, box 3; *McCormick Co. to Freudenreich* 1.31.1885. Mss 1x, LPCB 457; *McCormick Co. to Freudenreich* 8.6.1885. Mss 1x, LPCB 458.

55 See for example *Cyrus Jr. to McCormick Co.* 8.13.1887. Mss 3b, box 7; *Butler to Edward Ackerman* 8.18.1887. Mss 1x, LPCB 459; *Hornsby's to Lankester.* Mss 2c, box 112.

56 *Cyrus Jr. to Freudenreich* 5.29.1885. Mss 1x, LPCB 458.

57 *Lankester to Cyrus Jr.* 8.28.1888. Mss 2c, box 112.
In 1865 at the Paris International Exposition Massey had exhibited its machine for the first time in Europe. This event did not lead to foreign sales. Consequently, in 1886, according to Denison, the company made its second entree at the Indian and Colonial Exhibition in London, again without raising great interest in the audience. The Massey Company was still in 1887 a relatively small enterprise with an output of only 8851 machines. In spite of this modest beginning, the company sent its own representative in 1887 to London and next year was able to report sales in England, Scotland, Ireland, France, Germany, Belgium and France, consisting of 127 reapers, 30 binders, 58 mowers and 44 rakes. *Denison* 1949, 51-52, 100, 104-105.

58 *Freudenreich to Cyrus Jr.* 9.21.1888. Mss 2c, box 112.

59 *Lankester to Cyrus Jr.* 10.5. and 10.10.1888. Mss 2c, box 112.

The only major change during the 1880s, compared to the 1870s, was that the McCormick Co. had recouped in competition and that the Canadian Massey Company was increasing its impact on the trade. The same companies fought each other everywhere: in Russia, France, Germany, or even in small countries such as Sweden or Finland.[60]

Although European manufacturers copied American machines, they were not, except for the English makers, a real threat to the latter.[61] The U.S. consul for Belgium, Geo. C. Tanner, counted twenty-three imitations of American agricultural machines at an exhibition in Liege. In some cases, names of American companies were erased or painted over.[62] The situation was similar in France, where twelve of the sixteen competing machines were American and the remaining four more or less copies of them.[63] This situation remained very stable for years. In 1890 the consul for France reported a heavy demand for mowers, reapers and binders; the field was still totally in the hands of the Americans. At the field test of the French Department of Agriculture in Perigueux, only two of the competing machines were not American.[64] Even the English firms had to admit the superiority of their American competitors.[65]

---

60 See for example *Lankester to Cyrus Jr.* 7.22.1889. Mss 10c, box 10; *Butler to Lankester* 11.4.1890. Mss 1x, LPCB 460; *Rönnbäck 1883*, 214. Present at the eighth general meeting of Finnish agriculturists in 1881 were at least Hornsby & Son, Walter A. Wood and William Anson Wood; *Hällström 1889*, 283-286. In the corrsponding meeting at Viipuri in 1887 the Swedish Aktiebolaget Palmcranz & Co. and Vesterås mekaniska verkstad were awarded the first prize silver medal for their mowers, as was the Finnish Åbo Jernmanufakturbolag. The Swedish firm Överum's mowers also reaped the second prize bronze medal. The same honor was given to William Anson Wood's mower, which was the only American manufacturer to receive any offical attention. On the other hand, the question might be one of domestic jurors and powerful agents like Wictor Forselius or Francke & Hackman, who represented William Anson Wood for the Swedish companies.

61 R.B. Swift from the McCormick headquarters made bitter and very instructive comments on the English companies in his letter. According to him "his (Hornsby's) machine should be a good one. There are enough of our ideas upon it to make it a good one." *R.B. Swift to Lankester* 5.20.1889. Mss 1x, LPCB 459.

62 *Reports of the consuls of the United States.* No. 38. 1884. p. 555-556. Belgian firms took castings and made exact molds of them or took the machines to pieces, made drawings of them, and built as near a copy as possible. But when finished "it bears very near the same resemblance to the machine they are trying to steal as that a locomotive of fifty years ago bears to one of to-day". Ibid.

63 *Reports from the consuls of the United States. No. 72.* 1886. p. 558.

64 *Reports from the consuls of the United States. No.123.* 1890. p. 649. These two European machines were from France and from England but "were made after the fashion of patents that have expired in the United States" and "it is certain that no English or French mower or reaper could compete at present with the American models in beauty, workmanship, and, we believe, in price."

65 *Consular reports on commerce, manufactures, etc. No. 154.* 1893. p. 314-315.

In the long run, it was the German manufacturers who offered the most keen resistance to the American firms. In Russia, German makers outnumbered their American rivals in the sales of plows and harrows, but could not compete with the harvester companies.[66] The growing interest of the European manufacturers in harvesting machines reflects the increasing demand. They clearly anticipated profits in that business and wanted to have their share of it.

The growing productivity of American industry and adoption of new and more efficient production methods began to arouse fears in Europe, even in small and remote areas such as Finland. The United States was regarded as a giant which would in the long run destroy Europe's industry.[67] Consequently, tariff walls began to rise against American exports. In Russia, foreign companies were able to sell their products free of import charges until 1884, when a duty of fifty kopecks per pood,[68] effective in 1885, was placed upon imported agricultural machinery and implements made of iron and steel. In 1887 this duty was raised to seventy kopecks per pood, but was reduced in 1896 to fifty-two kopecks per pood.[69]

Foreign machinery or implements made principally of wood or cast iron imported into Germany were placed under a duty of 3 marks per 100 kilograms, and the same goods made of wrought iron 5 marks per 100 kilograms.[70]

Russian tariffs were raised to protect the country's own growing reaper industry. The "lobogreika" (forehead sweat) was a domestic Russian copy of the Walter A. Wood Company's handrake reaper that was produced in the 1860s. It was crude, simple and easy to

---

66 *Consular reports. Commerce, manufactures, etc. Vol. LV.* 1897. p. 270-277; *Consular reports. Commerce, manufactures, etc. Vol. LXIII.* Nos. 236, 237, 238, and 239. 1900. p. 73-74;

67 See for example *Tallqvist* 1906, 24-25, 30-35.

68 1 pood is equivalent to 16.4 kilograms or 36.4 pounds. Facta 2001 1984,vol. 11, 397.

69 *Queen 1942*, 144; *United States Consular Report No. 48*, 1884. p. 490, 493; *Consular reports. Commerce, manufactures, etc. No. 205.* 1897. p. 271.

70 *United States Consular Reports. No. 48*, 1884. p. 498, 518.
German tariff laws had many peculiar details. Although in principle tariff was based on the material of which the machine was principally composed, nevertheless, polish, painting and finish or outward decoration also had an effect upon the final duty. In practise it was difficult even for a professional importer to say what the duty for a new product would be until a sample machine had been imported. If a machine was composed of several materials, it was set in the class to which the greatest weight of its materials belonged to. However, the final rate of duty was dependent on the outward finish of the article. So plows which were rough and unpainted paid a duty of six marks, but if they were painted or polished the duty was ten marks. The German tariffs remained at the same level until 1899. *Consular reports. Commerce, manufactures, etc. Vol. LXI.* Nos. 228, 229, 230 and 231. 1899. p.124-125.

manufacture and repair, and what was most important, it was cheap. It was a real threat[71] to all foreign harvester companies in a country where only 8.9 percent of the peasantry could produce surplus grain to sell, and where a peasant's expenses could amount to fifty-five roubles, while his income was only thirty-three roubles. It was therefore no wonder that imports of American machines dropped in value from 417 000 roubles in 1884 to 28 000 roubles in 1886.[72] In 1896, there were 196 factories producing agricultural machinery, to the value of 9.6 million roubles. The high duties did not protect and develop domestic Russian manufacture of agricultural machines; instead they raised the prices of harvesters by 18 to 20 percent.[73]

To avoid paying the Russian duties, American firms began to repack their products in Germany to meet the demands of the Russian tariff administration. This led to another problem. When the German companies began to stamp their own names on American machines, the Russian agents could no longer distinguish American products from German ones.[74] This was not the case with the McCormick Company, however, which shipped its machines directly from the United States to South Russian ports, normally to Odessa.[75] Sometimes the English customs caused delays in shipments. Importers were in such cases almost totally powerless, and all they could do was wait for customs decisions.[76]

In addition to tariffs, all the American harvester companies were

---

71 The McCormick Company considered the situation but found it profitable to recommence production of its old handraking models. Modification of the reaper would swallow too much time and energy and finally the profit marginal for such a cheap machine was too narrow and further narrowed by Russian competition. *Cyrus Jr. to Freudenreich* 8.6.1885. Mss 1x, LPCB 458.

72 *Queen 1942*, 141-142, 144-145, 147-150; *Freudenreich to Cyrus Jr.* 9.21.1888. Mss 2c, box 112.

73 *Consular reports. Commerce, manufactures, etc. Vol. LV. No. 205. 1897. p. 270-271; Freudenreich to Cyrus Jr.* 9.21.1888. Mss 2c, box 112.

74 *Queen* 1942, 144-145. The case was similar with shipments through other countries, for example through Britain to France. *Reports from the consuls of the United States.* No. 3. 1881. p. 100.

75 *Emil Liphardt & Co. McCormick Co.* 2.15-17.1884. Mss 1x, LPCB 457; *McCormick Co. to J. Maszewski* 2.28.1884. Mss 1x, LPCB 457.
Moscow shipments were sent from London to Reval in the Baltic, which was the normal port for Moscow. *McCormick Co. to Emil Liphardt & Co.* 4.30.1884. 1x, LPCB 457.

76 *Lankester to Cyrus Jr.* 8.9.1888. Mss 2c, box 112. McCormick's machines to Budapest were stopped without any reason and Lankester expected at least 10 to 15 days' delay in transportation. That of course was a severe setback for the local agent, who was unable to supply the machines to his customers. Besides, there was always a threat that farmers might refuse to take the machines any more if they came too late for the harvest.

influenced in about the same way by changes in weather. This was equally evident in America and in Oceania. Crop failures at frequent intervals had especially dramatic influences in Russia. In four of the eight years between 1885-1892 Russia suffered serious shortfalls in crops. Consequently, the price for labor sank to a level where laborers could be hired for their board, which meant only black bread, cucumbers and an occasional drink of poor vodka.[77] The famines and poverty arising therefrom were reflected in harvester sales. In general, if the promise of a good harvest continued throughout the spring, agents' orders rose, in anticipation of large sales. Conversely, if any country was hit by a drought or heavy rains, machines were left unsold on agents' hands.[78]

7.2.2. Trials, services and pamphlets

By the 1880s, Europe had become an extension of the American harvester market, with the same factories struggling for the markets on both sides of the Atlantic. As has been noted before, the American harvester companies did not transfer their sales organization to Europe; only a few firms maintained what they called branches in Europe. Nor did they extend the system of hire purchase directly to farmers, but, as in McCormick's case, sold solely to agents f.o.b. in cash. In such circumstances, how could they sell their products to European farmers in competition with native producers during agricultural crises?

In 1884, the McCormick Company had to admit that, for the time being, it had lost its leading position in the Colonies to its English competitors. Defeats in England, on the other hand, were blamed on unfair domestic competition, aimed at keeping American firms out of Britain. Leaders of the Company tried to convince its agents of the good qualities of its machines: the machines were good, but competitors unfair and losses accidental.[79]

---

77 *Queen 1942*, 143-144, 149-150.

78 *McCormick Co. to Freudenreich* 6.5.1884. Mss 1x, LPCB 457; *E.K. Butler to Freudenreich* 6.22.1887. Mss 1x, LPCB 459; *E.K. Butler to Freudenreich* 1x, LPCB 460; Queen 1942, 143-144.
For example, in France after heavy rains in the fall, farmers even returned binders that had already been delivered. *Lankester to Cyrus Jr.* 8.2.1888. Mss 2c, box 112.

79 *E.K. Butler to Lankester & Co.* 1.8.1884. Mss 1x, LPCB 457. Butler tried to reassure Lankester by stating that "we stand further at the front to-day than we have ever heretofore, both in New Zealand and in Australia..."
Cyrus Jr. to Lankester & Co. 8.27.1884. Mss 1x, LPCB 457. The English companies that were victorious in the Shrewsbury show were Hornsby and Howard. Although Cyrus Jr. was not pleased with the results of the show, he was satisfied with Lankester's efforts and did not anticipate a dramatic impact on trade due to the late date of the trial.

Cyrus Jr. was not concerned without cause. All the companies were trying to obtain as much publicity for their victories as possible. Cyrus himself wrote immediately to his mother of their triumph in Hungary,[80] and urged the local expert in Budapest to write immediately, in full, about the jurors and how much it would be possible to increase sales thanks to that victory.[81] News of the event was also sent to Russia, and presumably also to other agents, to make use of.[82] Furthermore, Butler sent news of the victories in the Colonies to Lankaster and asked him to send immediately all positive accounts from England and from the Continent to South America and Australasia.[83]

The importance of these trials for trade becomes more evident during later years. After the successful test in France, Nettie McCormick decided to take an active part in the daily operations of the McCormick Company. Apparently General Manager E.K. Butler had not taken these victories seriously enough, for Nettie urged him to publish the results in the press with no more delay. Butler also received a lecture on the significance of these shows.[84]

This confirms the earlier statement, of how important it was not only to stay on the market and keep the market share, but also how the whole world was one market where news of the results of the tests spread rapidly from country to country.

Cyrus McCormick's report on the French trials also offers an opportunity to look behind the scenes of these tests. Cyrus Jr., together with Lankester and an expert mechanic, had made careful preparations and brought the newest specially finished machines to France for the Show. However, he soon noticed that "it was not to be a trial between machines but acquaintanceship, influence and smooth-talking". Walter A. Wood had been lucky enough to contract a prominent agent for his machines. Consequently, the result of the trial was already decided, according to Cyrus Jr., before the event itself. Cyrus was especially bitter about the results, since the Company's older wooden frame machine beat the new iron-frame binder. For him the Great National Government trial was "as complete a humbug as far as fair trial was concerned as I ever saw,

---

80 *Cyrus Jr. to Nettie McCormick* 7.9.1885. Mss 2b, box 31. Cyrus was especially happy that Hornsby was "at the bottom of the list."

81 *Cyrus Jr. to Lee Borrell* 7.9.1885. Mss 1x, LPCB 458.

82 *Cyrus Jr. Freudenreich* 8.6.1885. Mss 1x, LPCB 458.

83 *Butler to Lankester & Co.* 12.8.1885. Mss 1x, LPCB 458.

84 *Nettie McCormick to E.K. Butler* 8.9.1887. Mss 1b, box 24.

more fuss and talking and less action and work".[85] Cyrus's comments disclose how the outcome of the test depended on the jurors and the ability to affect them. On the other hand, it is surprising that he was not alredy aware of this, since he had been attending trials in Europe for ten years.

As became clear in McCormick's case, results of the trials were published as soon as possible, whereas the losers, on the other hand, tried to foil the winner's efforts. Percy Lankester had printed a circular immediately after a show in France, on the basis of information he had received from his French agent; the Hornsbys instantly attacked and claimed that the medal which the McCormick Co. had won in Goderville was for the agent's collection and not for the machine or machines. Lankester did not withdraw his circular, and the result was a lengthy debate that was put in the hands of attorneys, but finally was allowed to drop.[86]

Foreign exhibitions, such the Paris Exposition in 1889, were cases of special interest. Cyrus Jr. himself decided to attend this, and preparations for it were under way at least half a year before the event.[87]

Trials and agricultural shows also offered an opportunity to conquer new markets. The McCormick Company had appointed agents in Germany, Sweden, Norway and Denmark, and to help get them started gave either direct financial support, as in Sweden, or sent its experts to assist in trials, as in Germany.[88]

---

85 *Cyrus Jr. to McCormick Co.* 8.13.1887. Mss 3b, box 7.

86 *Hornsbys to Lankester* 7.24.1888. Mss 2c, box 112; *Lankester to Cyrus Jr.* 7.25., 7.30., 8.2., 8.9.1888. Mss 2c, box 112.

87 Cyrus once again anticipated jury tactics. This time all possible points were taken under consideration. By January, special exhibition machines were under construction; Lankester was asked to reserve two fine teams of horses, and to make arrangements so that Cyrus would be able to meet all the necessary people to bring influence to bear upon the result. Besides, Lankester had to reserve suitable accommodation where Cyrus could entertain, "as that is a pretty good way to reach a Frenchman's heart". McCormick's French agent, H.T. Mot & Co., was asked to look after the members of the jury. None of the men should be appointed who had supported the Wood Company in Mitry in 1887. *Cyrus Jr. to Lankester* 1.26. and 5.22.1889. Mss 1x, LPCB 459; *Cyrus Jr. to H.T. Mot & Co.* 6.10.1889. Mss 1x, LPCB 459.
The McCormick Company's efforts were fruitful: it won the Grand Prize and was selected as the most desirable machine. McCormick's delight was overshadowed by Walter A. Wood, who was also awarded the Grand Prize. *Butler to Freudenreich* 10.12.1889. Mss 1x, LPCB 460. *Awards to an American Genius. The Development of the Walter A. Wood Harvesting Machinery.* 1900. Mss 4z, box 25.

88 *Lankester to Cyrus Jr.* 7.22.1889. Mss 10c, box 10. Other competitors at the Hildesheim trial in addition to McCormick were Hornsby, the German firm Hennef and Zimmerman, Osborne, Wood, Howard, Massey, Samuelson and Johnston. Walter A. Wood won the race, followed by McCormick. *Butler to Lankester* 7.31.1889. Mss 1x, LPCB 460.
On the Gothenburg International Exhibition in 1891 see *Butler to Lankester* 11.4.1890. Mss 1x, LPCB 460.

While trials were a place to advertise and gain publicity for one's products, one of the key factors affecting actual sales was pricing policy. On the domestic field, the McCormick Company was reluctant to become involved in price wars, preferring to follow the general market trend but on average hold its prices at a higher level than its competitors. In Europe the same tendencies are visible as in America. Machines became cheaper under the pressure of competition.[89]

The figures show variation in the agents' net prices, which seem to have been elastic in response to local competition. Percy Lankester, for example, asked for a reduction of mower prices from $50 to at least $47.50 to fight the English makers, who sold at $50 f.o.b. in Paris when Lankester's price was f.o.b. in New York.[90] In Russia the Deering Company offered its machines at $175 f.o.b. New York, and Osborne's agent sold reapers at $82.50, but there were rumors of an English company that intended to retail reapers at $110. McCormick's Daisy reaper cost the agent from $95 to $90.25, and it was impossible to sell it with any profit under $140.[91]

During the 1880s, prices of harvesting machines on all the major markets were very unstable. The downward trend became possible not only as a result of competition, by also of falling production costs. While the prices referred to above in the mid-1880s were agents' prices in Europe, by the end of the decade some companies were selling their binders in America even to farmers at $100.[92]

---

89 In 1884 McCormick's agent cash price for a harvester and binder for the Algerian trade was $210, and for a mower $55. *Butler to Lankester & Co.* 4.15.1884. Mss 1x, LPCB 457. From the previous season there was a drop of $5 in binder prices. In Russia, J. Maszewski had to pay $215 and $55 for the same machines but his prices were dropped to $200 and $52.50 respectively to meet the competition. *Butler to Freudenreich* 11.17.1884. Mss 1x, LPCB 457. If the sales were made in two installments, the McCormick company added ten percent "to be settled for a good approved note". Besides, the buyer had to pay the lawful interest in Russia. Emil Liphardt & Co. of Moscow paid for a one-foot-wide binder $220 and for a mower $55. *Freudenreich to McCormick Company* 12.6-18.1884. Mss 3b, box 3. Liphardt's commission was five percent of the net price.

90 *Lankester & Co. to McCormick Company* 12.15.1884. Mss 3b, box 3.

91 *Freudenreich to McCormick Co.* 12.6-18.1884. Mss 3b, box 3. According to Freudenreich the dealer must have a "profit of $25 at least or else he cannot subsist". This price war started rumors of extremely low prices in America. McCormick had to assure Lankester that not even the poorest factory had sold reapers to farmers at $165, as had been reported. On the other hand, by 1885 McCormick's sold its binders to Lankester to be forwarded to Cosimini & Sons in Italy at $155, and in Russia, Freudenreich had tried to sell binders to his agents at $175, but without any success. *Butler to Lankester* 1.8.1885. Mss 1x, LPCB 457. *Cyrus Jr. to Freudenreich* 8.6.1885. Mss 1x, LPCB 458. Cyrus Jr. kept Freudenreich's price very low even in America. Other manufacturers also had difficulties in Russia during 1885. According to Cyrus Jr., Johnston, Hornsby and Deering would have a large number of machines left over in the Volga region.

92 *Butler to Lankester* 7.11.1888. Mss 1x, LPCB 459.

Severe competition forced companies to considerable price cuts in Europe too. In England, Wood and Hornsby were happy to get rid of their stock at almost any price, and Wood offered its machines to farmers at 25 percent discount from list prices.[93]

The price war was an aftermath of the knockout game between the harvester companies in America. In the early 1880s, they had tried to reach agreement on prices and production quotas, but discussions had failed. Although competition forced some companies to sell even under their production costs,[94] by the end of the decade prices began to stabilize. Net prices to Percy Lankester were dropped to $110 for binders and $32-35 for mowers; he was also given a discount of seven percent.[95] Normal agents had to pay around $125 for their binders.[96] When the drastic drop in the machine prices is taken into account, it is plausible to suppose that the severe competion among the American firms drove the new European manufacturers from the field. And it was even difficult for the existing ones to survive.

The McCormick Company's basic idea in its price policy was to make as big a profit per sold unit as possible, and the success of this approach can be seen from its distribution of dividends in Table 11. E.K. Butler's message to the Company's foreign agents in 1888, when he took charge of the foreign affairs, was clear: business must be profitable.[97] The other key idea was quality. From time to time Butler assured that McCormick machines "compete for merit, not for price";[98] he was "holding to the theory of merit and hard work rather than to that of low price and ease". Furthermore, "we never have, and never shall allow competitors to fix prices at which we must sell our goods".[99] Butler's strong comments give a picture of a hard-boiled manager but also of a company that regarded itself as a market leader that has the best machines available for sale.

David A. Hounshell has raised the question of whether the

---

93 *Lankester to Cyrus Jr.* 8.2.1888. Mss 2c, box 112.

94 *Butler to Lankester* 9.23.1890. Mss 1x, LPCB 460.

95 *Butler to Lankester* 12.27.1888. Mss 1x, LPCB 459; *Butler to Lankester* 9.23.1890. Mss 1x, LPCB 460.

96 *Butler to Lankester* 11.26.1890. Mss 1x, LPCB 460. The price referred to was for Mot & Co. in France.

97 Butler stated that though competitors had sold machines at very low prices "we have sold no machines to net us less than those we have shipped to you, and will average much better." *Butler to Lankester & Co.* 7.11.1888. Mss 1x, LPCB 459.

98 *Butler to Freudenreich* 5.11.1889. Mss 1x, LPCB 459.

99 *Butler to Lankester* 9.23.1890. Mss 1x, LPCB 460.

widely-held notion that American-made products succeeded in the market because they were cheaply made and low priced is correct. According to his findings, firms like Singer, McCormick, the Pope Manufacturing Company, and the Western Wheel Works, all leaders in their industries, also stand at the top of the prices list and were known for their quality.[100] Hounshell's statement coincides with the findings of the present work.

It has become evident that the McCormick Company was able to acquire a good share of the European harvesting machine market. If we consider its success at trials, Butler's words are based on solid ground. Every year McCormick's machines brought home a bunch of medals and other prizes. It is, nevertheless, another question how reliable and indicative of the competitive situation the tests and trials really were. In that respect there seem to be some discordant notes in the otherwise unanimous picture that is constructed from the McCormick Collection.

The McCormick Company's Russian agent, George Freudenreich, had struggled energetically to acquire the same conditions for his trade as obtained in America, with some success. Russia was the only place where McCormick sold machines on consignment;[101] neither did Freudenreich hesitate to express critical comments on defects in the construction of machines. During the peak year of 1884, he did not take the negative voices of his Russian customers too seriously; Freudenreich reported that they had simply expected too much of the binders, something that the machines could not perform.[102] Soon, however, Freudenreich began to demand modifications in the standard harvesters, which if carried out would mean extra work and expenses for the factory.[103] From the correspondence it can be read that McCormick machines, so victorious in the trials, had problems in practice.[104] Lankester

---

100 *Hounshell* 1987, 5-6, 9.

101 *Butler to Freudenreich* 11.17.1884. Mss 1x, LPCB 457.

102 *Freudenreich to McCormick Company* 12.6-18.1884. Mss 3b, box 3.

103 *Cyrus Jr. to Freudenreich* 1.31.1885. Mss 1x, LPCB 457. In spite of all the costs it caused, McCormick Co. began to construct a new model of the "Daisy" reaper adapted to Russian conditions. *Cyrus Jr. to Freudenreich* 5.29.1885. Mss 1x, LPCB 458.

104 The first notice of the complaints from Russia dates from 1883, when Butler wrote to Freudenreich that "your several favors of late have carried the impression that our machines were not altogether adapted for the Russian business". *Butler to Freudenreich* 1.3.1883. Mss 1x, LPCB 456; *Cyrus Jr. to Freudenreich* 8.6.1885. Mss 1x, LPCB 458. The negative news is normally in the form of short dispatches here and there; "we are glad to find that the "Daisy" of this year seems to meet with approval..."

reported machine breakages, and Mot & Co. in France were replacing McCormick mowers with Buckeyes.[105] McCormick's previous French agent, Paul Francey, had also persistently declined to handle any of the Company's "Daisy" reapers, but sold similar competing machines.[106] Besides, Freudenreich reported a growing dissatisfaction with the McCormick mowers.[107]

The explanation for the differences between the success in America and at trials and, on the other hand, complaints from the field may arise from the fact that the machines were not properly adapted to foreign conditions. Freudenreich, like other agents, frequently expressed his desire for modifications or for totally new machines to meet local conditions. Agents' interest in such changes was awakened because remodeling of the basic harvesting machines and the establishment of service and repair systems were key components in marketing.

Cyrus Hall McCormick's experiences during his first years of foreign trade had already shown the importance of modifying American machines for Europe. One of the basic technical problems was caused by differences in farming ideologies. In America, farmers normally did not collect straw for bedding, and consequently harvesting machines were constructed to cut the grain near the ears. An especially problematic task for the first harvesters was the long European rye: the platform of the American harvesting machines was too short, and McCormick was forced to adapt the deeper English model.[108] In this way, innovations also moved from Europe to America, and not only vice versa.

Minor changes in the construction of the basic models did not cause serious difficulties for the production line, and these modifications were normal practise in the factory.[109] If the restyling demanded new patterns, jigs and fittings, the factory began to make calculations about profitability. If the foreign trade's requirements collided with domestic business, normally the foreign had to give way. When Lankester asked for some modifications in McCormick's French mowers, Butler answered that "our pattern department has

---

105 *Butler to Lankester* 6.24.1889. Mss 1x, LPCB 459. The change of machines Butler could not understand because the McCormick mower, according to him, was superior in America. He supposed that the reason had to be in the users.

106 *Cyrus Jr. to Lankester* 1.26.1889. Mss 1x, LPCB 459.

107 *Butler to Freudenreich* 1.15.1890. Mss 1x, LPCB 460.

108 *Butler to Freudenreich* 1.31.1885. Mss 1x, LPCB 457.

109 *Butler to Lankester* 12.6.1889. Mss 1x, LPCB 460. Butler's answer to Lankester's requests is illustrative on this subject. Butler goes through all the principal machines item by item and lists the possible changes.

been driven to its utmost to take care of our regular work, which is of far more importance to us than this model". On the other hand, the factory was at the same time completing a special one-horse mower and a folding-bar reaper to meet demands in France; this was possible because they did not require demanding technical changes.[110]

The change from wire-binders to twine-binders was a big technical challenge for the harvester companies, and a question of survival. An equally trying test was the introduction of a new model on the market. Almost every year, harvester companies made modifications in their machines; farmers learned to expect this, and to demand the newest models or discounts on older ones.[111] After the invention of the twine-binder, perhaps the greatest change in the binders was when wood was replaced by iron. The McCormick Company made its first experiments with iron-frame machines in 1885, and already the next year produced 20 000 steel-frame machines and only 2500 wood-frames.[112]

A major change like this demanded significant investments, which created difficulties for smaller factories. Cyrus Jr. anticipated problems also for Hornsby. In America, according to his information, three or four smaller factories that already held patents for steel-frame machines were still building wood-frame harvesters because of the great risks involved in the changeover.[113] Since most of the minor yearly changes in models were merely marketing tricks to attract new customers, companies had to consider the situation carefully so as to avoid undermining sales of the previous year's models.[114]

Although it is clear that there were numerous faults in the machines, some of the problems were also caused by unpractised

---

110 *Butler to Lankester* 1.17.1889. Mss 1x, LPCB 459. In this case Butler, however, was ready to order a model machine from some model maker.
Making an entirely new pattern machine would have required an entirely new set of forms, jigs and patterns. The expense of the change was estimated at about $7000. *Butler to Lankester* 12.6.1889. 1x, LPCB 460.

111 See for example *Cyrus Jr. to Freudenreich* 5.29.1885. Mss 1x, LPCB 458; *McCormick Co. to Lankester* 5.20. 1889. Mss 1x, LPCB 459;

112 *Butler to Cyrus Jr.* 10.22.1885. 3b, box 4.
The change from wood to iron meant a complete change in the machinery of the factory. All the lathes, forms and presses were replaced. It caused delays in production and "thousands of dollars in new machinery." *McCormick Co. to Lankester* 5.20.1889. Mss 1x, LPCB 459.

113 *McCormick Co. to Lankester* 5.20.1889. Mss 1x, LPCB 459.

114 *Cyrus Jr. to Lankester* 1.16.1889. Mss 1x, LPCB 459; *Butler to Lankester* 12.6.1889. Mss 1x, LPCB 460.

users.[115] This fact forced factories to send travelling experts abroad
to set up and repair machines. A working machine was the best
advertisement and inducement for further sales. In Russia, the need
for experienced mechanics was so great that agents were prepared
to bear the cost themselves. Normally the McCormick Company tried

---

115 *McCormick Co. to F. Vogeler* (McCormick's agent in Prussia, Germany)
10.24.1888. Mss 1x, LPCB 459.

to share an expert's salary and expenses.[116] McCormick's travelling mechanics were men experienced on the fields, men who helped in sales work but were simultaneously also the Company's eyes and ears.[117] One of the many requirements for an expert in Europe was that he had to speak several languages. Travelling over a huge area extending from Russia to South America made work exhausting, and not all of them could stand the strain.[118]

In America, a central element of sales campaigns was advertising. As was stated earlier, factories published not only yearly catalogs but also numerous other articles: posters, showcards and flyers, and this approach was transferred to Europe too.

It is not clear when the McCormick Company began to send its own printed material to its European agents. This must have been before 1885, for in that year Cyrus Jr. informs his Russian agent that the Company is not able to furnish him with special pamphlets; he would have to manage with the repair catalogs and instructions which the Company had sent to him. The Champion Co., however, had sent its American catalog to Russia and added into it two other languages.[119] For the 1886 pamphlet, the McCormick Company asked about special features for the English trade, and for the first time thought about publishing a French edition.[120] The next year the Company's selection had expanded to folders, showcards and to its own magazine, Farmers Advance. German pamphlets were also published for the first time.[121] McCormick's business connections

---

116 *Emil Liphardt & Co. to McCormick Co.* 2.15-27.1884. Mss 1x, LPCB 457; *McCormick Co. to Freudenreich* 5.6.1884. Mss 1x, LPCB. The experts that McCormick normally sent abroad were Poppe and Borell. Ibid

117 *Poppe to McCormick Co.* 6.9.1888. Mss 2c, box 112.

118 *Butler to Lankester* 4.6.1889. Mss 1x, LPCB 459. Poppe left the McCormick Company after a dispute over his salary. Half a year later, however, Poppe was back in service and on his way from South America to Russia. *Butler to Freudenreich* 12.18.1889. Mss 1x, LPCB 460.
Borell's problem was his heavy drinking. Butler asked Lankester "to look after him a little in this regard". Borell's salary was $100 a month and the Company covered his travelling expenses. *Butler to Lankester* 6.4.1889. Mss 1x, LPCB 459.

119 *Cyrus Jr. to Freudenreich* 1.31.1885. Mss 1x, LPCB 457. Cyrus Jr., while not able to send company's catalogs to Freudenreich, boasted that they were better than their competitors'.

120 *McCormick Co. to Lankester & Co.* 8.29.1885. Mss 1x, LPCB 458.
The McCormick Company fulfilled its aims and printed a French pamphlet for 1886 and a poster. *McCormick Co. to Lankester & Co.* 4.11.1886. Mss 1x, LPCB 458.

121 *McCormick Co. to Freudenreich* 1.13.1887. Mss 1x, LPCB 458; *Butler to Emil Müller* 7.6.1887. Mss 1x, LPCB 459.
It was, however, curious that the McCormick Co. published printed instructions and directions for setting up and operating machines only in English. *McCormick Co. to Lankester & Co.* 4.14.1887. Mss 1x, LPCB 458; *McCormick Co. to F. Vogeler* 10.24.1888. Mss 1x, LPCB 459.

had expanded over such a large area that it caused the Company's officials great problems to adapt advertising material for different countries.[122] A major step forward in advertising was to send catalogs directly to the members of the agricultural societies, as in England.[123]

In addition to direct advertisements, reporters were also used to promote sales. In 1889, Percy Lankester met a contributor to the German agricultural press, who offered his services to the McCormick Company. For 20 shillings per article he was ready to write columns in the seven leading German agricultural magazines, dealing not only with McCormick's factory and its success in the trials but also McCormick's ideas. Although it is not a hundred percent sure if this proposal was carried out, it casts serious doubt over the reliability of the articles published in the agricultural newspapers and periodicals.[124]

---

122 *McCormick Co. to Freudenreich* 1.13.1887. Mss 1x, LPCB 458. Different conditions had forced the factory to make changes in the basic construction, which had to be taken into account when pamphlets and other material were published. At this phase the McCormick Company used the same pamphlets or catalogs both in America and in Europe: only the language was changed and the name of the agent was printed on the cover. *McCormick Co. to Lankester* 1.28. and 4.14.1890. Mss 1x, LPCB 460.

123 *McCormick Co. to Lankester* 4.14.1890. Mss 1x, LPCB 460. Lankester decided to send pamphlets to the members of the Royal Agricultural Society of England. He asked McCormick Co. to send 15 000 catalogs for that purpose.

124 *Lankester to Cyrus Jr.*7.16.1889. Mss 10c, box 10. Lankester asked authorization for his move but was ready to pay for all seven articles only five pounds, since he assumed that the German journalist was "somewhat in needy circumstances". Lankester saw a need for these articles, since the McCormick Company was in Germany more or less on the defensive and needed new wind for its sales.

VIII

■ The Foreign Trade Grows in Importance, 1890-1898

8.1. The new division of the home markets

8.1.1. The first merger attempt

In the hands of Cyrus McCormick Jr., the McCormick Harvesting Machine Company was transformed from a family enterprise into a managerially run large-scale corporation. In spite of its changing nature, the McCormick Company remained in many respects a family firm. Its energetic General Manager, E.K. Butler, was in charge of daily operations, but under the close eye of the President of the Company, Cyrus Jr., and especially of his mother, Nettie McCormick. Besides, the Company's shares were in the hands of family members. At the formation of the Company in 1879, one fourth of its stock was left to Leander McCormick, altogether 6250 shares. In 1880, when Leander and his son Hall were forced to withdraw from the active conduct of the business, they retained their stock; besides, Leander remained Vice-President of the Company and Hall a director.[1]

Although Leander and Hall were outside daily operations, their positions enabled them to follow the Company's decision making on the highest level. They were able to force distribution of the profits and made development and expansion of the business almost impossible. Of course it has to be taken into account that the material in the McCormick Collection is hostile to Leander's ideas and approach to the business. Not only had there been prolonged dispute and accumulated hatred between the two brothers, but also two totally different business ideologies; Cyrus demanding constant expansion of production, and Leander more cautiously attempting to restrict his brother's vigor.[2]

Continued friction between the two families finally led Cyrus Jr. to offer to buy Leander's share in the Company. After discussions between Cyrus Jr. and Hall McCormick, the two parties finally agreed

---

1 *Records of the Directors Meetings.* 2.4.1890. Mss M/I, box 4; *Opinion of John P. Wilson in connection with the purchase of L.J. McCormick stock.* 7.25.1918. Mss 6c, box 30. p. 1.

2 *Opinion of John P. Wilson.* 7.25.1918. Mss 6c, box 30. page 3-4.

on the terms of the purchase on December 28th, 1889. For Leander's shares Cyrus and Nettie McCormick paid $3 200 000, of which $600 000 was paid in cash at the time of the signing of the agreement. The remainder was paid over the five following years and was secured by collateral to the amount of $1 000 000. Leander and Hall, for their part, promised to entirely to give up the reaper business, including the goodwill. They were also forced to surrender their positions in the Company management.[3]

Nettie McCormick's already strong (actually crucial) position in the business was reinforced when her brother, Eldridge M. Fowler, was elected as the new Vice-President of the Company. They were to be the conservative voice on the Board of Directors. The division of the shares[4] and the duties and officers of the Company[5] were

---

3 *Cyrus Jr. to Leander McCormick* 12.16.1889. Mss 1a, box 108; *Opinion of John P. Wilson* 7.25.1918. Mss 6c, box 30. p. 1-5; *Copy of the Records of the Director's Meeting.* 2.4.1890. Mss 1a, box 109; *Meeting of the Board of Director's.* 2.4.1890. Mss M/I, box 4.

4 The estate of Cyrus H. McCormick was by his will to be divided on May 13th 1889 but was not in fact executed until Jan. 13th 1890. The shares of those under the age of 25 were to be held in trust for them until they reached the required age. At the time of the purchase of the stock 500 shares were set apart for Mrs. Blaine (Anita McCormick), Harold and Stanley McCormick respectively. Decades later the division of the shares led to an investigation of the matter which produced the report of John P. Wilson. Ibid. Mss 6c, box 30. p.1-2.

5 The division of the shares was:

	Nettie McCormick	5752	shares
	Cyrus H. McCormick	5317	"
Trustees of	Harold F. McCormick	3500	"
"	Stanley R. McCormick	3500	"
"	M. Virginia McCormick	300	"
	Anita McCormick Blaine	3499	"
	Emmons Blaine	1	"
	E.M. Fowler	20	"
	N.F. McCormick, Trustee	200	"
	E.K. Butler	30	"
	W.R. Selleck	1	"
	total	2500	shares

Stockholders' Records. Annual meeting of the stockholders of the McCormick Harvesting Machine Comopany. 12.30.1890. p. 34. Mss M/I, box 24. Vol. 36.

The value of the entire estate of Cyrus H. McCormick Sr. other than the stock in the Company was in 1889 valued at about $6 million. At the date of the purchase not less than three-fifths of the entire estate consisted of stock in the McCormick Company. *Opinion of John P. Wilson* 7.25.1918. Mss 6c, box 30.

The rate of the shares was $520 per share in 1889, at least to Nettie McCormick, who purchased 200 shares in that year. *Stockholders' Records. Annual meeting of the stockholders of the McCormick Harvesting Machine Comopany.* 12.30.1890. p. 41. Mss M/I, box 24. Vol. 36.

The Board of Directors was to consist of five persons, each of whom also had to be a stockholder. The Company would have a President, a Vice President, a Secretary, a Treasurer and a General Manager. The General Manager was to have "such powers and shall discharge such duties as may be delegated to him from time to time by the Board of Directors or the President."

Stockholders' Records. Annual meeting of the stockholders of the McCormick Harvesting Machine Comopany. 12.30.1890. p. 34-39. Mss M/I, box 24. Vol. 36.

confirmed. Even after the redistribution of shares, Nettie remained the largest shareholder in the Company.

The purchase of Leander's stock also explains why the McCormick Company paid no dividends in 1889, and the slow, and in some cases even reluctant, moves and decisions in foreign business. Leander had opposed foreign actions from the very beginning and had hardly changed his mind during the 1880s.

Cyrus Jr. also apparently had other questions on his mind when he began to bargain with Leander. Competition in the harvester business had increased both at home and abroad. Negotiations on the reduction of rivalry and fixed prices between the companies had ended in the early 1880s without any settlement. At the end of the decade, however, Colonel A.L. Conger, President of the Whitman & Barnes Manufacturing Company in Akron, Ohio started discreet new consultations with various parties to organize a harvester trust.[6] The reasons were apparent. The fierce rivalry had forced companies to extend their agencies even to the smallest country towns. A patent war had continued since the very beginning of the industry and still in the 1880s companies held overlapping patents and had suits pending against each other.[7] Besides, as has already been stated, competition led to constant price cutting, which was of course beneficial to farmers but ruinous to manufacturers.[8]

On November 19th, 1890, meetings of the leading harvester men during the fall of 1890 led to the formation of the American Harvester Company. Its capital stock was fixed at $35 000 000. It was the first large-scale merger in the American agricultural machine and implement industry. The companies to enter into the agreement were the McCormick Company of Chicago; the Walter A. Wood Company of Hoosic Falls, New York; the Warder, Bushnell and Glessner Company of Chicago; the Aultman & Miller Company of Akron; the William A. Deering & Company of Chicago; and the

---

6 Conger had obtained options on the plants of nearly 20 different harvester manufacturers, including the McCormick, Deering and Walter A. Wood. Whitman & Barnes was itself a combination of 13 mower and reaper knifemaking facturies that sold their products to larger companies. *The International Harvester Co* 1913, 57-58.

7 The McCormick Company paid $100 000 in 1884 for the patents of Marquis L. Gorham, which anticipated the Appleby twine-binding apparatus. These patents were added to a pool of patents and thereafter manufacturing rights sold to competing firms. *McCormick* 1931, 92.
In 1886 the McCormick Company had a suit pending against D.M. Osborne on infringement of Gorham patents. McCormick demanded from Osborne $25 000 for the machines already built and sold and $5 000 as a regular royalty per machine for future production. *Cyrus Jr. to D.M. Osborne* 1.6 and 1.7.1886. Mss 1a, box 103. Three months later the McCormick Company bought half of the rights for Sylvanus D. Locke's patents and production for $75 000. *Contract between Sylvanus D. Locke and the McCormick Harvesting Machine Company* 4.6.1886. Mss 1a, box 103.

8 *Benson* 1936, 5-6. Colonel Conger could remember more than eighty harvester companies that had gone bankrupt. Ibid, 3.

Whitman & Barnes Manufacturing Company of Akron.[9] The
McCormick Company subscribed for $11 000 000 and the Deering
Company for $8 000 000 of the total amount of the capital stock,
which gave them a majority of the entire capital stock. These two
companies were clearly the leading firms in the business.[10]

The formation of the American Harvester Company had raised
opposition among farmers and implement dealers, who were afraid
for their future. Amalgamation also met with resistance from within
the Company, for the old hostility between the companies did not
die overnight. Suspicions persisted, and were aggravated by
thoughtless acts by Company officers; not all of the parties engaged
in the enterprise were convinced of its benefits.[11] Besides, the

---

9 *Benson* 1936, 3-7; *The International Harvester Co.* 1913, 57-58.
Besides these "Big Six" the following smaller firms also merged into the American
Harvester Co.: Plano Manufacturing Company, Plano, Illinois; Milwaukee Harvester
Company, Milwaukee, Wisconsin; Easterly Harvesting Machine Company, Whitewater,
Wisconsin; Minneapolis Harvester Works, Minneapolis, Minnesota; Emerson, Talcott
& Company, Rockford, Illinois; J.F. Seiberling Company, Akron, Ohio; Seiberling,
Miller & Company, Boylestown, Ohio; Amos Whiteley & Company, Springfield, Ohio;
Hoover & Gamble, Miamisburg, Ohio; D.M. Osborne, Auburn, New York; Richardson
Manufacturing Company, Worcester, Massachusetts; Adriance, Platt & Company,
Poughkeepsie, New York; D.S. Morgan & Company, Brockport, New York; the
Johnston Harvester Company, Batavia, New York. Ibid. Altogether 19 of the 21
manufacturers of reapers and mowers took part in the consolidation. *Clark* 1929,
361.

10 *Benson* 1936, 8. The new giant elected Cyrus McCormick as its President, William
Deering as Chairman of the Board of Directors, Walter A. Wood as Vice-President,
E.K. Butler as General Manager and A.L. Conger as Secretary and Associate General
Manager. Others elected to the Board of Directors were Asa S. Bushnell, Lewis Miller
and William Goudy. The Board of Directors delegated its powers of management and
control to the Executive Committee. The general business of the Company was to
be divided into three principal departments: financial, manufacturing and sales.
Officers of the new concern decided to divide the markets of the United States into
three parts. E.K. Butler would manage the western division, A.L. Conger the middle
and Walter A. Wood the eastern division. Because one of the ideas behind the merger
was to eliminate overlapping functions between the companies, the American
Harvester Company gave notice of dismissal to some of its employees and began
moves to establish fixed retail prices for the agents. *Memorandum of agreement
between William Deering, Walter A. Wood, A.S. Bushnell, A.L. Conger and Cyrus
McCormick as subscribers for and owners of stock in the American Harvester
Company.* Mss 3b, box 11; *Cyrus to Harold McCormick* 12.8.1890. Mss 1a, box 109;
Benson 1936, 8; *Plan of organization for the American Harvester Company.* Mss
w, box 2; *The International Harvester Co.* 1913, 58.

11 Already about a month after the formation of the American Harvester Company,
someone (probably the Treasurer of the Deering Company) informed his superiors
of malpractices in the company. E.K. Butler of the McCormick Co., as the new General
Manager, had decided to put in charge of the general agencies old McCormick men,
except in Harrisburg where a Deering agent was appointed. Now the Deering clerk
wanted to warn his company of the possibility that Deering's name might vanish
from the market. What was more dangerous, if the McCormick men were in charge
of the agencies, the availability of extra parts could deteriorate and accordingly also
the value of uncollected notes, at the time being $4 875 000. Besides, there was a
threat that in the eyes of the public, the McCormick Company would control the
situation. *H.S. Shield (?) to William Deering & Co.* 12.23.1890. Mss W, box 1.
For the concerns of farmers and implement dealers see *Benson* 1936, 11.

Map 1. Harvesting and moving machine manufacturers of the U.S., 1890.

Source: Benson 1936, 3-7; The International Harvester Co. 1913, 57-58; Clark 1928, 361.

American Harvester Company was unable to finance its operations with the expected sale of bonds to the value of $15 000 000, since the United States Trust Company of New York declined to provide the necessary financing for the operation; and finally the American Harvester Company was dissolved on January 11th, 1891.[12] One reason for the dissolution was the opinion of Mrs. McCormick, who was afraid that her husband's name would disappear from the harvesting machine industry.[13]

The American Harvester Company was the first serious attempt to fuse the mutual interests of the largest harvesting manufacturers. It was also for the prevailing time a typical attempt to merge mutual interests. The merger was an expression of the strategy of horizontal combination and was aimed at maintaining profits by controlling the price and output of the operating units. The first steps in that direction were taken already in the 1870s by setting up nationwide associations to control price and production. By the 1880s these federations were a normal way of doing business, although they were found difficult to maintain. Nevertheless, only a few manufacturers moved from cartels to legal consolidation during the 1880s. The first real merger wave occurred between 1890 and 1893 as a result of the legal attack on combinations, the passage of the Sherman Act, and the revisions of the New Jersey law. It lasted until the depression of the 1890s.[14]

Although the benefits of the amalgamation were evident, the time was not yet ripe for its realization: the bonds that tied the reaper men to their own companies were still too tight. Most of them were inventors, and founders of their businesses. Mutual suspicions

---

12 *Benson* 1936, 9-11; *McCormick* 1931, 108-109. McCormick has also a version of his own of the end of the American Harvester Company. According to him, Cyrus Jr. and William Deering had travelled to see the New York bankers in January 1891 and having found the city bankers cold to their ideas were spending their night in the same hotel parlor. Late at night Deering had come to see McCormick and a conversation followed:
"McCormick, he said at last, "are these other fellows trying to make the two of us carry water for them?"
 "It looks that way to me!"
"All right, let's go home and call it off."
"I agree," said the younger man"... Ibid, 108-109.

13 *Benson* 1936, 12. Benson bases his statement on newspaper material, in this case the New York Herald. The Chicago Times also took part in the debate, quoting Cyrus McCormick's statements. Cyrus firmly denied the role of his mother in the case, was sorry for the outcome of the event and blamed the Sherman Antitrust Law for the dissolution. The Chicago Times found the real reason for the downfall to be the activities of the Deering Company. Deering wanted to benefit from the discontent among the farmers against the merger, and sent to its agents circulars, where it explained it would maintain its own prices and published its intention to withdraw from the combine. *Chicago Times* 1.10.1891. Mss 6x, box 1.

14 *Chandler* 1977, 315-318, 320, 331-332.

prevented, as in so many similar combinations before, the founding of an agreement on production quotas and prices. As long there was some profit available on the market, there was always somebody who was ready to benefit from it.

For research this episode is, nevertheless, very fruitful. Before the formation of the American Harvester Company a thorough investigation was made in all the major participants' production plants, which gives a detailed picture of the state of the various companies, their competitiveness and their position on the market.

The report reveals the reasons for the positions of the competing manufacturers. The largest and most aggressive firms had invested in technology and organization, while the smaller ones had trusted to outdated practises and processes. The Buckeye line machine shop of Aultman, Miller & Company had hardly any special-purpose machines, and used outmoded methods with inadequate inspection. The same situation prevailed at the Peerless Reaper Company, at the Amos Whiteley Harvester & Reaper Factory and at D.M. Osborne & Company. On the other hand, larger manufacturers had adopted specialized machinery, and had installed the latest technology in every department from foundry to machine shop. The companies in best shape were Warder, Bushnell & Glessner (producers of the Champion line), Walter A. Wood and Deering. The benchmark for comparison in the report was the McCormick Company. Both of the inspectors were McCormick employees and certainly not neutral; their report was, on the other hand, meant only for their superiors and can be rearded at least as suggestive.[15]

New machinery alone does not make production profitable, if the workforce is not properly organized. The largest companies had

---

15 *Tour of Inspection of the Different Factories of the American Harvester Company*. Report of B.A. Kennedy and H.B. Uttley. Submitted either at the end of 1890 or in the beginning of 1891. Mss 3b, box 11; *H.B. Uttley to the American Harvester Company* 12.17.1890. Mss w, box 1.

David Hounshell showed in his research how companies such as Singer and McCormick in the beginning of the 1880s still heavily relied on general-purpose machinery and handfitting, and had not adopted the New English manufacturing systems, the 'American Manufacturing System'. *Hounshell* 1987, 99, 105-109, 173-175, 182-186.

The hiring of Lewis Wilkinson and firing of Leander McCormick were the outset of a new era in the McCormick Company. That also becomes indirectly evident in Uttley's and Kennedy's report, when they compare the other manufacturing plants to the McCormick factory. In a decade the larger manufacturers had began to furnish their plants with special-purpose machines. The smaller factories still used in 1890 old-fashioned hand tools and "many of the tools in machine shops were out of date and not suitable for the economical production of Harvesting Machinery", as was the case at Aultman, Miller & Company. The same company had not changed from wooden-frame machines to steel harvesters either, which also indicates the shape of its business. At D.M. Osborne, the conclusion was the same, "nothing modern in the way of machinery, all work being done on ordinary tools such as were in use years ago". *Tour of Inspection...*Mss 3b, box 11.

arranged the machinery in their workshops and foundries to allow the best possible output. They had also adopted piece rates in general use, and controlled both the flow of raw materials and the entire process with timekeepers. Even the machines ran faster at their plants. The organization of the production process alone gave to these plants great advantages over the smaller manufacturers and increased their competitive ability. If we can trust the report, the McCormick Company was in this respect too a forerunner.[16]

The new technology, organization of work and close control of costs were reflected in production costs per machine. At Aultman, Miller & Co. the building of a harvester cost not less than $75 and a mower not less than $28. Production costs were at the same level at the Amos Whitley Harvester & Reaper Factory, namely not less than $75 to $80 for a harvester and not less than $34 to $36 for a mower.[17] These figures can only be taken as indicative, since there is no information about how they were calculated. From McCormick's we know that the estimated cost per harvester in 1891 was $54.75 and per mower $18.25 to $22.[18] In the American Harvester Company the average cost to make a harvester was set at

---

16 *Tour of Inspection*...Mss 3b, box 11.
Aultman, Miller & Co. used piece rates in their foundry, where average rates were $2.15 per ten hours work or 37 cents per hour, for laborers 14 cents per hour and for mechanics from 20 to 27.5 cents per hour. Seiberling & Company's general average in the foundry was about 27.5 cents per hour and the day labor rate was 13.5 cents. Rates were also about the same at the Amos Whiteley Malleable Iron foundry but considerably higher at Warder, Bushnell & Glessner; for an average work 17 cents per hour and 28 cents in the foundry. Inspectors noted that at the McCormick works piece rates were about half of these rates, labor costs being $2.14 for each mower compared with Warder's $5.12. Robert Ozanne has calculated the average labor cost per machine at the McCormick plant in 1890 to have been $6.73. *Ozanne* 1962, 369. According to the report the average salary at the Whitman & Barnes Manufacturing Company was about $1.90 per ten hours, which was the same as at McCormicks. There was also great variation in the organization of the work. Aultman, Miller & Co. had no regular timekeepers at all; at Warder, Bushnell & Glessner all the work, except in the foundry, was done by the day or hour and consequently "their system of day labor more thoroughly than ever demonstrated to me, from what I saw there, and my past experience, not to be the right way to economize and get the largest amount of work for the least money." At Aultman, Miller & Co. the lack of special devices, such as jigs and forms, meant that the work was done in a slow and expensive manner. Aultman also allowed their piece workers to leave their work at any time. At the McCormick works that habit had been terminated more than ten years earlier.

17 Ibid.

18 *Cost to Manufacture* 1891 Machines. Mss 2c, box 29.

$60 and for a mower at $22.[19]

The impact of the formation of the American Harvester Co. were not felt in Europe, but its dissolution was. Butler informed Percy Lankester in good time of the coming merger; he anticipated changes in the contracts and organization of the foreign trade, but instructed Lankester to wait and stay calm. For the time being McCormick's was willing to continue old contracts but declined to make new ones.[20] When the amalgamation was dissolved, Butler urged Lankester to move immediately to take over the business of Osborne, Johnston and Seymour & Morgan in Europe; he expected these firms to discontinue their foreign activities after the American Harvester episode. Lankester should react before Walter A. Wood realized the situation too. McCormick was even ready to invest extra money in the effort and hire a man from Hornsbys.[21] On the other hand, Osborne, Bradley, Morgan and Johnston faced difficulties, forcing them to lower their prices in order to get rid of the stock carried over from 1890 at any price. Butler even accused them of dumping their machines in France. McCormick had now decided to take up the markets in France, and Lankester was advised to reduce his prices at once by up to $10.[22] Butler was not worried about Wood without cause: in 1890 the value of its foreign business was about a million dollars, making it the McCormick Company's leading opponent. Deering's business at the end of the 1880s was still almost entirely domestic, foreign trade playing only a minor role.[23]

---

19 *Estimated cost of production for season of 1891*. American Harvester Co. Papers. Mss w, box 2.
Estimates of the volume of production for 1891 were:

	Harvester	Mower	Reaper
McCormick	24 000	29 000	1 800
Deering	21 000	26 000	800
Wood	11 500	26 000	2 100
Aultman	8 350	16 500	750
Warder	6 100	18 700	750
Total	70 950	116 300	6 200

The quotas show clearly how the McCormick and the Deering companies dominated the harvester trade in America.
In Canada the merger of the leading implement manufacturers was completed in 1891. The next year the Massey-Harris Company declared its output to be 41 474 units. *Denison* 1949, 135. It is unclear which implements that output included, but even as such it reveals that the Canadian company was gaining steam.

20 *Butler to Lankester* 11.26.1890. Mss 1x, LPCB 460.

21 *Butler to Lankester* 1.21.1891. Mss 1x, LPCB 460.

22 *Butler to Lankester* 2.17.1891. Mss 1x, LPCB 460.
According to Butler, the Johnston Harvester Co. was practically in the hands of a receiver and they were trying to find a buyer for the concern.

23 *Tour of Inspection...* Mss 3b, box 11; *Report of the state of Deering Company, 1890.* Mss w, box 2.

8.1.2. Only the strongest survive

The collapse of the American Harvester Company marked the beginning of renewed competition. The two giants, McCormick and Deering, increased their sales efforts and expanded their volume, and the weakest companies had to give way.[24]

The literature dealing with this period is full of stories of furious rivalry, even open warfare, between salesmen of the various companies. Cyrus McCormick, grandson of the inventor, tells in his book how the salesmen tried to outdo their rivals: donating funds for a new church, arranging picnic parties or special delivery days for the harvesters, where machines were lined up on the street, bands were playing, banners fluttered across the street and finally a grand dinner was served. Even more dramatic encounters took place when two or more firms fought for the customer. Even after the price was set as low as possible, and a deal made, competing agents could haul away the rival machine during the night and replace it with their own. Nor were fights uncommon. McCormick tells an example of the price war, how a farmer was so poor that he was unable to buy a hammer and a wrench. Eager agents sold him a binder even though he did not have grain to cut: the tools he needed were in the tool box. Equally illustrative is a story of price cutting between two rival agents to the point when the price of the machine had dropped to half, and the farmer decided to buy both of them.[25]

Cut-throat competition was also noted by the leaders of the harvesting business, such as Cyrus McCormick Jr. and John J. Glessner.[26] Yet was competition really as fierce as has been claimed, or just a myth constructed for the sake of the International Harvester Company to convince the court in the case where it was accused of constituting an unlawful trust?

It is certainly true that the change to branch-offices, long-term credit for farmers, and after-sale service involved heavy capital investment, which the smaller manufacturers could not meet.[27]

---

24 At least St. Paul, Seiberling, Winona and Easterly failed during the 1890s, and Morgan, Whitely and Wood encountered severe economic difficulties *The International Harvester Co.* 1913, 62; *Benson* 1936, 13-14; *Schonberger* 1964, 56; *Butler to Lankester* 5.11.1892. Mss 1x, LPCB 461.

25 *McCormick* 1931, 95-107. Cyrus McCormick's book is full of interesting stories of the old harvester men. These tales he had collected from interviews with the old agents. The reliability of these narratives is questionable, but however, numerous writers have cited his information as evidence of competition. See for example *Schonberger* 1964, 62-63 and *Eckles* 1953, 81-84.

26 *The International Harvester Co.* 1913, 50-62.

27 In the District Court of the United States for the District of Minnesota. *Hearing*

Prices of the McCormick Company's machines in the Commission Agency Contracts remained on the same stable level from 1893 to 1899.[28] Indeed, the stable price level points rather toward fixed prices agreed between the manufacturers, as the jurors in the International Harvester case surmised.[29] Similarly, the key figures for the McCormick Company in Table 11 show an increase in net sales, except in 1893. Profits also almost doubled from 1892 to 1899; consequently the Company was able to pay considerable dividends to its shareholders.

There undoubtedly was serious rivalry; whether it was harder than during previous decades, is another question. Companies tried to beat each other in field tests and trials, as before, but perhaps some intensification and transformation can be seen.[30] Even prominent firms like the McCormick Company had to extend credit to farmers by an extra year, but that was due to the crop failure in 1894 and not to unforeseen competition, as Schonberger claims.[31] Besides, the same method was also used in the foreign trade.[32] All the companies offered longer terms to their clients than before, and even sent out special circulars on how to handle difficult customers.[33]

In 1890, the Canadian Harris Company brought out an 'open-end'

Before Circuit Judges Sanborn, Hook and Smith, at St. Paul, Minn., Nov. 3-5, 1913. Oral argument of Wm. D. McHugh. 4-5.

28 In the District Court of the United States for the District of Minnesota. The United States of America, Petitioner, vs International Harvester Company et al. Defendants. Volume V. *Testimony of Witnesses for the Defendants.* p. 264-280.
For example the price of a 5-ft cut binder with a carrier was $91 in 1893 and in 1899 exactly the same. The fixed price for the customer in 1892 was $145. Since the agent's commission was the difference between the net price and the sale price there was in principle a good chance for profitable trade.

29 *The International Harvester Co.* 1913, 59-60.

30 Now one manufacturer could challenge its competitors to drive a machine against a pole, or lock the cutter bars at the outer ends with a chain and then drive the horses in opposite directions, as Champion did. *Attention Farmers.* On Mon. June 25 there will be a Mower Test at Aplington. Flyer 1900. Mss 5x, box 2; *Champion's Challenge to McCormick*, Oral, W. Va., July 16, 1900. Flyer. Mss 5x, box 2.
According to the McCormick Company, however, the importance of these trials was very low, and it paid only minimal attention to them. *E.K. Butler to Percy Lankester* 5.11.1892. Mss 1x, LPCB 461. Finally the leading companies declined to send their machines to fairs any more. *Warder, Bushnell & Glessner to McCormick Co.* 8.30.1898. Mss 2x, box 316.

31 *Schonberger* 1964, 62. Schonberger refers to the letter of E.K. Butler to M.D. Harter 1.11.1894. "This (extension of credit) prevents anyone from buying any other machine and insures the sale for the second year, thus putting the purchasers on the shelf for the two years away from the competition." Ibid.

32 *Butler to Lankester* 7.17.1893. Mss 1x, LPCB 461.

33 *William Deering Company.* Circular, 4.4.1892.

binder, which for the first time enabled the cutting of grain with any length of straw. Its value was not recognized in America, but it became a necessity in Europe where straw was used for bedding and even for feed.[34] The open-end binder, and the low-down binder of Adriance, Platt & Co. from 1891, forced the McCormick Company, as well as the others, to react,[35] as had happened before in the history of harvesting machinery.

There still remains the question of the failed firms. During the 1890s some smaller factories went bankrupt, and even the Walter A. Wood Company encountered difficulties after the founder of the business died in 1892. Should we blame acute competition for these failures, or were there other reasons? The episode of the American Harvester Company revealed the overall condition of the harvesting machinery factories: the largest had up-to-date machinery and organization in their plants, while the poorer ones had to rely on handfitting and other old-fashioned methods. When these factors are combined with the overall depression that hit the country in 1893 and to the crop failure the following year, we might find the solution for the bankruptcy of some of the factories, especially when we know that the demand for harvesting machines was so high during the early 1890s that the McCormick Company was not able to supply

---

34 *Denison* 1948, 117.

35 *E.K. Butler to Lankester* 7.8.1891. Mss 1x, LPCB 460. McCormick had obtained a low-down Adriance for experimentation and was developing an open-end machine too.
The Adriance low-down binder forced the McCormick Company to buy the patents for it and the Company began to produce "the Bindlochine", its own copy of the same. Consequently, Adriance, Platt & Co. brought a case against the McCormick Company and forced it to stop the production and foreign sales of the Bindlochine. Although the McCormick Co. was not able to produce a low-down machine of its own, it owned the patents and gained as royalties $5 for each machine that Adriance sold in foreign countries. Besides, it planned to open a new plant in Canada to make the Bindlochine, but because of the limited demand abandoned the idea. The Bindlochine is not mentioned in the pamphlet of 1894. *Butler to Ackerman* 11.5.1891. Mss 1x, LPCB 460; *The Open Elevator. McCormick Pamphlet*, 1894. Mss 5x, box 1; *Butler to Lankester* 4.25.1892, 5.30.1892. Mss 1x, LPCB 461; *Butler to Freudenreich* 12.12.1892. Mss 1x, LPCB 461; *Butler to Lankester* 4.28.1893. Mss 1x, LPCB 461; *Butler to H.T. Mot & Co.* 4.17.1893. Mss 1x, LPCB 461; *Butler to Lankester* 5.31.1893, 7.17.1893, 8.12.1893, 10.24.1893, 11.7.1893. Mss 1x, LPCB 461; *The Farm Implement News.* Vol. XIV. No. 13. 3.13.1893. p.25.
Experiments with the open-end binder began in 1891. *Butler to Lankester* 7.8.1891. Mss 1x, LPCB 460. Development of the new binder took more time than was anticipated and was also costly to the factory. The new machine demanded the making of from 150 to 175 new patterns and the Company was forced to build "the whole number without the forms, jigs and dies such as we use in turning out machines regularly". *Butler to Lankester* 1.19.1892, 4.25.1893, 7.15.1893. Mss 1x, LPCB 460. The new open-end binder was introduced finally in 1894. *The Open Elevator. McCormick Pamphlet*, 1894. Mss 5x, box 1

all the machines ordered.[36]

It was, however, clear that the large companies could stand the hard years better than the smaller ones, and when the Wood Company ran into problems, after the mid-1890s there remained only two major firms in the field.[37] Even these firms held serious discussions on a merger of their interests.[38]

The division of the harvester industry into two categories became more and more evident during the 1890s. The McCormick and the Deering companies wrestled in their own class; behind them the

---

36 *Butler to Freudenreich* 6.22.1891 and *Butler to Lankester* 7.8.1891. Mss 1x, LPCB 460.
Butler boasted that the McCormick Company in 1891 had 40 000 binders on the market and would have sold 60 000 if they had had them in time. Ibid.   The lucrative situation in 1891 and 1892 tempted the McCormick Company to produce a record number of machines the next year, in the midst of the depression. Even such a prominent firm was unable to borrow money, either from America or in the international financial markets. The impact of the recession were soon felt in the McCormick Company too. Its sales and profits dropped temporarily in two successive years, as can be seen in Table 11. It also had to curtail its production by about 30 000 machines from 1893 to 1894, and there still remained over 15 000 machines unsold from 1893. By 1894, however, the McCormick Company was able to sell about a thousand machines more than it manufactured. *Schonberger* 1964, 41-44; *Statements showing machines manufactured and sold, 1884-1898. inc.* Mss M/I, box 18.

37 See for example *the International Harvester Co.* 1913, table 6, page 86. Of the capital stock of the International Harvester Co. $26 321 657 went to McCormick, $21 362 555 to Deering and to the next largest, Champion, only $3 372 186.

38 Negotiations on a possible sale of the Deering company opened in July 1897. These dealings also throw some new light on the market situation. The capitalization of the McCormick Company was estimated at $17 million. It became evident that McCormick really was holding its prices over the other companies. McCormick's $135 for time sales against Deering's $100. *Harold McCormick to Nettie McCormick* 7.13.1896. Mss 3b, box 17.
In August Cyrus Jr. asked Nettie's approval for the purchase of the Deering company. The idea was to increase the capital stock of the McCormick Co. by selling the new stock to a syndicate to be formed. The capital stock was planned to be divided between the following persons:

Nettie McCormick	5752
Cyrus Jr.	6978
Harold McCormick	5160
Stanley McCormick	5160
R (Rockefeller?)	4980
Anita McCormick	3500
Mary Virginia McCormick	3000
E.K. Butler	230
Eldridge Fowler	200

R possibly means John D. Rockefeller, whose daughter Edith was married to Harold McCormick. *Cyrus Jr. to Nettie McCormick* 8.4.1897. Mss 3b, box 17.
Nettie McCormick, nevertheless, took a conservative view on the matter. She suggested careful moves and expressed concern how "we can *meet them* in *economy* in *making reapers*. By paying *too much* we *lose* our *justification for buying.*" (spacing Nettie McCormick). *Nettie McCormick to Butler* 9.22.1897. Mss 1b, box 24.
Although Cyrus McCormick III explains the collapse of the plan by the difficulties to find financing for the operation, it is obvious that the real reason was Nettie McCormick's resistance. *McCormick* 1931, 109.

Milwaukee Company, the Champion line, the Walter A. Wood Company, the Plano, Adriance, Platt & Co., the Johnston Harvester Co. and D.M. Osborne fought for survival.[39]

From this perspective the American harvester markets had actually stabilized from the 1880s to the 1890s. This view does not deny the competition which certainly existed, but the fight for supremacy was transferred to the new level of a battle between two virtually equal rivals.[40]

On the northern horizon the Canadian Massey and Harris companies, which had fought a bitter struggle over supremacy in Canada, united their forces and merged in 1890.[41] In Europe and in Australasia, Massey had already shown its teeth. The World Columbian Exposition, held in Chicago from May to October 1893, opened a new channel for Massey-Harris to demonstrate its strength right before the eyes of its American competitors. It secured the largest floor area, for carloads of machinery. The Company's central office was panelled with fifteen Canadian soft and hardwoods and decorated with the flags of over fifty countries where Massey-Harris machinery was in use. Despite all these efforts, the Massey-Harris Company failed to win the awards, because of active and even scandalous maneuvers by the McCormick Company, which wanted to secure the leading position at any price in its home town.[42]

---

39 *Estimates of the key figures of the members of the American Harvester Co.* Mss w, box 2; *The International Harvester Co.* 1913, 126-132.

40 *G.L. Keith to Nettie McCormick* 5.16.1891. Mss 3b, box 12.
Philipps distinguishes three effects of the stabilization of the markets: It decreased the importance of spatial competition so that producers were freed from searching for a favorable location in relation to the market; Second, many small firms which had manufactured on contract for local markets disappeared when the means of transportation were improved; Third, companies began to tidy and develop their organizations. *Phillips* 1956, 11-12.

41 *Phillips* 1956, 12 chart 1, 43; *Kuuse* 1974, 275.
The origins of harvester making in Canada were based on American ideas and patents. Canadian manufacturers bought patent rights or simply copied the American machines outright. They could produce for their own protected markets and expand their territory without foreign competition. On the other hand, the American companies reaped home royalties and escaped paying the imports duties imposed on the machines. *Phillips* 1956, 39-40.

42 *Denison* 1948, 131-133. Howard Schonberger gives a more detailed picture of the events at the Chicago fair. *Schonberger* 1964, 66-82. The Chairman of the Executive Committee of Awards at the Columbian Exposition, John Boyd Thacher, decided before the World Fair that each exhibit was to be examined on the floor of the implement annex and awards would be given without field trials. Suddenly, two months after the opening of the fair he notified the harvester companies of field trials to be held in five days at Wayne, Illinois. All the leading companies, except McCormick Co., refused to go to tests at such short notice. The foreign companies were exempted from the field trials. The exemption gave rise to vigorous protests from the American exhibitors. The struggle expanded to a quarrel over the invention of the orginal reaper.

Closely connected to the threat from the Massey-Harris Company was the question of tariffs. Canada had protected its implement industry with high tariffs since 1847. Initially, duties were 10 percent, but were raised in 1858 to 20 percent and further to 35 percent in 1883, which was the highest level the implement tariff ever reached, practically closing Canada to foreign competition.[43]

The U.S. raised tariff walls at least as high, to safeguard its own growing industry. The Morrill Tariff Act of 1861 raised the average tariff rate on dutiable imports to 47 percent and it remained above 40 percent until the First World War.[44] Under the McKinley Tariff of 1890, the duty on imported agricultural implements was set at 45 percent. The proposed Wilson Tariff of 1893 placed agricultural machinery on the free list, however, because it was able to meet foreign competition with superior quality and low prices. This action raised a wave of objection from the American manufacturers, who demanded a reciprocity proviso in the bill, with the intention of excluding the Canadian makers from the American markets. This action was directed towards the Massey-Harris Company; the American companies were afraid that the Canadian giant would use the United States as dumping ground for its surplus machines.[45]

The American manufacturers were not on the alert for Massey-Harris without cause. It acquired a controlling interest in the Johnston Harvester Company at Batavia, New York, in 1910 and also in the Deyo-Macey Engine Company at about the same time,[46] whereas U.S. companies did not open factories in Canada before the turn of the century because of difficulties in obtaining adequate patent protection, Canadian tariffs, and because of uncertainty regarding the permanence of the Canadian market. At the height of the quarrel over the low-down binder with Adriance, Platt & Co., the McCormick Company planned to open a factory of its own in Canada to circumvent the possible negative outcome; that plan was never fulfilled, but it did establish in 1892 in Winnipeg a warehouse of its own to handle the increased trade.[47] The Deering Company

---

43 *Phillips* 1956, 10-11, 42-43.

44 *Hughes* 1987, 369-370.

45 *Schonberger* 1964, 90-95.

46 In the District Court of the United States for the District of Minnesota. The United States of America, Petitioner, vs. International Harvester Company et al., Defendants. Volume XIII. *Testimony of the Witnesses for the Defendants.* p. 184; The *International Harvester Co.* 1913, 49; *Phillips* 1956, 12, chart 1

47 *Cyrus Jr. to Nettie McCormick* 11.22.1892. Mss 3b, box 13; *The Wilkinson Plough Co. to E.K. Butler* 10.9.1893. Mss 3b, box 14. The Wilkinson Plough Company offered to make from 500 to 1000 Bindlochines for the sum of $40 each. This letter shows

was the first to buy a site for a factory in Canada, in 1900, which later became the location of the International Harvester Co. of Canada.[48]

8.2. From Russia to Iceland

8.2.1. The giant knows no boundaries

Researchers have often seen the depression of the 1890s, which began in 1893, as a turning-point in the foreign trade in harvesting machines. According to these explanations, because of the depression of the 1890s, firms had large numbers of unsold machines, and acute competition forced them to seek foreign markets.[49]

Mira Wilkins, on the other hand, found the Sherman Antitrust Act to be an indirect reason for the extension to foreign lands. This Act forbade agreement between companies, but not mergers. Consequently, firms which survived the depression and competition grew powerful, and expanded into giant corporations which had good prospects for foreign business too.[50]

The explanations that have attracted researchers were already mentioned in the contemporary press, which has perhaps affected later interpretations, although in the implement magazines there was no sharp increase in the number of articles.[51] It is understandable

---

how far-reaching plans the McCormick Company had made for the production of the low-down binder in Canada. If they licensed a Canadian company to build Bindlochines, it would keep the patents out of the hands of the Massey-Harris Company. *Butler to Lankester* 10.7.1893. Mss 1x, LPCB 461.

48 *Phillips* 1956, 42; *Schonberger* 1964, 99.

49 *Manning* 1961, 10; *Schonberger* 1964, 106, 114-117; *Carstensen* 1984, 115. Carstensen is more careful in his statement, saying that McCormick "presumably was ready to look at undeveloped markets".

50 *Wilkins* 1970, 71-72.

51 The following excerpts are titles from the Farm Implement News:
"How can foreign trade be established?" 5.11.1893. No.19.
"The condition of the American farm implement trade on the continent of Europe may be reviewed as follows:". 9.28.1893. No.39.
"Now is the opportune time for pushing foreign trade". 11.29.1893. No. 47.
"Review of the agricultural machinery trade in Russia". 3.29.1894. No. 13
"Increasing interest in foreign trade". 4.5.1894. No. 14.
Common to all the articles is their instructive style. They advised exporters to avoid unreliable European agents, to investigate the situation on the ground, and be readier for business. Equally striking is the very literary nature of the articles compared with the everyday nature of the foreign trade for example of the McCormick Company.

that the press tried to find solutions for the hard times, and encouraged firms into foreign business; but in the harvester business there was no dramatic turn abroad. Most of the American harvester companies had tried their wings on foreign fields for years before the depression of the 1890s. Foreign trade was a long process, which needed careful investigation of the market, the kinds of machines needed, and familiarity with the agents and the legal environment. There was no room for hasty decisions if one wanted to succeed. Besides, competition at home prevented the industry from investing capital on overseas operations, as Benson has shown.[52]

For some of the smaller companies, Europe had, according to the McCormick Company, become a dumping ground where they sold their unsold stock of machines. The McCormick Company did not move in that direction; it had far-reaching plans for Europe and did not want to spoil its chances with such tricks.[53] McCormick was continuously receiving new applications for agencies all over Europe,[54] but moved cautiously. Butler continued his yearly trips to Europe and tightened his grip on Mot & Co., the French agent, who was also selling Johnston machines.[55] McCormick had established agencies in all the Scandinavian countries except Finland,[56] and

---

52 *Benson* 1936, 14.

53 *Butler to Maszewski* 2.28.1891. Mss 1x, LPCB 460. Butler stated strongly to the Company's Russian dealer that it would not follow Wood or other companies in price cuts; moreover, "We will venture to say that neither Wood nor his representative had the slightest margin of profit when they were through the season. Furthermore, they have established a low price, which they cannot well raise."
Butler to Lankester 2.3.1892. Mss 1x, LPCB. The price question was repeated in several letters. In this letter Butler expressed his ideology in his characteristic manner: "It matters not what Wood's people choose to do about fixing prices on their machines. We cannot allow their figures to govern us."

54 See for example *Butler to Lankester* 3.3.1891 and 8.24.1891. Mss 1x, LPCB 460. S.C.A. Holth from Norway applies for an agency, as does A.O. Wolthuis from Groningen in Holland.

55 *Butler to Cyrus Jr.* 10.5.1891. Mss 8c, box 11.

56 In 1892 Adolf Petersson from Helsinki applied for an agency. His letter was transferred to the care of Percy Lankester instead of George Freudenreich in Odessa, who otherwise was in charge of Russia. Butler thought that Lankester had better possibilities to take care of Finland. *Butler to Adolf Petersson* 2.2.1892 and *Butler to Lankester* 2.2.1892. Mss 1x, LPCB 460.
In 1893 also Johannes Preetzmann from Albo (Åbo) wanted McCormick machines for sale. *Butler to Lankester* and *to Johannes Preetzmann* 8.25.1893. Mss 1x, LPCB 461. A similar inquiry came from Emil Rehnberg in Helsinki. *Butler to Emil Rehnberg* and *Butler to Lankester* 10.19.1893. Mss 1x, LPCB 461.
Although McCormick's machines were not sold in Finland, it was not an unknown brand name. In his article on the Chicago World Fair in 1893, Edv. Björkenheim reported in detail on the latest models of the Company, but noted that it was not represented in Finland, unlike Walter A. Wood. *Teknikern* No:87, 1994.

continued these operations.[57] The significance of European trade was growing, although its economic value was not yet very important, in such a measure that the McCormick Company in 1892, amidst its domestic hustles, was prepared to develop special models and the new open-end binder for the European markets.[58]

The McCormick Company saw great prospects in Germany. From the very beginning, Germany was excluded from Lankester & Co.'s own sales area, and operated by McCormick with Lankester as salaried manager. In this role, either in 1891 or early in 1892 Lankester appointed Max Paulsen as salaried representative to Germany.[59] After his trip through Europe, Butler planned to open shipping facilities and storage space in Hamburg, where machines could be supplied to various agents; he was on the look-out for a possible enlargement of business in Germany, and this was achieved during the next spring, when Paulsen was installed in temporary headquarters at Bremen.[60] In 1893, the German trade gained further promotion when headquarters sent a special German-speaking salaried agent to assist in sales.[61]

After his first full year as a McCormick's agent in Germany, Max Paulsen wanted to handle the trade on his own account. This

---

57 See for example *Butler to Lankester* 12.3.1891. Mss 1x, LPCB 460.

58 *Butler to Lankester* 4.25.1892. Mss 1x, LPCB 461.
"...we have laid very much more important matters aside to rush this through, to keep you to the front and in the lead of all competitors. It is a matter of pride with us, rather than money".
In the hectic years of 1891 and 1892, the home markets were much more important to the McCormick Co. than foreign fields.

59 Germany was a virgin land to the McCormick Co. and Butler was very careful with all aspects. This comes out for example in the case of a German emigrant farmer who wanted to buy and send a binder to his brother in Germany. This would be a chance to conquer new territory in the Coblenz area. *Butler to Lankester* 2.13.1892. Mss 1x, LPCB 460.
Butler to Lankester 1.26.1892. Mss 1x, LPCB 460. Butler used in this letter the name Max Polson (sic).

60 *Butler to Lankester* 12.13.1892. Mss 1x, LPCB 461. This plan was one of the first steps to open the old problem of extra machines and repairs.
The key position of Germany in McCormick's European activities becomes evident from the following passage in Butler's letter:"All your remarks concerning the pushing of this German trade has been duly noted and approved. We only wish we had a better man now, ready to go there at once." *Butler to Lankester* 12.28.1892. Mss 1x, LPCB 461.
For the temporary headquarters at Bremen see *Butler to Lankester* 5.3.1893. Mss 1x, LPCB 461.

61 *Butler to Lankester* 5.22.1893, *to Max Paulsen* 6.1.1893 and *to C.H. Fincke* 6.30.1893. Mss 1x, LPCB 461.
Expert was C.H. Finke's. His salary was set at $100 a month. His purpose was not merely to assist the trade but to build it up. He was the man Butler had sought some months earlier. Fincke repeated his trip to Germany the next year. *Lankester to McCormick Co.* 5.7.1894. Mss 2x, box 243.

Lankester was not ready to accept. He regarded the German business as very slow, and explained that the demand still needed to be created and farmers had to be taught the advantages of the machines. This case also illustrates the basic approach of E.K. Butler toward expansion in new areas; the Company should sell in the beginning only as many machines as it was able to set up, service and to deliver spare parts for. For future sales, it was imperative to show the real abilities of the machines to farmers and make them satisfied. Butler was equally strict concerning the agents. The McCormick Co. would not sell to any agent unless his intentions and background were known. Butler accepted the move of the agency from Bremen to Hamburg, however, which was both a waterway and railroad junction.[62]

The same careful but steady extension of the European business is typical of the McCormick Co. throughout the 1890s. In France it promptly exploited the economic difficulties of Mot & Co., and confirmed a sole contract with the agent. Mot & Co. was a rare example of an agent which was able for some time to fight successfully with the manufacturers.[63] In Hungary, Butler also proposed a salaried representative to replace Müller & Weitz, whose reputation and standing were sinking.[64] Operations were extended to Bohemia in 1893, where Pomerath & Co. of Prague began to represent McCormick.[65] McCormick machines also found their way to Switzerland and inquiries for them came even from Iceland.[66] For

---

62 *Lankester to McCormick Co.* 9.9.1893. Mss 2x, box 227.
Paulsen was not satisfied with his salary of $900 and wanted it raised to $1200, but had to be satisfied with $1100. Ibid and *Butler to Lankester* 9.29.1893 and 10.24.1893. Mss 1x, LPCB 461.
Paulsen was regarded as an energetic young man. This description seems to have been accurate, for he contracted with a new subagent in Schleswig for the sale of 40 machines. *Butler to Lankester* 11.21.1893. Mss 1x, LPCB 461.

63 *Butler to Cyrus Jr.* 10.5.1891. Mss 8c, box 11; *Butler to Lankester* 1.7.1892. Mss 1x, LPCB 460; *Butler to Lankester* 3.4.1893 and 11.14.1893. Mss 1x, LPCB 461.

64 *Butler to Lankester* 12.13.1892. Mss 1x, LPCB 461.

65 *Butler to Lankester* 5.19.1893, 9.29.1893 and 11.7.1893. Mss 1x, LPCB 461; *Lankester to Butler* 9.9.1893. Mss 2x, box 227. Lankester also held negotiations with Umrath & Co. of Prague, but was not satisfied with their small order.

66 the case of Iceland the initiative came from McCormick's side. It had sent a questionnaire to Bendix about sales possibilities in Iceland. Bendix was not positive, but promised to investigate, and half a year later Stefan Johnson from Iceland was ready to take an agency there. Butler, however, decided to sell there through Bendix. *Brodr.Bendix to McCormick Co.* 5.10.1894. Mss 2x, box 235; *Butler to Stefan Johnson* 12.5.1894. Mss 1x, LPCB 462.
In Switzerland Lankester was forced to sell machines outright to a farmers's co-operative to prevent Deering from getting the sale. Nevertheless, McCormick also had an agent in that country. *Lankester to McCormick Co.* 6.30.1894. Mss 2x, box 235.

the Scandinavian countries the McCormick Company had made contracts at the turn of the decade. Bröder Bendix continued in Denmark, and Andersson and Mattson in Sweden, but in Norway the agency was transferred from Werner to the Amerikanske Maskin Compagniet.[67]

Finland was a totally new territory for McCormick: the Company had received numerous inquiries, and some machines had been sold to a Finnish retailer. In 1893, Andersson & Mattson extended operations to Finland, but Bröder Bendix was also ready to expand to Finland. Only in 1896 was the McCormick Co. ready to make a contract for agencies in Finland, a further illustration of McCormick's cautious foreign trade policy.[68]

In Russia, George Freudenreich had for some time disappointed Butler with his sharp criticism, obstinacy and inefficiency. From time to time Butler sent Freudenreich bitter comments over his doings, and demanded that he follows directions from Chicago.[69] In 1892, Butler expressed for the first time the idea of replacing Freudenreich with a younger and more energetic agent.[70]

By the end of 1893 the McCormick Company had fulfilled its strategy of expansion in western Europe. It had appointed agencies in practically every European country, including Iceland. Now the time was ripe to turn its sights on Russia. In 1894 Butler extended

---

67 *Butler to Lankester* 12.28.1892. Mss 1x, LPCB 461; *Lankester to McCormick Co.* 9.14.1893. Mss 2x, box 227; *Butler to Lankester* 10.7.1893. Mss 1x, LPCB 461; *Lankester to McCormick Co.* 10.25.1893 and 12.19.1893. Mss 2x, box 227; *Lankester to McCormick Co.* 10.27.1894. Mss 2x, box 235.

68 *McCormick Co. to George Freudenreich* 3.14.1892. Mss 1x, LPCB 461. The first two mowers were bought by P. Sidorow from Helsinki for the sum of $73.80; *Lankester to McCormick Co.* 9.30.1893 and 12.14.1893. Mss 2x, box 227. Lankester made his investigations in the normal manner and found all the Finnish agents economically in very poor condition. "The parties we are in negotiation with are all financially weak, and we have had to pay for our experience of dealing with those kind of people in such remote places, for if they do not pay you, it is only throwing good money after bad to sue them at such a distance"; *Butler to Lankester* 11.7.1893. Mss 1x, LPCB 461; *Lankester to Butler* 10.25.1893. Mss 2x, box 227; A.W. Jakobson from Viipuri asked for McCormick machines for sale but the Company was ready to give the agency only if Jakobson would be ready to buy at a carload, that is 40 machines. *Butler to A.W. Jakobson* 12.10.1894. Mss 1x, LPCB 462; In 1896 Edward Ackerman finally appointed Francke & Hackman from Viipuri to represent the McCormick Co. in the eastern part of Finland. *Ackerman to McCormick Co.* 10.6.1896. Mss 2x, box 282.

69 *Butler to Freudenreich* 1.21.1891. Mss 1x, LPCB 460. An illustrative dispatch of the relations between Odessa and Chicago is the following remark of Butler:"You are there, and paid to fight our battles for us... If you are going to represent us in that country you will have to brace up and help us out of difficulties rather than to lie down when your agents make complaints...The like never occurred from any known land other than under your management." Ibid. *Butler to Freudenreich* 12.8.1891. Mss 1x, LPCB 460; *Butler to Freudenreich* 1.5.1893. Mss 1x, LPCB 461.

70 *Butler to Lankester* 12.13.1892. Mss 1x, LPCB 461.

his annual trip for the first time to Russia. After extensive traveling and a detailed survey of the potential of the market, Butler decided to open a new warehouse in Odessa. Thus far, Freudenreich had conducted the trade under his own name; Butler now wanted to manage it directly under McCormick's name. He decided to send a young and aggressive assistant, George W. Tracy, to help Freudenreich in Odessa.[71] The new warehouse was the answer to Freudenreich's repeated requests. Finally he had the possibility to carry a stock of machines and spare parts. This warehouse was also the start for McCormick's own retail business in Russia and was its first foreign capital investment.

In the beginning of the 1890s, the McCormick Company had taken very decided steps to expand its foreign business. Nevertheless, there was no sudden rush abroad after the depression of 1893 and the crop failure of 1894. Consequently, it is evident that the decisions on and preparations for expansion had been made before the depression. In the foreign correspondence of the McCormick Company, the only remark concerning the effects of the depression was Butler's remarks on shortage of storage space;[72] neither did McCormick's competitors take any drastic actions to enlarge their foreign operations. One new move came in 1892, when the Deering Company sent its representative Charles H. Haney to Europe to find responsible agents and outlets for the Company's goods. During several succeeding trips, Haney also inspected possibilities on the other continents.[73]

As a result of the steady and determined development of the foreign organization, France was transferred in 1895 from Mot & Co. to the more energetic Wallut & Hoffman. In Germany, Max Paulsen's status changed from that of salaried agent to an independent agent

---

71 *Manning* 1961, 10; *Schonberger* 1964, 115; *Carstensen* 1984, 125-127.
Before Butler's journey, Freudenreich had a plan to form a syndicate together with Richard Garret & Sons of England, the Eckert Manufacturing Co. of Germany and Weissmann & Co. of Russia to sell agricultural machinery of various but not overlapping lines in Russia. *Freudenreich to McCormick Co.* 2.3-15.1894. Mss 2x, box 235. This plan is an illustrative example of how far Freudenreich had slid from the McCormick ideology.
Butler was very positive on the future prospects in Russia;"There is a better opportunity there than here, or anywhere in the world, so fas as our judgement goes, for an increased profitable business. Everything is ripe for it. The country is no more uncertain than ours." *Butler to Tracy* 12.13.1894. Mss 1x, LPCB 462.

72 *Butler to Lankester* 11.7.1894. Mss 1x, LPCB 462. "Meantime there is no special hurry on our account, except the fact that we shall be badly in need of storage after the 1st of January."

73 The District court of the United States for the District of Minnesota...*Testimony of Witnesses for the Defendants.* p.134-135.

working on his own account.[74] In Russia, too, the McCormick Company tightened its grip on the business. George A. Freudenreich, who had worked in Russia since 1880 as McCormick's salaried agent, was now replaced by his assistant, George Tracy, and the Russian agency resembled the American branch houses, with its salaried manager and a warehouse.[75]

Although General Manager E.K. Butler had kept a keen eye on the foreign affairs throughout his career, he clearly put more energy into this field after the American Harvester Co. episode. Nomination of the new agents and the replacement of Freudenreich were only the beginning of a larger change. The pros and cons of the jobbing business have already been discussed; as the McCormick Co. intensified its foreign business, the negative sides became increasingly evident.

For some time, Butler had received requests to open McCormick's own branch houses in Europe. The jobbers would not risk ordering large numbers of machines, and consequently they often fell short of them during the season. In the same way they were short of spare parts. In the face of intensifying competition, the cost of the machines was one more factor working against jobbers. After his journey through the Scandinavian countries, Edward Ackerman was convinced that "the margin on each machine does not admit of a middleman's profit". The only way to avert loss would be to deal directly with each country. Ackerman also raised doubts concerning Percy Lankester and his real intentions.[76]

---

74 *Lankester to McCormick Co.* 1.22.1895. Mss 2x, box 263. Mot & Co. had contracted with the English Hornsbys to sell the rest of their French stock while Hornsby was closing their French house; *Butler to Lankester* 3.11.1895. Mss 1x, LPCB 462; *Max Paulsen to McCormick Co.* 8.9.1895. Mss 2x, box 263; *Lankester to McCormick Co.* 12.23.1895. Mss 2x, box 263; *Butler to Lankester* 3.23.1897. Mss 1x, LPCB 463.

75 Dismissal of Freudenreich was a lengthy process. The McCormick Company did not want to lose Freudenreich's experience, and requested Tracy to keep him in a good humor and meanwhile to learn Russian and all possible aspects concerning the business. Power of attorney had to be transferred from Freudenreich to the McCormick Co. in the name of Tracy. The transfer of power occurred fairly well, although Butler was prepared to send Lankester to assist in the case of open conflict. Freudenreich left Odessa in fall 1895. *Butler to Tracy* 1.31.1895, 3.8.1895 and 4.10. 1895. Mss 1x, LPCB 462; *Butler to Freudenreich* 3.13.1895. Mss 1x, LPCB 462; *Carstensen* 1984, 127. Butler was not satisfied with the doings of Tracy either. He sent the Company's reliable foreign traveler Edward Ackerman to put some more "stamina" into him. *Butler to Lankester* 2.7.1895. LCBP 462; *Lankester to McCormick Co.* 3.2.1895. Mss 2x, box 263.
In the agreement between Tracy and McCormick his salary was in 1898 set at $2400 plus traveling expenses. The contract would be canceled "for just cause by the said Cyrus H. McCormick". *Agreement between Cyrus H. McCormick and George W. Tracy.* 3.17.1898. Mss 2x, box 299.

76 *Emil Müller to Butler* 7.2.1895. Mss 2x, box 263; Bröder Bendix to McCormick Co. 6.15.1895. Mss 2x, box 263; *Max Paulsen to McCormick Co.* 8.9.1895. Mss 2x,

The first moves toward establishing McCormick's own direct presentation in Europe were taken in Hungary, where the sales and reputation of the Company had been falling for some time.[77] At the end of 1897, Butler proposed setting up McCormick's own office in Budapest the next spring, and also communicated this idea to Lankester.[78] The following year McCormick sent its representative to take over the Hungarian business. Due to his late arrival, he decided to retain the Company's old representative for the coming season, during which he could familiarize himself with the local conditions and agents.[79]

The greatest change in the McCormick Company's organization was triggered by the announcement by E.K. Butler in the beginning of 1898 of his intention to retire from the McCormick Co. Cyrus Jr. saw the timing of Butler's retirement as most opportune for McCormick's. The years of the great fights within and outside the Company were over, and McCormick's organization and factory were in good shape. The place where Butler's loss would be most felt was in the European business. Butler's plans to leave the Company did not prevent him from conducting the sales as heretofore. He made his annual trip to Europe, settled his accounts, and left the Company on November 1st, 1898.[80]

---

box 263; *Edward Ackerman to McCormick Co.* 10.6.1896. Mss 2x, box 282; *Ackerman to McCormick Co.* 5.3.1897. Mss 2x, box 287. Ackerman found out that Lankester had bought spare parts from some other manufacturer and delivered them to his agents as "will fit McCormick machines". Ackerman repeated his demand for direct sales in 1897 after his round trip among the European agents.

77 *Emil Müller to Butler* 7.2.1895. Mss 2x, box 263; *Lankester to McCormick Co.* 12.16.1896. Mss 2x, box 283; *Butler to Ackerman* 5.18.1897. Mss 1x, LPCB 463. Butler instructed Ackerman that during 1898 something was going to happen in Hungary. Until then, Lankester had time to unload the machines; *Ackerman to McCormick Co.* 10.4.1897 and 10.6.1897. Mss 2x, box 287. Ackerman worked directly under Butler in his journeys throughout the world. He helped to set up the machines, handled them in trials and was the Company's eyes and ears.

78 *Butler to Ackerman* 10.22.1897, *to Lankester* 10.23.1897, 11.16.1897, 12.11.1897 and *to Ackerman* 11.2.1897. Mss 1x, LPCB 463. One of the biggest reasons for taking the business in their own hands was too high prices. In this connection Butler also anticipated the future of the German trade and planned to make the Hungarian office the main office for nearly all of Europe. Ackerman did not agree with Butler over the site of the general agency. He advised Butler to place it either in Hamburg or in Bremen. According to him, all the other companies had their own houses in Germany, from which they delivered their machines. Besides, he warned Butler not to depend anymore on Lankester in the Hungarian question, since he had "done his best to put obstacles in our way..." Lankester had lost both Finland and Hungary. *Ackerman to McCormick Co.* 11.28.1897. Mss 2x, box 287.

79 The man sent to Budapest was W.J. Stillman. *Stillman to McCormick Co.* 2.4.1898, 5.27.1898. Mss 2x, box 314.

80 *Cyrus Jr. to Nettie McCormick* 1.13.1898. Mss 3b, box 19. Cyrus immediately

The retirement of E.K. Butler opened the way for a reorganization of the Company. The post of General Manager was discontinued, and his tasks were divided between the McCormick brothers and new departmental managers. Cyrus McCormick continued as the President of the Company, while his brother Harold McCormick replaced E.M. Fowler as Vice-President and Secretary; the youngest brother, Stanley McCormick, was elected a director and the Assistant Secretary of the Company. One of the new departments was the Foreign Department under William C. Mundt. The factory stayed under the Superintendency of H.B. Utley.[81]

Many of the basic features in the development of American business life which Alfred Chandler has described, became apparent in the McCormick Company too. Although it remained a family owned enterprise, and its top managers were family members even after the reforms in organization, the middle managers, on the other hand, were clearly professionals. Directors of the various departments took part in the decision making and planning of the corporation's future. In that sense the McCormick Company did not differ from other large companies[82]. Only its top managers also owned the stock which was becoming uncommon as enterprises grew and diversified their activities.

In the foreign field, Butler's resignation had immediate effects. The old cautious policy was put aside and McCormick's began to seek new growth on the foreign market too. The Company had become increasingly dissatisfied with its long-serving European manager, Percy Lankester, who had represented the Company but also handled his own machine trade in parts of Europe as a jobber. In November 1898, Cyrus Jr. informed Lankester of the retirement of Butler and of the changes in the foreign organization: the Company would not continue Lankester's contract as a salaried agent after the end of his current contract period; he could, however, carry on his

informed his mother of Butler's plans. The reason for Butler's retirement was probably partly burnout, under the high pressure during the long years in the McCormick ranks since 1879. Partly, the reasons are also to be found from his family. He wanted to give more time to his sons, who were approaching their adult years. Cyrus explained to his mother the various possibilities of how the situation should be handled. The resignation was, nevertheless, peaceful and Butler even at this moment was a McCormick man. He did not want to cause any harm to the company; *Manning* 1961, 15.

81 *President's annual report to the stockholders*, July 13th, 1899. Mss M/I, box 18; *Organization of the McCormick Reaper Works*, "staff". Mss 2c, box 29; *Manning* 1961, 15-16.

82 *Chandler* 1977, 9-10, 145-148.

own jobbing business in England.[83]

During the 1890s there is a visible change in the McCormick Company's approach to its foreign business. After the failed merger

---

83 *Cyrus Jr. to Lankester* 11.4.1898. Mss 1x, LPCB 464. Cyrus justified his decision on extension of the foreign business and the increased expenditures resulting thereof; *Lankester to Cyrus Jr.* 11.19.1898. Mss 2x, box 306. Lankester only asked from what date Cyrus wanted to discontinue the existing arrangements. He also asked if the company could inform the European agents that "the expansion of the British trade requires my undivided attention"; The day of termination of Lankester's services was settled to be April 1st, 1899. Cyrus, nevertheless, asked for Lankester's support also in the future, especially because he planned to open an office in Hamburg. *Cyrus Jr. to Lankester* 12.12.1898. Mss 1x, LPCB 464.

attempt the American harvester market was divided into two groups, the two giants fighting in the first class for supremacy. That fight was soon extended also to overseas and intensified in the middle of the 1890s. The McCormick Company extended its agencies to nearly every possible grain and dairy producing country, although mostly the incentive came from the agents themselves. McCormick's strategy began to change hand in hand with the growth of the foreign sales. The old jobbing house system proved to be unsuitable in the new circumstances and the foundation for the Company's own presence in Europe was set up at the end of the decade. On the other hand, the key business ideas remained unchanged; the best advertisement was the satisfied customer. It did not expand to such areas where it was not able to handle the setting up of machines, service and spare parts. The McCormick Co. continued its strict price policy. It did not trust bad customers and tried to minimize thereby the possible risks. On the other hand, the structures of agriculture began to change at the turn of the century. In western and northern Europe the supply of labor declined and wages began to grow[84]. When machine prices simultaneously sank[85], there were incentives for increased machine demand.

8.2.2. The fight over the European markets

During the 1880s the McCormick Company had fought over the foreign markets mostly with the Wood, Osborne and Champion companies and with a couple of English firms. At the end of the decade, the Canadian Massey Co. cast its nets in the same waters. The Deering Company entered the game with new strength from 1892 onwards. The vigor of the American manufacturers to extend business in Europe is understandable, since in most west and north European countries mechanization of agriculture was under way or was beginning.[86]

In France, both Whitman & Barnes and Aultman, Miller & Co. opened their own houses in 1892. Butler, nevertheless, did not expect any great harm to result, and was satisfied with the Company's situation. There was good reason for his satisfaction, since the French agent had been able to sell almost all of his mowers.

---

84 For Britain see *Collins* 1989, 213-214, for Finland information can be obtained in *Maanviljelyshallituksen kertomus* 1898, 157-160.

85 *Heikkonen* 1989, 280 table 24.

86 *Heikkonen* 1983, 26-27, 81-83, 85-86.

In the Scandinavian countries and in Germany, reaper sales were now also expanding. Butler assumed that the Wood Company was losing its grip in Germany.[87] In 1895 the trend in Central Europe remained the same: McCormick was taking the lead, while the Wood Co. was losing ground. In Germany, the business was still divided between McCormick, Deering and Wood.[88] Competition centering around the Deering and the McCormick companies became more apparent after the mid-1890s,[89] and in 1896 Lankester was no longer sure if Paulsen would be able to fight the Deerings to such an extent as he claimed.[90]

In spite of increasing sales, the Scandinavian countries were a difficult area for McCormick's. Especially Deering, Johnston and Morgan had in Denmark a good standing, which was reinforced by the defeat of McCormick's at some trials.[91] In Norway the same competitors had a strong hold on the business, in addition to Aultman and Adriance. According to Lankester's information, Wood had sold less than 400 machines against over 800 the previous year, Johnston 110 against 150 the previous year and Aultman about 25. McCormick's own order was for 300 mowers and 20 binders,[92] which meant that its share of the business had developed quite well, if we keep in mind that McCormick's had established itself in Norway only three years earlier. In Sweden the field was in the hands of the Wood Company.[93] The next year Lankester reported matters to be in satisfactory shape in that country. In 1895 he had to ask for a reduction in freights because of keen competition from

---

87 *Butler to Lankester* 1.7.1892. Mss 1x, LPCB 460. Butler was sure that neither of the houses would get rich; "...if they get enough profit to pay his (manager) salary alone and the rent of the building they will be in great luck."; *Butler to Lankester & Co.* 7.15.1892. Mss 1x, LPCB 461; *Butler to Lankester* 7.18.1892. Mss 1x, LPCB 461.

88 *Louis Kypke to McCormick Co.* 6.18.1895. Mss 2x, box 263. Louis Kypke was one of the McCormick Company's foreign experts; Max Paulsen confirmed Kypkes details and reported he had won new agents from Wood, Deering, Osborne and Adriance. Also the Milwaukee company had emerged in the German business. *Paulsen to McCormick Co.* 6.21.1895. Mss 2x, box 263; Deering had sold some 300 machines in Germany during 1895. *Lankester to McCormick Co.* 12.23.1895. Mss 2x, box 263.

89 *Paulsen to McCormick Co.* 3.10.1896. Mss 2x, box 270.

90 *Lankester to McCormick Co.* 12.16.1896. Mss 2x, box 283; Also McCormick's agent, D. Wachtel from Breslau informed Lankester of the increasing pressure of Deering and Osborne but also reminded him that the Wood company still had its good old connections in the continent. *D. Wachtel to Lankester* 4.29.1897. Mss 2x, box 288.

91 *Lankester to McCormick Co.* 9.14.1893. Mss 2x, box 227.

92 *Lankester to McCormick Co.* 9.22.1893. Mss 2x, box 227.

93 *Lankester to McCormick Co.* 10.25.1893. 2x, box 227.

Deering,[94] which was also tightening its grip on Norway.[95]

When McCormick's travelling foreign agent Edward Ackerman was sent to inspect the Scandinavian countries, his comments were crushing. According to him, neither in Denmark nor in Sweden had the McCormick Company reached its share of the business, which was still in the hands of the Wood Company.[96] In Norway the local McCormick agent, Amerikanske Maskincompagniet, had to admit that Deering was outselling it.[97] In fact, Ackerman stated that only in a few places was competition harder than in Scandinavia, where it was reaching the point of survival. He blamed the Company itself, but also the lazy agents, for a lack of energy in the situation.[98]

The difference between the reports from Ackerman and Lankester throws more light on the relations between the agents and the Chicago administration. Ackerman reported in a very sharp manner on the state of the business in Scandinavia, whereas according to Lankester, the same situation was satisfactory. He could state that Andersson & Mattson had sold almost all of their machines and "matters are all in good shape in Europe notwithstanding the drought..."[99] For Lankester, it was enough when the agents sold all the ordered machines; Ackerman, on the contrary, as a Company man, was fighting for supremacy of the markets.

In Finland, the McCormick Company had done only minor trade before 1896. When Ackerman visited Sweden, he noticed that a considerable number of machines were sold in Finland each year, but he complained that McCormick's was not getting its share. He therefore asked for time off from Chicago to make a visit to Finland,[100] which he carried out during the fall of 1896. Finland was a great surprise to him. McCormick was hardly known in the country, while Deering had sold over 600 mowers in the coastal town of Vaasa alone. Besides, the Wood Company had also done considerable trade, and a number of others were represented.

---

94 *Lankester to McCormick Co.* 10.27.1894. Mss 2x, box 235; *Lankester to McCormick Co.* 1.22.1895. Mss 2x, box 263.

95 *Lankester to McCormick Co.* 3.2.1895. Mss 2x, box 263.

96 *Ackerman to McCormick Co.* 8.21.1895. Mss 2x, box 268; Still in 1898 Wood was holding its strong position in Sweden. *Lankester to McCormick Co.* 4.1.1898. Mss 2x, box 306.

97 *Amerikanske Maskincompagniet to McCormick Co.* 2.6.1896. Mss 2x, box 270.

98 *Ackerman to McCormick Co.* 5.15. and 6.13.1896. Mss 2x, box 282. In his letters Ackerman warned of Deering Co. which was doing all it could to capture the trade.

99 *Lankester to McCormick Co.* 6.17.1896. Mss 2x, box 283.

100 *Ackerman to McCormick Co.* 5.15.1896. Mss 2x, box 282.

Competition was as acute as anywhere in Europe and did not admit of middlemen's profits. Lankester, on the other hand, had not mentioned anything about Finnish prospects or difficulties.[101]

In Austria-Hungary McCormick's still had a promising business in 1894. The Wood and the Adriance companies were the only ones whose trade amounted to anything at all. Lankester could boast of selling in these countries more than all the other makers put together.[102] McCormick's problems with its Hungarian agent was soon reflected in its sales too, and consequently Deering and Osborne became more active.[103] Osborne's increasing activity was felt all over Europe: it opened new branch houses and advertised its products very vigorously.[104] In Romania both Osborne and Milwaukee encountered a severe setback when their agent went bankrupt in 1898.[105]

In Russia the picture was not as bright, in spite of the Company having its own agent. McCormick's travelling expert, A.C. Danner, reported during his trip that there were large regions where McCormick's name was totally unknown, while the Wood Company had done a large amount of business. With a proper drive, he anticipated almost unlimited prospects in Russia.[106]

Competition among the American harvester companies in the European harvesting machine market was going through the same process as in the United States. Smaller manufacturers tried to find room for their production in Europe, but were wiped out step by step by the McCormick and Deering companies, fighting for supremacy there as at home. Overall, American harvesting machinery had almost totally conquered the European field. During the 1890s there was a continuous demand for mowers, reapers and binders. Consular reports also confirm the trend that has become visible in the case of McCormick's. Companies sent abroad experienced men

---

101 *Ackerman to McCormick Co.* 10.6.1896. Mss 2x, box 282. Ackerman noticed also that Osborne's machine was copied and manufactured in quite large numbers; *Lankester to McCormick Co.* 10.10.1896. Mss 2x, box 283.

102 *Lankester to McCormick Co.* 5.7.1894. Mss 2x, box 243,

103 *Emil Müller to McCormick Co.* 7.2.1895. Mss 2x, box 263. In Hungary were represented, in addition to McCormick's, at least Wood, Harrison, Samuelson, Johnston, Adriance, Brantford, Massey, Deering, Osborne, Hofherr and Zimmerman & Cervinka.

104 *Lankester to McCormick Co.* 4.14.1896. Mss 2x, box 272. Lankester figured that Osborne's program incurred a good deal of expence and it would need a large trade to pay it.

105 *W. Staadecker to McCormick Co.* 2.20.1898. Mss 2x, box 314.

106 *A.C. Danner to McCormick Co.* 6.21.1895. Mss 2x, box 263.

to establish agencies and experts to assist the agents and farmers, but the long-term credit which was seen as essential for trade, was still missing.[107]

This overall picture of the market situation in Europe, which has been put together from the fragmentary McCormick correspondence, can be defined more precisely using accounts material of the same Company supplemented by official statistical material.

Figure 3. Exports of the main agricultural implements and machines from the U.S., 1870-1904.

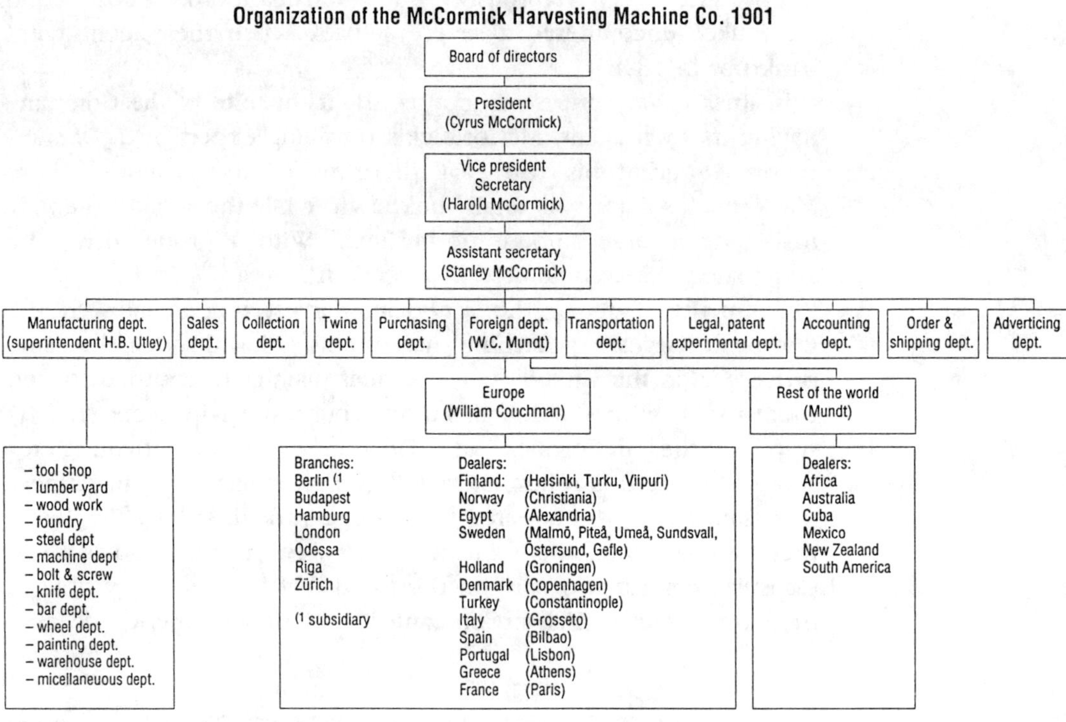

Source: The Foreign Commerce and Navigation of the U.S., 1870-1915.

The value of exports of agricultural machines never reached two percent of the total exports of the United States before 1902. From that perspective, the importance of agricultural machine exports was minimal. If the same phenomenon is compared with total production of agricultural machinery in the U.S., exports increased from 4.5 percent in 1889 to 11.9 percent in 1899 and reached 20.5 percent

---

107 Reports from the Consuls of the United States. No. 123. 1890. p. 649; *Consular reports. Commerce, manufactures, etc.* No. 154. 1893. p. 314-315; *Consular reports. Commerce, manufactures, etc.* No. 195. 1896. p. 609-611.

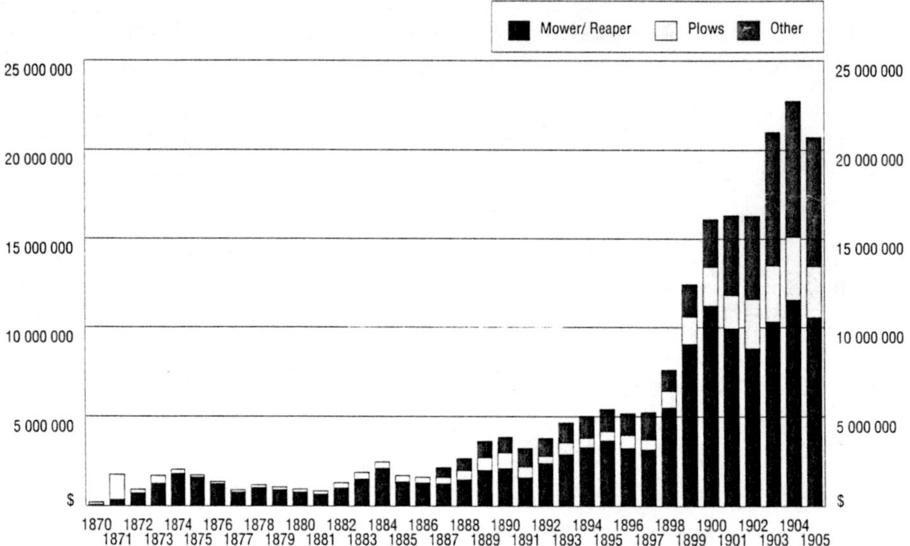

in 1904.[108] That meant a considerable growth, and a clear shift of interest towards foreign trade in the whole line of business. The largest proportion of exports consisted of harvesting machinery and plows, as can be seen in Figure 3. Harvesting machinery, however, increased its share during the 1890s and at the end of the decade clearly formed the largest section of agricultural machinery exports.[109] Machines were mostly sold to Europe, where England, France, Germany and Russia were the central sales areas.[110]

Figure 4. U.S. exports of harvesting machines, by continent, 1890-1905.

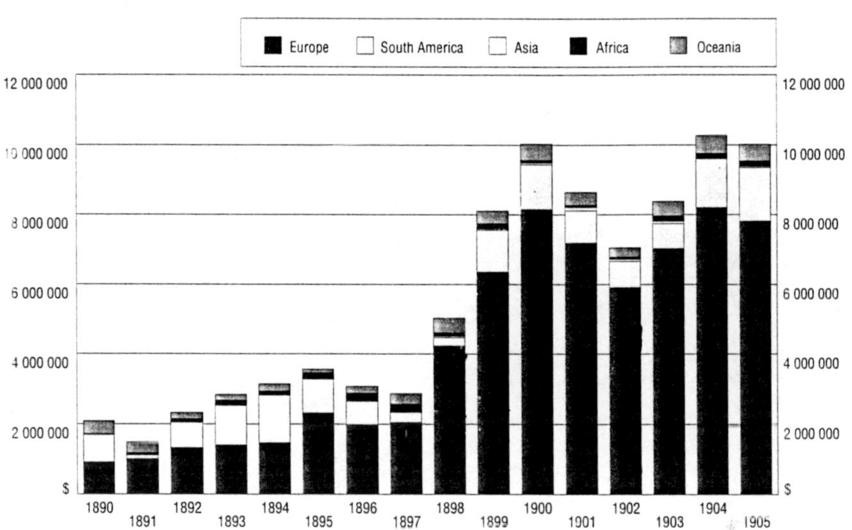

Source: The Foreign Commerce and Navigation of the United States, 1890-1905.

If these general data are compared with the McCormick Company's activities, they coincide very closely. Table 8 shows that between 1884-1898 over 19 percent of the Company's foreign business was done in Russia, with France and England coming next with 16 and 13 percent respectively.

---

108 *The Foreign Commerce and Navigation of the U.S.*, 1869-1915; *Statistical Abstract of the U.S.* 1897-1917; *Heikkonen* 1989, 134-135.

109 The Moline Plow Company, which was one of the largest plow makers in the U.S., after its first careful attempts in the 1870s, began to export its products to Argentina and Denmark in 1892 without any great success. *Thomas* 1976, 72-73.

110 *Statistical Abstract of the United States*, 1870-1917; *The Foreign Commerce and Navigation of the United States*, 1891-1915; *Heikkonen* 1989, 136-138.

in 1904.[108] That meant a considerable growth, and a clear shift of interest towards foreign trade in the whole line of business. The largest proportion of exports consisted of harvesting machinery and plows, as can be seen in Figure 3. Harvesting machinery, however, increased its share during the 1890s and at the end of the decade clearly formed the largest section of agricultural machinery exports.[109] Machines were mostly sold to Europe, where England, France, Germany and Russia were the central sales areas.[110]

Figure 4. U.S. exports of harvesting machines, by continent, 1890-1905.

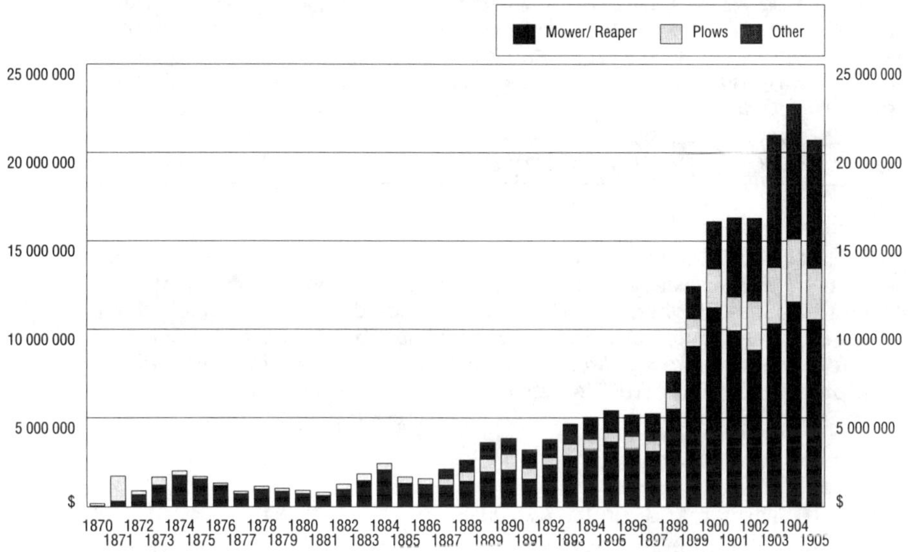

Source: The Foreign Commerce and Navigation of the United States, 1890-1905.

If these general data are compared with the McCormick Company's activities, they coincide very closely. Table 8 shows that between 1884-1898 over 19 percent of the Company's foreign business was done in Russia, with France and England coming next with 16 and 13 percent respectively.

---

108 *The Foreign Commerce and Navigation of the U.S.*, 1869-1915; *Statistical Abstract of the U.S.* 1897-1917; *Heikkonen* 1989, 134-135.

109 The Moline Plow Company, which was one of the largest plow makers in the U.S., after its first careful attempts in the 1870s, began to export its products to Argentina and Denmark in 1892 without any great success. *Thomas* 1976, 72-73.

110 *Statistical Abstract of the United States*, 1870-1917; *The Foreign Commerce and Navigation of the United States*, 1891-1915; *Heikkonen* 1989, 136-138.

Table 16. Total exports of harvesting machines, U.S., and the McCormick Company, 1891-1902.

Year	Total export of the U.S. ($)	Total sales of McCormick ($)	Total sales of McCormick (machines)	McCormick's foreign sales (machines)	Exports as % of total sales
1891	1 579 976	7 386 102	170 666	4 243	3
1892	2 372 938	8 787 887	202 350	..	..
1893	2 873 897	7 598 245	179 643	..	..
1894	3 261 892	6 002 714	143 143 (72 465)[a]	.. (5 156)[a]	.. (7)[a]
1895	3 659 735	8 580 123	206 488 (99 542)	10 851 (8 843)	5 (9)
1896	3 212 423	8 126 821	129 100 (98 773)	15 200 (10 191)	12 (10)
1897	3 127 415	8 945 504	151 885 (122 956)	.. (13 915)	.. (11)
1898	5 500 665	12 511 934	208 346 (161 713)	30 217 (23 885)	15 (15)
1899	9 053 830	13 799 537	266 849 (213 808)	46 359 (33 806)	17 (16)
1900	11 243 763	15 554 122	312 128	48 923	16
1901	9 943 680	18 580 458	402 362	62 813	16
1902	8 818 370	18 650 498	432 100	71 294	17

Source: Statistical Abstract of the U.S. 1891-1903; Machine Records. Mss 3x, box 26. [a] Figures in parentheses are from an unnamed statement of 11.16.1899. Mss 3b, box 21. They do not include machine attachments, which explains the difference between the sources. The normal practise of the McCormick Company was to calculate them as machines. Exceptions have produced a number of materials which have been difficult to include in this study, but, as in this case, offer valuable new information.

Table 16 shows a definite increase in the foreign trade of the McCormick Company. This confirms the picture that has been drawn on the basis of the McCormick correspondence. The significance of foreign sales was visible, reaching in 1899 about 16 or 17 percent of total sales, equivalent to over $2 million.[111]

---

111 The figures in Table 16 do not agree with the information in Table 13, but this difference can be explained by the use of different source material. At the same time it demonstrates how difficult it is to construct a reliable picture of McCormick's foreign trade.

Table 17. Foreign sales of the McCormick Company by main market areas, 1891-1902.

Number

Year	Russia	France	Germany	England	S.Am.	Oceania	S.Afr.	Others	Total
1891	1 231	187	231	1 044	371	371	135	673	4 243
1895	2 741	1 434	997	1 483	1 171	497	648	1 880	10 851
1896	4 784	1 728	1 529	1 655	715	2 604	382	1 803	15 200
1898	3 690	9 358	3 506	3 430	3 618	2 648	478	3 489	30 217
1899	5 870	16 752	6 098	5 059	4 782	1 570	377	5 851	46 359
1900	9 522	19 616	7 183	722	2 872	2 143	368	6 497	48 923
1901	13 327	15 907	16 066	4 993	2 183	1 287	221	8 829	62 813
1902	10 896	16 286	14 069	6 277	4 892	2 900	621	15 353	71 294

%

Year	Russia	France	Germany	England	S.Am	Oceania	S.Afr.	Others	Total
1891	29	4	5	25	9	9	3	16	100
1895	25	13	9	14	11	5	6	17	100
1896	31	11	10	11	5	17	3	12	100
1898	12	31	12	11	12	9	2	1	100
1899	13	36	13	11	10	3	1	13	100
1900	19	40	15	2	6	4	1	13	100
190	21	25	26	8	4	2	0	14	100
1902	15	23	20	9	7	4	1	21	100

Source: Machine Records. Mss 3x, box 26.

Table 17 confirms that in the 1890s Europe was the most important foreign territory.[112] Within Europe, England, the first country where the McCormick Company had begun to sell its machines, was still in the forefront of business, but was losing its position to Russia, France and Germany. Trade in France and Germany, in particular, grew considerably towards the turn of the century.

The development of the harvesting machine trade can also be examined by looking at what kinds of machines were sold in different countries. In the early 1890s the agriculture in England was, if we can trust the McCormick data, the most advanced. In 1891, McCormick still sold more binders there than in all the other European countries together. Its share of binders stayed at the same level throughout the decade; only a few reapers were sold, but the

---

112 The Machine Records of the McCormick Company are not quite consistent with the figures in the *Statement No.5. Profits on foreign machine business.* Mss M/I, box 18, which are used later on in this study. The difference is not, however, very notable and follows the overall trend. In the case of France the difference is largest. Machine Records inform of 16,752 machines sold in 1899 and Statement No.5 of only 10,883 pieces. The growth rates are, nevertheless considerably close to each other, about 56 percent in the former and about 60 percent in the latter material.

sales of mowers increased, which reflects the changes in the structure of agricultural production. On the other hand, in Russia, reapers were clearly the machines most in demand, although mower sales also increased at the turn of the century. The Northern European countries were a mower region. Especially in Scandinavia, binder trade was minimal, and only at the end of the 1890s did growth begin in Denmark. In Germany the same phenomenon can be seen on a larger scale.[113] In this way machine sales followed and directly reflected the state and demands of agriculture in various parts of the world. In Scandinavian countries, production was directed towards dairy products. This was visible also in machine sales; most of the machines sold in that area were mowers. France, eastern Germany and Russia were main grain growing regions and the harvesting machine trade was concentrated in these areas.

Nearly the only way to test the reliability of the Machine Records of the McCormick Company is through its correspondence with its agents. In 1893, machine sales in France seem to have been on the same low level as two years before, only 195 machines altogether, but for 1894, Mot & Co. ordered 768 machines, of which 575 were mowers. Max Paulsen started in Germany with 185 machines, Bröder Bendix in Denmark ordered 258 and Amerikanske Maskincompagniet 400 mowers for Norway, which means that at that phase Denmark and Norway had overtaken Germany, where business was still only beginning.[114]

The Machine Records and the machine orders found in the McCormick correspondence are consistent, although they do not exactly coincide. The realized sales of reapers and mowers in 1895 in Denmark were 400 and 150 machines respectively, and the advance order for the same country 300 and 125. Max Paulsen ordered 725 machines for 1895, of which 125 were binders. That exceeds the figures found in the Machine Records.[115] The difference in the figures is on average not significant, and may be explainable by the duration of the fiscal year, which ended in the McCormick Co. in May.[116] If dealers ordered more machines after that, they were

---

113 *Machine Records.* Mss 3x, box 26.

114 *Butler to Mot & Co.* 4.17.1893. Mss 1x, LPCB 461; *Lankester to McCormick Co.* 9.9., 14.9., 10.25. and 11.14.1893. Mss 2x, box 227

115 *Machine Records,* 1895. Mss 3x, box 26; *Bröder Bendix to McCormick Co.* 9.29.1894. Mss 2x, box 242; *Lankester to McCormick Co.* 10.27.1894. Mss 2x, box 235.
In the case of Russia see for example *Funch, Edye & Co. to McCormick Co.* 3.16.1896. Mss 2x, box 276. For Sweden see *Andersson & Mattson to McCormick Co.* 12.18.1896. Mss 2x, box 270.

116 Collections of the McCormick Co. year ending May 1st 1892. Mss 2c, box 29.

not included in that year's sales. Besides, the Machine Records record only materialized actual sales; an agent might order a larger lot of machines than he was was able to sell.

The McCormick correspondence fills in gaps and at the same time confirms the Company's official accounts for 1897, which are missing from the records. For Denmark, Bendix ordered 1000 machines, Wallut & Co. for France 3250.[117] According to these two examples, the trend in machine sales shows continued growth. In France the effects of the change of dealer can also be seen.

According to the calculations produced by the McCormick Company, the total output of harvesting machines (binders, reapers and mowers) in the world in 1899 was 503 000 of which U.S. manufacturers furnished 443 000 machines; that left only 13.5 percent for the rest of the world. The competition of English companies had diminished to only 13 000 machines, and the largest makers outside the U.S. came from Canada.[118] This information has to be approached with some caution, at least concerning the world data. Estimates made in the McCormick Company with regard to the situation in the United States seem to follow the figures collected for the case against the International Harvester Company.[119]

In the United States McCormick was unquestionably the largest manufacturer of harvesting machines. Its total output was about 160 000 machines, compared to Deering's 132 000. Consequently, McCormick's net profits in 1899 were $4 677 733, and Deering's $3 375 294. The value of McCormick's production plant (land, buildings and machinery) was estimated probably at the end of 1900 at $4 087 266, and Deering's at $3 99 868. Behind the two giants came Champion, Plano, Milwaukee, Wood and Osborne with 37 000, 32 500, 24 000, 20 000 and 17 500 machines respectively. The division of the American harvester industry into two categories had become more visible throughout the 1890s. This division was also transferred to the world markets, of which the McCormick Company controlled about 32 percent and the Deering Company about 26 percent.[120]

---

117 *Bröder Bendix to McCormick Co.* 1.18.1897. Mss 2x, box 294, *Butler to Ackerman* 6.28.1897. Mss 1x, LPCB 463.

118 *Undated estimate of the harvesting machine output of the world found in the McCormick collection.* Mss 1a, box 113.

119 *The Internationa Harvester Co.* 1913, 64-65. The offical figures give 62,203 binders as McCormick's production against 65,000 by the company estimates.

120 *Undated estimate of the harvesting machine output of the world found in the McCormick collection.* Mss 1a, box 113; Comparative statement showing the value of the McCormick and Deering Companies. Mss M/I, box 1.

Table 18. Profits of foreign trade of the McCormick Company in its central trading areas in 1899 ($).

Country	Proceeds of machine sales	Cost of manuf.		Expense selling		Profits on machines	
	$	$	%	$	%	$	%
Russia	284 199	104 717	37	46 504	16	132 978	47
Germany	213 232	95 493	45	14 402	7	103 337	48
France	542 071	261 676	48	50 427	9	229 967	42
England	168 372	90 383	54	15 126	9	62 863	37
Denmark	78 747	35 927	46	5 196	7	37 623	47
Total	1 754 288	796 297	45	159 128	9	799 030	46

Source: Statement No.5. Profits on foreign machine business. Mss M/I, box 18.

Table 18 shows the main grain growth areas in Europe. Denmark is taken as an example of a small Scandinavian country. Although the figures displayed are only for one year, it is evident that foreign business was profitable for the McCormick Company and in all probability for other manufacturers too.[121] Profits from the business seem to have been higher than in the domestic trade. In 1899 domestic sales yielded 38 percent profit, compared with 46 percent from foreign business. On the other hand, profits from the home field were 42 and 41 percent in the two earlier years. However, foreign trade seems to have offered good possibilities at the turn of the century, if we can rely on the surviving data.

Machines for the English markets seem to have been the most expensive ones and for Russia the cheapest to produce. The difference between these two extremes can be explained by the more sophisticated machinery sold in England. In Russia reapers and mowers were the machines most in demand, while in England

Information concerning the situation in the U.S. is confirmed by the McCormick correspondence. Butler notified Lankester that the Osborne company "do not cut much figure in the Reaper and Binder world, in the States" and the Aultman & Miller company has even no organization in America. They, however, had started their factory but in a very feeble way. Of the Wood company Butler was not able to give any details. *Butler to Lankester* 1.14.1898. Mss 1x, LPCB 463.

121 Value of Deering Company's sales in 1902 in Europe was $1,261,789. Total expences were $139,497 or 11.05 percent on sales. If this figure is compared with McCormick's expences in Table 18, it looks like the McCormick Company was more efficient than its archrival. *Selling in Foreign Market* 1919, 147.

binders and reapers were the main articles.[122] On the other hand, selling expenses were considerably higher in Russia than in other European countries. Distances in Russia were huge and demanded extensive travelling. Besides, farmers were, on average, uneducated and experts were needed to set up and advise on the use of the machinery. However, total profits from Russia were above the average profit rate, while England remained clearly below. One factor for the situation in England may have been the contract with Percy Lankester, who was a salaried European main agent for the McCormick Company, but still kept the British Isles on his own account.

8.2.3. Conducting the European trade

The daily functions and difficulties of an exporting company were very varied. To limit a study of an emerging multinational corporation to the level of the organizational development would belie the real size and range of its everyday operations.

Although the McCormick Company, like its competitors, sent special agents and experts to assist the European trade, the fight in practise was lost or won by the agents. It was crucial to find reliable and active agents who were ready to push the machines on the market. On the other hand, they must follow instructions from the manufacturer, which in many cases caused problems. One of the most common complaints concerned prices. Competitors sold their machines well under McCormick's prices, and agents wanted to lower them. In this respect the McCormick Company was firm, as has been stated before: it was unwilling to fight with prices. [123]

The Deering and Wood companies were repeatedly accused of cutting prices; Lankester even maintained that their object was to undersell everybody else and force their machines out by price alone.[124] The Plano and Milwaukee companies also resorted to the same policy, at least in Hungary.[125] At least in one case the other

---

122 The new foreign manager of the McCormick Company, William C. Mundt stated that "the machines for Russia embody, as you know, fewer special features than any other foreign lots..." *Mundt to Couchman* 4.10.1899. Mss 1x, LPCB 465.

123 See for example *Butler to Lankester* 2.3.1892. Mss 1x, LPCB 460; *Bröder Bendix to McCormick Co.* 6.15.1895. Mss 2x, box 263; *Lankester to McCormick Co.* 11.16.1896. Mss 2x, box 272; Still in 1898 Butler repeated his message to Lankester: "They (competitors) appear to think it is necessary to make the prices do business. We find it is necessary to keep the price where there is a little money in it for ourselves..." *Butler to Lankester* 1.14.1898. Mss 1x, LPCB 463.

124 *Lankester to McCormick Co.* 6.18.1897. Mss 2x, box 288.

125 *Stillman to McCormick Co.* 3.16.1898. Mss 2x, box 314.

companies allied against the Wood Company, which had reduced its prices:[126] the agents of the competing machines held meetings where they tried to agree over the prices, but were unable to come to any conclusion.[127] One side of the question was the dumping of old machine stock in the foreign countries at extremely low prices. The McCormick Company accused the Wood Company, especially, of such activities.[128] Verification of this statement is difficult and at least information from Norway

Table 19. Unit prices of the mowing machines of Walter A. Wood Co. in Norway (kroner), 1873-1900.

Year	1873	1875	1885	1898	1900
1-horse	380	–	260	200	210
2-horse	440	440	280	230	230

Source: Tveite 1980, 26.

does not give any direct answer to this question in the absence of comparative material. Machine prices of the Wood Co. certainly sank, but that could have been merely the result of a similar development as in the U.S., where prices fell rapidly during the 1880s because of streamlining in the production systems and increased competition. Prices in general in this period were also falling.

Table 20 shows the development of mower prices of various companies in Finland. Although this material reveals information only from one country, where McCormick,s began to sell its machine relatively late, it does not reveal any dramatic differentiation as to prices. These data have been gathered from agents' catalogs, however, prior to bargaining and therefore do not necessarily show final consumer prices.

---

126 *Lankester to McCormick Co.* 6.22.1896. Mss 2x, box 283. The other makers had an informal meeting where the opinion was that Woods should be left to pursue their own tactics. Only Hornsby did not follow the decision.

127 *Lankester to McCormick Co.* 12.13.1899. Mss 2x, box 344.

128 *Butler to Agar, Gross & Co.* 2.8.1897. Mss 1x, LPCB 463.

Table 20. Mower prices of different companies in Finland, 1890-1902. (In Finnish marks)

Manufacturer		1890	1891	1894	1895	1896	1897	1898	1899	1900	1901
Turun Buckeye	(2)	450	450[a]	360[a]	..	..	..	..	..	..	..
		..	..	..	..	..	..	..	..	..	
"	(1)	375	350[a]	280[a]	..	..	..	..	..	..	..
Turun Palmcranz	(2)	440	440[a]	270[a]	..	..	..	..	..	..	..
"	(1)	350	350[a]	..	..	..	..	..	..	..	..
Vesteråsin Buckeye	(2)	450	..	..	..	..	..	..	..	..	..
"	(1)	375	..	..	..	..	..	..	..	..	..
Buckeye	(2)	..	450[a]	360[a]	..	350[a]	350[a]	..	..	..	..
"	(1)	..	375[a]	300[a]	..	..	280	..	..	..	..
Osborne	(2)	..	..	..	340	340	345	..	340[d]	..	..
"	(1)	..	..	..	270	265	275	..	270[d]	..	..
Deering	(2)	..	..	..	350[b]	345	345	..	..	340[f]	300[g]
"	(1)	..	..	..	280[b]	275	275	..	..	270[f]	275[g]
Karkkila Osborne	(2)	..	..	..	..	..	340	..	..	..	..
"	(1)	..	..	..	..	..	265	..	..	..	..
Wood	(2)	..	..	400[a]	..	360[a]	..	..	..	..	345[g]
"	(1)	..	..	300[a]	..	275[c]	..	..	..	..	290[g]
Thor (Swedish)	(1)	..	..	..	..	280[a]	280[a]	..	..	..	..
McCormick	(2)	..	..	..	340[b]	360[c]	..	..	340[d]	..	..
"	(1)	..	..	..	270[b]	275[c]	..	..	270[d]	..	..
Adriance,Platt	(2)	..	..	..	..	..	360[b]	..	..	..	..
"	(1)	..	..	..	..	..	275[b]	..	..	..	..
Stoddard	(2)	..	..	..	..	325[c]	..	300[c]	..	..	..
Aultman	(2)	..	..	..	..	..	..	..	340[d]	..	..
"	(1)	..	..	..	..	..	..	..	..	270[d]	..
Herkules(Swedish)	(2)	..	..	..	..	..	..	..	340[d]	..	..
"	(1)	..	..	..	..	..	..	..	270[d]	265[f]	..
Toveri(amer.)	(2)	..	..	..	..	..	..	..	340[d]	..	..
"	(1)	..	..	..	..	..	..	..	270[d]	..	..
Plano	(2)	..	..	..	..	345[c]	..	..	..	..	..
"	(1)	..	..	..	..	275[c]	..	..	..	..	..

Mowers are either one-horse (1) or two-horse (2) machines. One-horse mowers are with a 3.5 foot cutter bar and two-horse machines with a 4.5 foot cutter bar. The following symbols refer to various Finnish machine dealers: a=Victor Forselius, b=Carl Jacobsen & Co., c=Johannes Preetzmann, d=Suomalainen Maanviljelyskauppa-Osakeyhtiö, e=Emil Björkell, f=Laurell & Åkerberg, g=Oy Agros, and unmarked = P. Sidorow.

Source: Sidorow 1890, 1895, 1896, 1897; Forselius 1891, 1894, 1896, 1897; Jacobsen 1895, 1897; Preetzmann 1896, 1898; Suomalainen Maanviljelyskauppa-Osakeyhtiö 1899; Björkell 1896; Laurell & Åkerberg 1900; Oy Agros 1901.

The question of prices was complex. A jobber negotiated wholesale prices for himself as low as possible, and the factory for its part lowered them to allow the jobber to lower his prices to his own agents or to farmers. In Germany and in Denmark this policy did not work out, and angry customers sent their complaints even directly to the Chicago headquarters.[129] In Denmark, Bröder Bendix allowed their agents so small a commission that it did not pay to promote the trade; moreover, their mower prices were 15 to 25 Kr. higher than their competitors'.[130] High prices were one of the factors that forced McCormick to consider its own branch house in Hungary and to send there its salaried representative.[131]

In their relations with the foreign jobbing houses, factories had to be very tactful.[132] The McCormick Company normally made only one-year contracts,[133] since agents were independent jobbing houses which represented many lines of machines, and in the case of Mot & Co. in France for some time even other harvester companies. Signing a contract and keeping good relations was more problematic with some agents than with others.[134] According to McCormick's business ideology, if an agent was not satisfied, it was better to let

---

129 Max Paulsen in Germany was claimed to hold his prices too high. *Lankester to McCormick Co.* 12.23.1895. Mss 2x, box 263; this claim was repeated many time during the following years. High prices placed the subagents at a disadvantage towards Deering Co. *Lankester to McCormick Co.* 12.16.1896. Mss 2x, box 283; *D. Wachtel to Lankester* 4.29.1897. Mss 2x, box 288; Also K. Martin from Offenburg in Baden complained of the high prices of Paulsen which made his business diffult, while the same machines were sold considerably cheaper both in France and in Switzerland. *K. Martin to McCormick Co.* 12.11.1897. Mss 2x, box 294. In his later letter Martin asked if it would be possible to buy the machines through Lankester if Paulsen would not lower his prices. *K. Martin to McCormick Co.* 1.14.1898. Mss 2x, box 314.

130 *Ackerman to McCormick Co.* 6.13.1896. Mss 2x, box 282; On the other hand, Bröder Bendix stated that Deering and Osborne are doing all they can to gain territory from them and sell their machines $10-15 lower than they were able to do. *Bröder Bendix to McCormick Co.* 1.18.1897. Mss 2x, box 294. In these two testimonies the dilemma between the views of an agent and the manufacturer becomes clear. Both wanted to gain business but in some cases in a different way. Ackerman wanted to increase McCormick's sales to Denmark because it was his task to do so. Bendix, on the other hand, wanted to have the biggest possible coverage for his business.

131 *Butler to Ackerman* 11.2.1897. Mss 1x, LPCB 463.

132 Relations between the McCormick Co. and its agents are reviewed from the side of the company to see how a giant corporation managed its business relations. This approach certainly does not give a neutral view but reveals new information.

133 The McCormick Company's ideology saved it from many problems. For example, in Norway when the agent retired and negotiations with his successor were becoming complex, the one year contract left the company's hands open to make a deal with a totally new partner. *Lankester to McCormick Co.* 10.25.1893. Mss 2x, box 227.

134 The Danish Bröder Bendix reserved a considerable amount of McCormick's time with its continuing complaints over prices, freights, repairs or some points in the machines. See for example *Lankester to McCormick Co.* 10.27.1894. Mss 2x, box 235; *Lankester to McCormick Co.* 5.15.1896. Mss 2x, box 283; *Ackerman to McCormick Co.* 5.24.1896. Mss 2x, box 282; Bendix also began to sell Aultman's Buckeyes in Sweden when McCormick denyed to add also Sweden to their territory. *Lankester to McCormick Co.* 4.14.1896. Mss 2x, box 272; *Lankester to McCormick Co.* 5.6.1897. Mss 2x, box 288.

him go: a dissatisfied agent merely makes every one else dissatisfied.[135]

To make an agent loyal and active, he should have so many machines
for sale that he would not think about competing brands; on the other
hand, the manufacturer should make sure that he would not need to
carry over machines to the next year.[136] Under the pressure of
competition, however, it was sometimes better to carry over some
machines than let competitors conquer new areas: "our business is a
chance business. We have got to keep you supplied with machines to
take care of a good crop, to prevent any getting away from us; if there
comes a poor crop we must carry them over".[137]

Closely connected to this question was the problem of distribution. The McCormick Company did not have its own warehouse
in Europe. This caused its agents many problems, as has already
been seen; it was a disadvantage in competition, since some other
manufacturers, like Deering and Johnston, had established depots in
their European general agencies.[138]

If a jobber did not fulfil his terms or was not efficient enough, he
would soon be reminded and could even be dismissed.[139] The
McCormick Company kept a close eye on its foreign affairs. Its
General Manager made yearly visits in Europe. Its European manager
monitored and managed daily activities, and could also test agents'
information;[140] but behind his back worked the travelling foreign

135 *Butler to Lankester* 9.29.1893. Mss 1x, LPCB 461.

136 *Butler to Lankester* 5.3. and 7.17.1893. Mss 1x, LPCB 461.

137 *Butler to Tracy* 2.8.1897. Mss 1x, LPCB 463; Also Edward Ackerman underlined
the necessity to have a large stock of machines. This way the shortage of machines
would not open doors to competitors; *Lankester to McCormick Co.* 7.20.1898. Mss
2x, box 306.

138 *Bröder Bendix to McCormick Co.* 6.15.1895. Mss 2x, box 263; In 1898 the
McCormick Co. sent 850 machines and one lot of repairs to Hamburg in the care of
Max Paulsen to be left in the Government warehouse until wanted by the European
agents. Each of the dealers took care of himself by his contract with the company.
The only matter Paulsen had to take care of was to handle advance freight and
handling and storage charges. *Butler to Paulsen* 4.28.1898. Mss 1x, LPCB 464; Butler
informed the agents of these extra machines which were to be used when an agent
had sold out all of his own stock. *Butler to Bendix* 4.28.1898. Mss 1x, LPCB 464.

139 Mot & Co. in France had caused the McCormick Co. over the years a considerable
amount of harm and extra work. *Butler to Lankester* 10.24.1893. Mss 1x, LPCB 461;
Ackerman to McCormick Co. 7.6.1896. Mss 2x, box 282. After the continued
complaints and difficulties with Bröder Bendix in Denmark Ackerman began to keep
an eye on other parties too; *Butler to Agar, Cross & Co.* 5.11.1897. Mss 1x, LPCB
463.

140 *Lankester to McCormick Co.* 6.22.1896. Mss 2x, box 283. Wallut & Co. from
France had instructed Lankester that a number of machines would remain unsold.
Lankester asked Wallut to spare him some of them but hurriedly got a negative
answer.

agent, Edward Ackerman, who reported on his investigations directly to the Chicago headquarters.[141] The salaried agents, such as George Freudenreich, were also under close supervision: if they no longer provided good service, they were dismissed. "Nothing personal can be allowed to enter into McCormick's affairs either on that side of the water or on this.."[142] Step by step the Chicago management also began to incorporate foreign trade as an organic part of the Company. For this purpose officials sent to Lankester the same forms as the general agents used in America, and sales accounts were put in line too.[143]

The terms of the contracts were different for each agent. Those who were able to sell more also received larger discounts. Percy Lankester was especially privileged among the agents.[144]

Another sensitive question for the manufacturer related to quarrels over sales territories. Every now and then an agent could not resist the temptation to sell machines in another's area;[145] sometimes, as in Germany, even the main agent might ignore the border.[146] Manufacturers also had to be very careful in situations where two dealers were fighting over representation, as was the case in Finland.[147]

---

141 See for example *Ackerman to McCormick Co.* 11.28.1897. Mss 2x, box 287.

142 *Butler to Freudenreich* 3.13.1895. Mss 1x, LPCB 462.

143 *McCormick Co. to Lankester* 9.30.1893. Mss 1x, LPCB 461.

144 *Butler to Lankester* 11.2.1892. Mss 1x, LPCB 461. Lankester payed in 1892 for a binder $95 and for a mower $30. For the normal repairs he got a 40% discount. Goods were to be payed one-half in three months and the other half in six months. Max Paulsen had to pay for the same machines in 1893 $105 and $35 respectively. *Lankester to McCormick Co.* 9.9.1893. Mss 2x, box 227; To Bröder Bendix, the price was already considerably higher, $115 and $36. Besides it must be kept in mind that at that time Paulsen was still a salaried agent of the McCormick Co. *Lankester to McCormick Co.* 9.14.1893. Mss 2x, box 227; In Norway prices were still higher, $125 and $38. *Lankester to McCormick Co.* 10.25.1893. Mss 2x, box 227.

145 *Lankester to McCormick Co.* 12.23.1895. Mss 2x, box 263. Dufour of Metz reported that Mot & Co. had sold 15 machines into his territory and demanded a 50 Fcs indemnity per machine;

146 *D. Wachtel to Lankester* 4.29.1897. Mss 2x, box 288. D. Wachtel claimed that Max Paulsen had nominated two new agents on his territory in Silesia; Franz Richter in Döbeln and Krätzig & Söhne in Jauer. He also stated that most of the agents with whom Paulsen had made contracts were only second class men; Lankester, however, did not take too seriously Wachtel's protest. To him Paulsen was a larger and more prominent agent and that was more important. Lankester to McCormick Co. 8.4.1897. Mss 2x, box 288.

147 *Lankester to McCormick Co.* 10.1. and 11.30.1897. Mss 2x, box 288; *Ackerman to McCormick Co.* 10.4., 10.13. and 10.16.1897. Mss 2x, box 287. Preetzman from Turku and Francke & Hackman from Viipuri had represented McCormick's in Finland but because of the continual quarrels the McCormick Co. decided to give the agency to Carl Jacobsen & Co. from Helsinki for 1898. See for more details *Heikkonen* 1989, 272-275; Preetzman had also violated Francke & Hackman's territory which aroused Lankester's disapproval. *Lankester to McCormick Co.* 6.8.1897. Mss 2x, box 288.

The crucial point for successful business was the machine itself. It had to satisfy both the customer and the agent. There were of course many ways to affect sales and to make one's own product more alluring. In America shows and tests were losing their significance, but in Europe they maintained their role.[148] The nature of the trials did not change; their results were widely publicized; judges in the tests were bribed and results decided beforehand[149]. The true significance of the trials becomes clear in Norway, where the local Wood agent threatened the McCormick dealer with a lawsuit for wrongly advertising the results of a machine test at Utah Agricultural Station in the United States![150] An equally illustrative example is the way Cyrus Jr. informed his mother of success in trials.[151] Manufacturers did not send their ordinary implements to trials, but special machines built for the show. McCormick's even went so far that it sent to one test their normal mower which was, however, constructed with special bearings and other mechanisms to minimize the friction.[152]

To boost their sales all the companies published posters, showcards, catalogs and newspaper advertisements. It was of great importance that catalogs were translated into various foreign languages.[153] Foreign-language material was used also in the U.S.A.,

---

148 In 1900 manufacturers agreed not to exhibit their machines at fairs of any kind in the United States and not to furnish their agents with machines to be exhibited or printed matter to be circulated at such fairs. The agreement was signed by Warder, Bushnell & Glessner Co., McCormick Co., the Plano Co., the Milwaukee Co., the Aultman, Miller & Co., the Walter A. Wood Co., the Adriance, Platt & Co., Deering Co., the Osborne Co. and the Johnston Co. January 1900. Mss M/I, box 6.

149 Results of the trials in Europe were regarded as so important that Cyrus Jr. himself wrote about them to agents. *Cyrus Jr. to Lankester* 9.30.1891 and to Koefoed & Haugberg's Maskinudsalg 10.10.1891. Mss 1x, LPCB 460.
Ackerman to McCormick Co. 7.6.1896. Mss 2x, box 282. This example comes from Norway where the first prize was intended for the Massey Harris Co. but their machine could not finish its part of the field and consequently lost the competition; *Maskincompagniet to Edward Ackerman* 9.2.1897. Mss 2x, box 294. *Ackerman to McCormick Co.* 9.14.1897. Mss 2x, box 287. In the machine trial in Norway also in 1897 the McCormick Co. was victorious. The second prize went to Osborn, the third to Massey, and Adriance, Deering and Plano came next in consecutive order.

150 *Lankester to McCormick Co.* 6.25.1892. Mss 2x, box 220; The suit was decided in favor of the McCormick's agent. *Butler to Lankester* 1.29.1894. Mss 1x, LPCB 462.

151 See for example *Cyrus Jr. to Nettie McCormick* 7.9.1895. Mss 1b, box 31. In this case the question was of a trial in Budapest where McCormick won the first prize, Wood the second and Deering the third.

152 See for example *Lankester to McCormick Co.* 10.27.1894. Mss 2x, box 235 and 4.1.1898. Mss 2x, box 306; There were many classes of fair finishing like State Fair finished and County Fair finished machines. *Lankester to McCormick Co.* 6.18.1897. Mss 2x, box 288.

153 *Bröder Bendix to McCormick Co.* 4.17.1896. Mss 2x, box 270; With catalogs

especially in the Mid-Western states, where there was demand for Norwegian, Swedish and German printed material.[154] The McCormick Company also used direct personal letters to farmers in their advertising. The 'personal' quality of the letters was questionable, for they were printed in different colors and typefaces to prevent farmers from the same locality from recognizing the system. Even the agent's name was ready signed below.[155] Extensive advertising was also a means to hide one's own problems.[156] In normal cases manufacturers furnished their agents with catalogs, but Wallut & Co. and Bröder Bendix, at least, printed their own pamphlets for which the McCormick Company provided only the covers.[157] The English copies of the catalogs were first sent to European dealers for translation, and then printed in the United States.[158]

The message in the catalogs was obvious. Superb American machinery was conquering the world like the "Ship of Progress" on McCormick's catalog cover in 1897, steaming with the flags of the United States and the McCormick Company at the mast and the latest model of a binder as the figurehead. Likewise, on the cover of Deering's pamphlet, its harvesting machines were running "Around the World on a Harvester" and on the back cover a family is looking at heaven where Deering's machines are running in clouds resembling a group of angels.[159]

Especially during the 1890s the American hegemony and missionary approach towards the foreign countries become visible. American machines were presented in exotic environments. The Deering "Pony" reaper was depicted in Finland drawn by reindeer,

---

there seems to have been variation as to the place of printing. In many cases the McCormick Co. took care of the whole process but in some cases it sent only catalog covers to the agents. *Butler to Wallut & Co.* 11.30.1897. Mss 1x, LPCB 463; *D. Wachtel to Lankester* 4.29.1897. Mss 2x, box 288.

154 *W.S. Krebs to McCormick Co.* 2.2. and 2.5.1898. Mss 2x, box 313. Krebs was the general agent in Minnesota.

155 *McCormick Co. to Paulsen* 3.30.1898. Mss 1x, LPCB 463; *Mundt to Carl Jacobsen & Co.* 1.20.1899. Mss 1x, LPCB 464.

156 *Lankester to McCormick Co.* 2.25.1896. Mss 2x, box 272. Lankester connected Wood's broad advertising campaign with its difficulties in the U.S. which the company wanted to keep out of the publicity.

157 *McCormick Company to W.V. Couchman* 2.16.1899. Mss 1x, LPCB 464.

158 *Couchman to McCormick Co.* 9.19.1900. Mss 1a, box 115.

159 *McCormick Company. Catalog 1897.* Mss 5x, box 3; *Deering Company. Catalog 1895.* Mss 4z, box 3.

in Argentina binders were attached to oxen and in Asia Minor to camels. The primitive Finnish brush harrow and a skeletal horse were contrasted against the achievements of modern technology. The same tendency can also be found in the travel books of American industrialists. Frank A. VANDERLIP, in his experiences from his journey around the world originally published in 1902, describes in the same manner superior American locomotives in Europe or steamdiggers in Siberia. In Europe, Vanderlip saw endless possibilities for American commercial and technical genius.[160]

Another appealing technique was to use in catalogs the testimonies of well-known farmers or pictures of famous persons. In its 1898 catalog, Deering showed Prince Otto von Bismarck examining his fields and a Deering binder. Lord Salisbury, Premier and Foreign Secretary of England, was presented in the same manner.[161]

In most cases, in order to suit foreign fields machines had to be adjusted or totally new kinds of machines built. When Massey-Harris introduced an open-end binder for the European trade, all the other manufacturers had to follow suit. Something similar happened when Deering began to sell its machines furnished with roller bearings[162] and McCormick introduced its right-hand cut machine.[163] Competitors' machines were kept all the time under a close eye,

---

160 *Deering Company. Catalog 1895.* Mss 4z, box 3; *Saving the World from Starvation.* The Miracle of Modern Farm Machinery. Pioneered by Cyrus McCormick and Perfected Now by the Worldwide "Harvester" Organization. International Harvester Company of America. s.a. Pam Ca-392. State Historical Society of Wisconsin Library collections.
In 1899 the McCormick Company ordered its agent in Russia to arrange photographs of its machines with "men of note" and of interesting harvesting scenes. Besides, Tracy should secure photos of primitive native methods of harvesting. The importance of the question emerges from the following comment:"We want to secure these photographs this year without fail, even though it may be necessary to make repeated attempts, and at a considerable expense." *McCormick Co. to Tracy* 4.12.1899. Mss 1x, LPCB 465; *Vanderlip* 1976, 18-19, 31, 37-38, 91. Keijo Virtanen has in his study found similar attitudes; a positive approach towards western and northern Europe but air of patronage towards eastern and southern Europe. *Virtanen* 1988, 341, 345, 348, 357.

161 *Deering Company. Catalog 1898.* Mss 4z, box 3.

162 Also the Massey-Harris Co. was in the game with its binder with roller bearings. *Lankester to McCormick Co.* 12.16.1896. Mss 2x, box 283; *Bröder Bendix to McCormick Co.* 9.1.1897. Mss 2x, box 302; *Maskincompagniet to McCormick Co.* 4.8.1898. Mss 2x, box 314.

163 *Lankester to McCormick Co.* 2.25.1896. Mss 2x, box 272. In the right hand machine the cutter bar was on the right side of the machine. In harvesters it had been traditionally on the left side but in mowers on the right side. McCormick's machine met with strong resistance from its competitors but also some of its dealers did not like it; *Lankester to McCormick Co.* 10.10.1896. Mss 2x, box 283. The Massey-Harris Co. offered also both left and right hand machines.

even tried out in practise, and practicable parts could be copied.[164] This was, however, a delicate matter. The Massey-Brantford open-end binder infringed McCormick's knotter patent; the Company let Massey carry the machines over to England, and brought a case there to prevent their sale, having first caused Massey considerable work and expense.[165] McCormick faced a similar action from Adriance & Co.[166]

McCormick was not, however, afraid of copying in Europe by the small factories. As has been stated before, it took patents in only a few countries, and did not spend too much energy on the rest.[167] There were, nevertheless, a considerable number of small machine shops that either imitated the original machines or used the brand names.[168]

The encounter with Adriance reveals the difficulties raised by the introduction of a totally new kind of machine. A small company had great difficulties to convince customers of the benefits of the machine, under pressure from the other makers.[169]

In addition to these major changes, agents sent in a yearly list of new requirements for their machines.[170] Sometimes this information

----

164 *Butler to Lankester* 7.8.1891. Mss 1x, LPCB 460. Wood's single apron binder was considered as a failure but Adriance's machine was found interesting; *Butler to Lankester* 11.14.1893. Mss 1x, LPCB 461; *Butler to Lankester* 1.29.1894. Mss 1x, LPCB 462; Deering's binder was doomed in the same class as the Wood's machine:"neither grand success nor an entire failure...sufficient to keep them pegging along and determined to make it go whether it will or not..." *Butler to Lankester* 3.11.1895. Mss 1x, LPCB 462.

165 *Butler to Lankester & Co.* 2.24.1892. Mss 1x, LPCB 461. Butler regarded this action as "a pretty good joke".

166 *Butler to Freudenreich* 12.12.1892 and 4.3.1893. Mss 1x, LPCB 461

167 *Lankester to McCormick Co.* 12.23.1895. Mss 2x, box 263. Lankester warned the McCormick Co. not to make an agency contract with the Erste Maschinenfabrik in Budabest who was known to begin to manufacture reapers and mowers by itself.

168 *Lankester to McCormick Co.* 2.14.1896. Mss 2x, box 272. In Switzerland one manufacturer advertised his mower as the "Cormick" mower. This action had alarmed McCormick's Swiss agent; Umrath & Co. of Prague was also thought to imitate imported machines. *D. Wachtel to McCormick Co.* 4.29.1897. Mss 2x, box 288; In Germany and in Austria-Hungary Lankester had to register the trade mark "Daisy" to prevent its use and still more important prevent other makers registering it and using it against the McCormick Co. *Lankester to McCormick Co.* 10.27.1894. Mss 2x, box 235.

169 *Butler to Lankester* 10.24.1893. Mss 1x, LPCB 461. "Adriance & Platt will make slower progress in fighting the battles alone on this Bindlo type of machine than they would if we were priviledged to go in there and help them along."

170 *R. Wallut & Cie. to McCormick Co.* 10.2.1896. Mss 2x, box 270; *Lankester to McCormick Co.* 11.16.1896. Mss 2x, box 272; *Bröder Bendix to McCormick Co.* 9.24.1897. Mss 2x, box 294.

could be alarming. McCormick's Norwegian agent criticized the Company because its mowers were too heavy, and frustrated farmers from Russia sent instead of a testimonial a long letter of complaints.[171] To assist in fixing and setting up the implements, all the companies sent an increasing number of experts to Europe.[172] It was equally important to furnish agents with a sufficient stock of spare parts; McCormick's simply refused to send its machines without an accompanying repair set.[173]

The McCormick Company was as strict in its credit policy as over prices. It favored cash sales and especially warned George Tracy in Russia against long-term credit. Everything should be due and payable not later than October and November.[174] Agents complained about McCormick's inflexibility, and asked for longer payments[175],

---

An example of the difficulties which harvester companies encountered abroad was the draft animals. In many places in Europe, like in France, oxen were still used and the McCormick Company had to modify its machines to suit oxens slow movements. *Mundt to Wallut & Co.* 4.25.1899. Mss 1x, LPCB 465.

171 *Amerikanske Maskincompagni to McCormick Co.* 2.6.1896. Mss 2x, box 270; *Benjamin Rednopp, Abraham Fehr, Bernhard Martens, Franz Rednopp, Julius Kasper, Peter Fahr and Jakob Friesen to the McCormic Co.* 4.11.1900. Mss 1a, box 115. Russian farmers complained that due to various and continuous breakages they had after six years finally put their McCormick binders aside and gone to work in the old way.
Complaints came also from other places in Russia. *A.Birnbaum to McCormick Co.* 7.29.1900 and *Itschkoff to McCormick Co.* 4.14.1900. Mss 1a, box 120.

172 *Butler to Lankester* 3.3.1891. Mss 1x, LPCB 460; *Lankester to McCormick Co.* 7.9.1896. Mss 2x, box 283; *Lankester to McCormick Co.* 8.4.1897. Mss 2x, box 288; *J.A. Palmer to the Aultman, Miller & Co.* 9.16.1897. Mss 2x, box 297.
To find good experts was a big problem. They were expensive to send and results seem to be meager, while to find a man who could speak European languages other than English and at the same time would be able to handle both the sales and repairs of the machines was not easy. *Butler to Freudenreich* 12.12.1892. Mss 1x, LPCB 461.

173 *Butler to McCormick Co.* 3.23.1897. Mss 1x, LPCB 463; Also Johannes Preetzman from Turku in Finland tried to buy machines without repairs which were, however, sent to Hamburg to wait for his order, with all the extra charges incurred on same added. *Lankester to McCormick Co.* 6.1.1897. Mss 2x, box 288.

174 *Butler to Tracy* 1.31.1895. Mss 1x, LPCB 462.

175 *Bröder Bendix to McCormick Co.* 6.15.1895. Mss 2x, box 263. Bendix stated that Deering and Osborne allowed even from two to three years' credits to farmers. Terms were not the same for all agents; for example, Andersson & Mattson of Sweden made their payments on September 1st, November 1st and January 1st, and received a a 35 percent reduction on spare parts. The Finnish agent, Carl Jacobsen & Co., had to pay for his goods 1/4 at sight, 1/4 on July 1st, 1/4 on August 1st and 1/4 on September 1st, and was granted only 33 1/3 percent off repairs. Bigger houses like Wallut & Co. in France were allowed a 50 percent discount on repairs, and paid for their machines 1/4 on September 1st, 1/4 on October 1st, 1/4 on November 1st and 1/4 on December 1st. If an extension in payments was needed, the McCormick Company charged interest for that period. *Lankester to McCormick Co.* 4.14.1896. Mss 2x, box 272; *Couchman to Stanley McCormick* 5.16.1900. Mss 2c, box 84. *Lankester to McCormick Co.* 5.15.1896. Mss 2x, box 283.

but for the Company it was only continuation of its domestic policy also to overseas businesses. Although it had extended its credits to farmers, McCormick's actively tried to minimize credit sales and preferred cash sales.

One of the almost daily activities which caused constant problems to McCormick was the shipments. Even in the early 1890s, foreign agents themselves or Percy Lankester still usually took care of their own freights through shipping agents.[176] The basic rule of the McCormick Co. for freights followed the same lines as in credits: "...we do not deliver our machines in foreign countries. We deliver them in New York, and when so delivered they are the property of the consignee, and must be insured at his risk, not ours."[177]

The shipping business deserves its own investigation, and consequently in this study it is discussed only insofar as it directly concerned McCormick's. One such instance concerns disputes over freight rates.[178] Shipping costs were a crucial factor in competition between dealers. If one could obtain his machines cheaper than the other he could either reduce his prices that much or put the difference in his own pocket.[179] On the other hand, if the machines were late, a dealer could lose his customers and consequently he might cancel his order to the manufacturer due to the delays.[180]

---

176 *McCormick Co. to Lankester & Co.* 1.13.1891. Mss 1x, LPCB 460. One of the shipbrokers normally used was Henry W. Peabody & Co. in New York; *Lankester to McCormick Co.* 3.5.1894. Mss 2x, box 243.

177 *Butler to Freudenreich* 6.22.1891. Mss 1x, LPCB 460. Nevertheless, the Company had to forward the goods to the ports and see that they reached the agents. Normally the machines were sent through New York, but also Boston and Newport News were used. Step by step McCormick's began to deal directly with shipping agents to get better contracts and cheaper rates. *McCormick Co. to Lankester* 2.16.1893. Mss 1x, LPCB 461. *Henry W. Peabody & Co. to McCormick Co.* 1.20.1894. Mss 2x, box 240; *Butler to Lankester* 11.7.1894. Mss 1x, LPCB 462; *Butler to Wallut & Co.* 3.9.1897. Mss 1x, LPCB 463.

178 If a shipping company noticed that it was necessary to send the goods via a particular route, they were very stiff in their rates, e.g. when the McCormick Company tried to send cargo by direct steamer to Copenhagen. Competition between harvester manufacturers over freights raised the prices in some years to a high level. Rates were naturally also higher to more remote harbors. To Copenhagen and Christiania (Oslo) charges were the same but to Malmö in Sweden already more. To Copenhagen and Christiania rates were 15/- for 40 cubic feet and to Malmö 6/- or 7/- more. Likewise to Odessa in Russia rates were 16/- and to Taganrog 18/6. *Henry W. Peabody to McCormick Co.* 1.27.1894. Mss 2x, box 240. The alternative route was to send the goods via Hull to Copenhagen. That would have taken only a few days longer. *Henry Peabody to McCormick Co.* 2.16.1894. Mss 2x, box 240. *Henry Peabody & Co. to McCormick Co.* 12.5.1894. Mss 2x, box 235.

179 *Lankester to McCormick Co.* 1.22.1895. Mss 2x, box 263.

180 *Carl Jacobsen & Co. to McCormick Co.* 5.18.1899. Mss 2x, box 336. The late shipments had caused Jacobsen & Co. considerable loss and it refused to take the ordered machines because they would arrive after the selling season.

One of the practical problems was the timing of shipments. Often the goods needed to arrive in harbor at a specified moment, because of transhipments or navigation problems, e.g. in the northern ports.[181] If the machines reached the port warehouse too soon, charges were incurred.[182] Mostly, the shipments were too late. Railroad companies in America might have failed in their responsibilities, or a dispute with a shipping agent could delay the machines. If the competitors delivered their machines earlier, it was a great advantage, as was the case in France.[183] Wallut & Co. finally tired of the endless delays in freights and recommended that McCormick's should send its own agent to New York to take care of its foreign shipments.[184]

Insurance of the cargo was normally undertaken by the manufacturer in the name of the receiver. Sometimes agents wanted to insure their machines themselves. Normally insurance was taken "to cover all goods shipped by them or for their account". The goods

---

Bröder Bendix did not entirely decline to take the machines coming too late for the season but expressed its dislike in a very sharp way. *Bendix to McCormick Co.* 2x, box 336. McCormick's tried to reduce expenses by finding shorter and cheaper routes: in Germany some of the machines were shipped to Hamburg, but those which were intended for the South of Germany went to Rotterdam and thence up the Rhine. Similarly, shipments to Finland were forwarded through Hamburg instead of Hull. *Lankester to McCormick Co.* 1.22.1895. Mss 2x, box 263. *Lankester to McCormick Co.* 12.16.1896. Mss 2x, box 283. Jos. Spiero of Hamburg offered a rate of 23/- from New York via Hamburg to Turku.

181 *Lankester to McCormick Co.* 12.23.1896. Mss 2x, box 283. For example the Finnish navigation opened in 1896 not until the middle of April.

182 *Lankester to McCormick Co.* 12.31.1896. Mss 2x, box 283; *Lankester to McCormick Co.* 3.23.1898. Mss 2x, box 306.

183 *Lankester to McCormick Co.*7.9.1896. Mss 2x, box 283; *Butler to Wallut & Co.* 3.9.1897. Mss 1x, LPCB 463; *Wallut & Co. McCormick Co.* 2.25.1898. Mss 2x, box 304; The same complaint came from Max Paulsen in Germany. *Paulsen to McCormick Co.* 2.25.1898. Mss 2x, box 313.

184 *Wallut to McCormick Co.* 4.22.1898. Mss 2x, box 304; The McCormick Company's shipping department was incable of handling all the foreign orders and in 1898 the company was forced to ask the agents to make their orders earlier. Local agents should be ready to take the machines as early as in December and January. *McCormick Co. to Lankester 10.21.1898 and to Paulsen 10.25.1898.* Mss 1x, LPCB 464.
It remains unclear, when exactly the company's own shipping agent was sent to New York. Eugen Manning explains in his seminar work that C.F. Gregory, who was the agent, was sent during the winter of 1895-1896 for the first time and after the third year was located there permanently. Manning has got his information from William C. Mundt's reminiscences for Cyrus III's "the Century of the Reaper". *Manning* 1961, 12; Wallut & Co., however, still in 1898 asked the McCormick Co. to send its representative to New York and the question was raised again during the next year by the new European manager, William Couchman. *Couchman to McCormick Co.* 5.2.1899. Mss 2x, box 343.

were valued at invoice and 10 % added until declared.[185] The broken packings also caused the agents trouble, when machines arrived in their hands in bad shape, often with many parts missing. It was bad advertising for the Company, which was responsible for breakages.[186] Increasing complaints forced the McCormick Company in 1899 to organize a special department for packing foreign machines.[187]

One of the normal points in the foreign trade were tariffs and customs regulations. Customs authorities could cause unpleasant surprises. McCormick's machines were held up by the English customs officers because according to the English directives every English name had to be followed by the place of origin, and on the binder frame arm the word "McCormick" was not identified as from Chicago, while in the mowers it was cast on the frame.[188]

These examples describe minor troubles caused by individual officers which, nevertheless, demanded energy and time. The obstacles put up by governments were in a totally different class. Earlier in this study, we have already seen how countries tried to protect their production with tariffs. In Sweden, duties were on an ad valorem basis: on all harvesting machines, at 10 percent of the net invoice value. In Denmark they were levied on weight and were $4.50 on mowers, $6.00 on reapers and $8.50 on binders.[189]

In Russia, in the 1880s, the government passed quite severe tariffs to protect domestic agricultural implement manufactures. These ordinances did not lead to the desired outcomes and in 1897 the convention of Russian landowners recommended that certain types of agricultural implements should be placed on the list of duty-free articles. This recommendation was put into effect, including harvesting machines, from September 1st, 1898 to December 31st, 1903.[190] In practise, the interpretation of the rules by the Russian Custom House depended largely "upon the whims of the officials".[191]

---

185 *Funch, Edye & Co. to McCormick Co.* 3.5.1896. Mss 2x, box 276.

186 *W. Staadecker to McCormick Co.* 6.23.1898. Mss 2x, box 321; *Lankester to McCormick Co.* 7.20.1898. Mss 2x, box 306; *G.I. Zappoff to McCormick Co.* 5.2.1900 and *Agababoff and Tochoff to McCormick Co.* 7.12.1900. Mss 1a, box 120.

187 *McCormick Company to W.V. Couchman* 4.10.1899. Mss 1x, LPCB 465.

188 *Lankester to McCormick Co.* 6.30.1894. Mss 2x, box 235. Although this rule sounds quite curious, it restricted import of the machines until the name of McCormick was obliterated from the binder arm of every machine.

189 *Couchman to McCormick Co.* "in reference to our business in Sweden" 6.17.1899 and "in reference to our business in Denmark 6.17.1899. Mss 2x, box 343.

190 *.Consular reports. Commerce, manufactures, etc.* No. 205. 1897. p. 271-273; *Queen* 1941, 155.

President McKinley was an advocate of freer trade with foreign countries. In 1899 he submitted reciprocity treaties with seven European countries to the Senate for ratification. The treaty with France provoked serious discussion and the ratification was postponed because of opposition in the Foreign Relations Committee. Because of the opposition, the tariff question was dropped during the administration of President Theodore Roosevelt, who was elected after the assassination of McKinley, and not taken up again until Franklin Roosevelt's administration.[192]

191 *McCormick Company to W.V. Couchman* 4.10.1899. Mss 1x, LPCB 465.

192 *Schonberger* 1964, 193-209. The French tariff was crusial also to the McCormick Company while the duty for the agricultural machinery was roughly $30 per long ton (1016 kilograms) on 15 francs per kilogram. The negotiated reciprocity treaty would have lowered the rate to 9 francs per kilogram. The difference in one binder which weighed about 1500 pounds would have been $9. This difference was in the favor of the European manufacturers who enjoyed the lower rate. Therefore it was natural that also the McCormick Company sent its most able lawyers to assist the fulfilling of the reciprocity treaties. Even Cyrus McCormick Jr. himself with other manufacturers personally visited the Senate. Ibid, 194-195, 205-206.

IX

■ Emergence of a Multinational

9.1. The Company in transition

9.1.1. Establishment of branch houses in Europe

The wave of change began to sweep with all its strength at the turn of the century over the McCormick Company's foreign organization. The European general agent, Percy Lankester, was the first to encounter it. His appointment was discontinued and he continued as a normal jobber. The McCormick Company's new touch in the foreign trade was also felt in Hungary, where it sent its own representative. The biggest changes, however, were still to come.

In December 1898, Cyrus Jr. informed Lankester that the Company was going to open its own office in Hamburg. He also asked for Lankester's support for the new European manager, William V. Couchman. Nominally Couchman's duties began on January 3rd, 1899, but he did not leave America until the beginning of February. All the European agents were informed of the change, and requested to communicate directly with him instead of the home office.[1]

Couchman's position was from the very beginning totally different from Lankester's. He was not only responsible for the whole of Europe, including Russia, but he was also provided with the necessary powers.[2] Couchman was sent to Europe to make a profit and he must have felt the pressure on his shoulders. When he came to Hamburg he found a letter waiting for him from Cyrus McCormick, with instructions to inspect the Odessa agency and to see if it was time to replace Tracy.[3] By April, Couchman had visited Russia, and sent a positive report on Tracy, and arranged for all

---

1 *Cyrus Jr. to Lankester* 1x, LPCB 464; *President's annual report to the stockholders.* 7.13.1899. Mss 3b, box 21 and Mss M/I, box 18. Couchman had formerly been McCormick's general agent in Marshalltown, Iowa; *Cyrus Jr. to Carl Jacobsen & Co.* 1.25.1899 and *to Wallut & Co.* 1.26.1899. Mss 1x, LPCB 464.

2 *Cyrus Jr. to Tracy* 2.4.1899. Mss 1x, LPCB 464.

3 *Cyrus Jr. to Couchman* 2.4.1899. Mss 1x, LPCB 464. Although Cyrus Jr. admitted Russia to be one of the most difficult agencies in Europe he noted, however, that "it has appeared to us that he (Tracy) has not shown marked ability or efficiency in the management of the Odessa agency..."

matters to go through his hands in Hamburg.[4]

Couchman continued his energetic management of the business with a trip to Copenhagen, Malmö, Stockholm, Turku and Helsinki. His journey reflects the potential these areas were expected to offer. Nothing could escape Couchman's critical eye. For the first time McCormick's European trade was in the hands of a professional manager used to large-scale business. Couchman was dissatisfied with the Swedish sales, and wanted to open a new agency in Stockholm. In Denmark he noticed the good standing of Bröder Bendix. In Finland the intentions of Russia to incorporate Finland and take away its liberties, together with poor crop prospects, were causing severe problems for the business.[5]

After his numerous travels to visit all the main agents of the McCormick Company,[6] Couchman had evidently made up his mind how to continue the European business. Jobbers had introduced McCormick's machines in Europe, but were becoming less effective. Now was the opportune time to take the business into the Company's own hands. The first plans for supplanting Lankester were made by July.[7] In Lankester's new contract for the coming season his territory covered only Great Britain.[8] In April 1900, Couchman sent the would-be British general agent C.H. Burlingame to London, and asked him to inform Lankester of the intentions of the McCormick Company to open its own branch house in Britain and to start negotiations over the terms of the change.[9] Lankester

---

4 *Mundt to Couchman* 4.10.1899. Mss 1x, LPCB 465.

5 *Couchman to McCormick Co.* 6.16.1899, 6.17.1899 in reference to our business in Denmark, 6.17.1899 in reference to the trade in Finland and 6.17.1899 in reference to our business in Sweden. Mss 2x, box 343.

6 Couchman was again in Scandinavia in September and traveled from there at once to France. He obviously wanted personally to get to know all the details and problems of the agents and the various countries. *Couchman to McCormick Co.* 9.5.1899. Mss 2x, box 343.

7 *Cyrus Jr. to Couchman* 7.3.1899. Mss M/I, box 5. The first possibility was to raise Lankester's prices to the same level as the other agents' and urge him to push harder. Couchman could have taken the British business under his control or a new man could be sent to England. The last alternative was to appoint some other firm to continue Lankester's business. Cyrus decided, however, to send a special traveler to investigate matters without letting Lankester know.

8 *Couchman to McCormick Co.* 10.3.1899. Mss 2x, box 343.

9 *Couchman to Burlingame* 4.1.1900. Mss 2x, box 362. The McCormick Company was very determined in its decision to open its own branch but, on the other hand, wanted to be as frank as possible with its long-time agent.
Lankester demanded as compensation for his goodwill an amount representing one and a half year's average net income, for the stock in hand $5000, for repairs $8000 and rights over the leasehold of the warehouse. *Lankester to Couchman* 3.29.1900. Mss 2x, box 362.

had no real chance in the negotiations. McCormick was ready to pay $10 000 for his goodwill instead of the $30 000 that Lankester asked for. An agreement was made on McCormick's terms and Lankester's entire business was transferred to the McCormick Company.[10]

E.K. Butler had already sent a Company representative to open McCormick's own branch in Budapest. He was unable to accomplish his plans until 1899, when the power of attorney was transferred to him.[11] Plans were made for the rest of Europe too. In fact by the fall of 1899 the McCormick Company had a scheme ready for drastic reorganization and had even listed the names of the men who were to be placed in the new branch houses.[12]

The question of the German representation rose to the surface when a German manufacturer proposed opening a joint venture for the production of mowers in Germany. The invitation was politely refused, but it opened their eyes to the growing significance of Germany in the machine trade.[13] In June 1900, the company's representative was already picking up information in regard to the German business and was in fact looking for desirable new agents for the McCormick Company. Soon thereafter Couchman decided to notify the McCormick Company's German agent Max Paulsen that his contract would be discontinued, and asked Cyrus Jr., who was visiting the Paris Exhibition, to come to Berlin to decide the location for the German office. Couchman raised for the first time the question of establishing a limited-liability subsidiary company in Germany.[14]

Burlingame had previously worked for the Osborne Company in South Africa. He was presumably Osborne's traveling representative in the same manner as Edward Ackerman was for the McCormick Company. *Mundt to Postin* 12.7.1899. Mss 1x, LPCB 466.

10 *McCormick Co. to Couchman* 4.28.1900. Mss 1a, box 118. *Couchman to McCormick Co.* 5.5.1900. Mss M/I, box 6. Couchman and Burlingame had investigated all the potential warehouses in London but had to acknowledge Lankester's to be the best possible. In the agreement Lankester had to promise not to engage in the agricultural machine business for five years, and be willing to give his assistance to Burlingame.

11 *Couchman to Harold McCormick* 10.5.1899, *Stillman to Couchman* 9.23.1899 and 10.2.1899. Mss 2c, box 84.

12 *Harold McCormick to Couchman* 10.26.1899. Mss M/I, box 5.

13 *Andreas Schilli & Co. to McCormick Co.* 5.1.1900. Mss 1a, box 119; *The McCormick Co to Andreas Schilli & Co. and to Couchman* 5.12.1900. Mss 1a, box 118.

14 *Couchman to McCormick Co.* 6.8.1900. Mss 2x, box 367; *Cyrus Jr. to McCormick Co.* 6.19.1900. Mss M/I, box 7; *Couchman to Max Paulsen* 6.23.1900. Mss 1a, box 115.
Couchman paid $6000 to Paulsen for all claims and damages under the contract, and

The founding of a subsidiary was Couchman's idea. One of his main arguments was to make McCormick's name more visible than before and at the same time to evade the control of the German authorities and taxation. For these reasons, most of the American companies were incorporated as limited companies in Germany. Taking the business into McCormick's own hands also demanded planning the methods of running the organization in the future. Couchman wanted to continue business with big dealers, but he also sought direct contacts with small local agents. The American system of doing direct business with farmers he was not ready to introduce in Europe, nor did he anticipate it in the near future.[15] The McCormick Harvesting Machine Co. m.b.H. was incorporated on September 7th, 1900, with a capital stock of 200 000 marks.[16]

Cyrus Jr. had traveled to Berlin as requested to make the final decision over the location and formation of the Company. When the remaining questions in Germany were settled, Cyrus continued his journey to Russia, where he wanted to hold discussions with the Company's long-established agent in Moscow, Liphart & Co., and to examine the prospects for opening another branch in Russia. After

took back from Paulsen his repair stock and twine. Thereafter the company's own agent, Mr. Hutmacher, had exclusive control over McCormick's business in Germany. *Couchman to Hutmacher* 10.9.1900. Mss 2x, box 367.

15 *Couchman to McCormick Co.* 6.29. and 7.13.1900. Mss 1a, box 115; *Couchman to Cyrus Jr.* 8.20.1900. Mss M/I, box 7. The other possibility would have been to register as a branch or 'filiale', which would have meant filing the yearly balance sheet of the American corporation with German authorities in addition to a separate sheet for the German branch. The income tax would then be calculated upon the profits shown by the American corporation on the basis of the proportion of profits made by the German company. Besides, it was totally at the German officials' discretion whether they would grant a license to do business.
The business could also have been run in the name of a German representative, leaving McCormick's name aside. The third possibility was to organize a joint stock company under German law. In that case the McCormick Company should appoint a board of directors and council, and submit the books of the corporation for inspection by the German officials.
The limited stock company or Gesellschaft mit Beschrenkter Haftung was the simplest method for the foreigner. It was not taxed on its profits, but individual stockholders were taxed on their holdings of stock. For foreigners there was practically no taxation because the tax law predated the laws authorizing the formation of limited companies and consequently, no provision had been made in the law to cover these companies. It was also easy to organize. The articles of the incorporation had to be drawn up, stating the purpose, name and domicile of the company, the amount of capital stock, holders of the stock and proportion of their interest, and the name of the responsible manager of the business.
According to Couchman, of the American companies at least Luxor Prism, National Cash Register, American Bicycle, Columbia Phonograph, American Radiator and Photographic Supply were incorporated as limited stock companies.

16 *Draft for the articles of incorporation the McCormick Harvester Machine Co. m.b.H.* Mss M/I, box 7; *The International Harvester Co.* 1913, 168.

investigating the matter, Cyrus planned to establish a general agency at Samara on the Volga some time in the future. Agricultural prospects and resources in Russia made an immense impact on Cyrus, but he also noticed its weaknesses, the inefficient native labor and the risky credits.[17] The Russian question was settled by establishing another branch house at Riga for the Baltic Provinces. Riga was selected because of its lively port.[18]

The agents in Switzerland did not please Couchman either, and Couchman resolved to take the Swiss trade into McCormick's hands, and placed the man intended for Riga there instead. He had found his agent not to be the right man to handle the large northern Russian region. Couchman's decision meant that he let Russia wait for another year, and allowed Tracy to take care of all of Russia while he was looking for a suitable man for Riga.[19]

While Cyrus McCormick visited Russia, his younger brother Stanley made a trip through Germany, Denmark, Norway and Sweden. In Denmark, Bröder Bendix made a good impression on Stanley, with their active appearance. Bendix claimed to be the leaders of the business, with Deering coming next. His pleasure changed to distress in Norway and in Sweden. McCormick's machines could not compete with their rivals' lighter machines, and consequently in Norway McCormick was in third place after Deering and Wood. In Sweden the situation was about the same, except that here the local manufacturers were also becoming a real threat to the American companies.[20]

---

17 McCormick's had not been satisfied with Liphart & Co. and consequently Tracy had curtailed its territory by appointing new agents. Liphart & Co. was, however, the strongest name in the agricultural world in Russia and the McCormick Company could not afford to let them go. Another difficulty in Russia was caused by the Siberian trade. It had been allocated to Storvell & Co. at Veronesh, but practically all the other agents also sold in Siberia whenever possible. The dilemma with Liphart & Co. was resolved in a new meeting at Berlin, attended by Cyrus Jr., Emil Liphart, Couchman and Tracy. Emil Liphart was not allowed to have back the areas which he had lost, but he was compensated for the loss with one percent more on the business of the new agents. *Cyrus to Stanley McCormick* 7.31.1900. Mss 2c, box 84; *Cyrus to McCormick Co.* 8.23.1900. Mss 2x, box 362. *Cyrus to McCormick Co.* 8.23.1900. Mss 1a, box 117.

18 *Cyrus to McCormick Co.* 8.23.1900. Mss 2x, box 362.

19 *Couchman to Stanley McCormick* 9.19.1900. Mss 2c, box 84. August Mury, in German-speaking Switzerland, had done good work, but Bucher-Manz was selling at too low prices and was therefore "giving our machine the wrong reputation in Switzerland." *Couchman to McCormick Co.* 9.21.1900. Mss 1a, box 115.

20 *Stanley McCormick to Cyrus Jr.* 9.3.1900. Mss 1a, box 119 and Mss 2c, box 84. In Norway, from the estimated number of 2210 machines sold in 1900 Deering's share was 600, Wood's 450 and McCormick's 350. In Sweden the total business was about 10 000 machines, of which Deering and Wood sold about 2500 each, McCormick coming next with about 1200. The Swedish companies sold together about 2000 machines.

Couchman sought new efficiency for the sales by appointing new agents. He opened negotiations for an agency in Holland and made a contract with Boeke & Huidekoper in Groningen. New agents were also appointed in Greece, in Alexandria for Egypt and Sudan, in Spain and in Constantinople. Couchman wanted to establish several agencies, for he anticipated "that they will have a tendency to create a demand for machines and that we will be known to the trade when that time comes".[21] In accordance with his principles, Couchman curtailed in Sweden the territory of the Company's present agent, Andersson & Mattson, due to their inefficiency, and nominated three new agents.[22]

The McCormick Company's sudden moves raised anxiety among the agents, who could no longer be sure of their status. Couchman began to use a technique of alternately cajoling and enticing the agents. In Sweden he had used hard measures to improve sales. In Denmark his travelling agent noted the good shape of the business, but Couchman, nevertheless, let them know of his actions in Germany to make it clear to the Danish agent that McCormick agents had to push the business in order to keep their contract. In Italy the nervous dealer wanted to make a five-year contract, but had to be content with the normal one-year one.[23]

New efficiency was needed, and it was sought for by sending new travelers and experts to Europe to fight over markets in highly competitive regions such as Scandinavia and England. It was, nevertheless, not too easy to find able men, and when found they were either reluctant to move abroad or their families refused to follow them. In some cases the American general agents did not let their best men go.[24] The status of the foreign trade had clearly

---

21 *Mundt to Couchman* 10.3.1899. Mss 1x, LPCB 465; *Couchman to Stanley McCormick* 5.16.1900. Mss 2c, box 84 also 1a, box 115; *A Few Points in Connection With Our Foreign Business.* 6.4.1900. Mss 3b, box 24 also 3b, box 24. These two reports contain Couchman's estimation of the standing of the various agents. A detail worth mentioning was that either in May or in June 1900 Edward Ackerman erected the first binders ever set up in Spain.

22 *Couchman to McCormick Co.* 7.10.1900. Mss 2x, box 367; Four letters from *Couchman to McCormick Co.* concerning Olsson & Larsson of Gefle, Eric Björkland of Stockholm, Joseph Carlander of Sundsvall and Andersson & Mattson 10.12.1900. Mss 2x, box 367.

23 *Shapiro* 1958, 13-15; *Couchman to McCormick Co.* 6.8.1900. Mss 2x, box 367.

24 *Mundt to Couchman* 11.4.1899. Mss 1x, LPCB 465; *Couchman to McCormick Co.* 2.6.1900. Mss 2x, box 363. Pressure of competitors was felt hard in Scandinavia, where Edward Ackerman so far had been the only traveler to visit. Now a totally new approach to the business was to sweep over the northern countries too. Couchman informed Lankester that he was going to send some help from the manufacturer, as the Deering Company had done for some time. *Couchman to Lankester* 2.23.1900. Mss 2x, box 362.

changed among the decision makers of the McCormick Company.

Couchman also began negotiations with the home office to stop holding a stock of machines in Hamburg. At first sight his demand seems strange, since reserve machines had been for many years one of the demands of the agents. Couchman stated, however, that when a large stock of machines was held, there remained always some machines unsold, since machines had to be adapted for various countries and it was impossible to know just what would be wanted. Now that he was on the spot, he could keep in close touch with agents and could send exact orders to the factory.[25] Accounting systems and better control over the production which had been extended to Europe too offered Couchman tools to follow the demand and supply accurately.

Adapting the machines to conditions in various parts of Europe had been discussed at the highest level in the McCormick Company for years. It looks as if in spite of frequent complaints no concrete changes had occurred, since so far there had been nobody in the organization to look after them. Foreign business had been under E.K. Butler, together with a thousand other things. Couchman returned to the question with new energy. He immediately understood how essential it was for the business to provide machines appropriate to local conditions. For the first time the McCormick Company received direct and reliable information on the state of agriculture in the European countries. Couchman also investigated the complaints over the quality of the machines, and found them in many cases well-founded. Branch managers and general agents were also asked to send a list of complaints. McCormick's machine inspection department had been under too heavy pressure or simply had not performed its functions properly. Couchman was able to list numerous defaults: in the "Daisy" reaper the rake arms hit together at the top, the chain tightener crank bent

---

In the letter of *Mundt to Couchman* 3.13.1900. Mss 1x, LPCB 466 is a list of the state of travelers and experts to some countries. The question is continued in *Couchman to McCormick Co.* 3.24.1900. Mss 2x, box 362, in which Couchman evaluates the traveler candidates.

In June 1900 Couchman had under him 24 men in Europe, either branch house managers, travelers, or experts, who covered practically the whole of Europe. *A Few Points in Connection With Our Foreign Business.* 6.4.1900. Mss 3b, box 24; *Couchman to McCormick Co.* 6.8.1900. Mss 2x, box 367; *Mundt to Couchman* 4.5.1901. Mss 1x, LPCB 467.

25 *Couchman to McCormick Co.* 11.15.1899. Mss 2x, box 343. There was one more reason not to hold machines in stock in Hamburg: some European countries (France especially) levied an extra charge on machines imported from America through Germany.

The McCormick Company accepted Couchman's proposal without any delay. *Mundt to Couchman* 11.28.1899. Mss 1x, LPCB 466.

and the spoons on the pitman were too weak owing to softness of the malleable iron. The mower, too, suffered from the spoons breaking. Besides, painting of the machines should have been more attractive.[26]

After his first full year as the European manager, William Couchman was able to report amazing progress in sales. The German branch alone had made a net profit of $100 000,[27] and the new management had shown its efficiency compared with the jobbing houses. In 1899, Cyrus Jr. expressed his satisfaction at having sent Couchman to Europe, and in the next year's President's annual report he was able to tell of a steady growth of the European business.[28]

In three years the McCormick Company had almost eliminated jobbing houses from its European business, and absorbed their profits. It had sought for an aggressive new drive in its sales by appointing new agents and increasing the number of its experts and travelers. Through these moves the McCormick Company was able to increase its share of the European markets. The elimination of big dealers[29] also provided an opportunity for direct contacts with local agents and thereby for closer control of the business. On the other hand, branch houses also considerably raised the costs of conducting the foreign business.[30]

---

26 *Couchman to McCormick Co.* 7.8.1900. Mss 1a, box 115; *Couchman to McCormick Co.* 8.13.1900. Mss 1a, box 117; *Hutmacher to Couchman* 9.8.1900. Mss 1a, box 116.
Couchman told, for example, how in Germany land was cultivated in ridges which gave a corrugated effect to the fields and was very difficult for working the machines. In Poland the situation was still worse and one of the first questions was to begin to educate farmers to make their fields suitable for the machines. This comment correlates with the analyses of Paul A. David on the situation of agriculture in Britain. *David* 1971, 145-175.

27 *Couchman to Cyrus Jr.* 12.8.1900. Mss M/I, box 7.

28 *President's annual report to the stockholders*, 1899. Mss 3b, box 21; *President's annual report* , June 1901. Mss M/I, box 18.
Sales of harvesting machines (binders, mowers, reapers, headers, shredders, corn harvesters and rakes) in 1899 was 29 417 the next year 36 913; and in 1901 already 47 375 machines.

29 Not all the big wholesalers were eliminated. In France R. Wallut & Co. continued its jobbing business and in 1902 placed a large order for machines for 1903, about 16 000 altogether. *The Implement Age* 1902, vol. XX.

30 *Cyrus Jr. to McCormick Co.* 6.191.1900. Mss M/I, box 7; *Couchman to Cyrus Jr.* 12.8.1900. Mss M/I, box 7. Show rooms, offices and repair facilities in the Berlin office alone cost about $4000 per year. The high costs soon paid themselves back in growing orders for machines. Couchman told Cyrus how two local German dealers had been very reluctant to order new machines; after visiting McCormick's new warehouse their eyes had been opened to the real size of its business and both gentlemen were thereafter more than eager to renew their contracts.

Map 2. Branch houses and main agents of the McCormick Company in Europe, 1901.

DEALERS

DENMARK
Bröder Bendix, Copenhagen

FINLAND
Suomalainen Maatalouskauppa Oy, Viipuri
Carl Jacobsen, Helsinki
Victor Forselius, Turku

FRANCE
R. Wallut & Co., Paris

GREECE
Aristote Tsakonas, Athens

HOLLAND
Boeke & Huidekoper, Groningen

ITALY
A. Cosimini & Figli, Grosseto

NORWAY
Aktieselskabet Maskinkompagniet,
Christiania (Oslo)

PORTUGAL
J.M. Sumner, Lisbon

SPAIN
Cortiez Hermanos Yermo & Co., Bilbao

SWEDEN
M.J. Backlund, Umeå
Ax. Colberg, Piteå
Joseph Carlander, Sundsvall
Olsson & Larsson, Gävle
Andersson & Mattson, Malmö
Hard & Forsberg, Östersund

TURKEY
Geo. Chisnell & Son, Constantinoble

BRANCHES

AUSTRO-HUNGARIAN
EMPIRE
William Stillman,
Budapest

GERMAN EMPIRE
Hutmacher, Berlin
Couchman, Hamburg

GREAT BRITAIN
C.H. Burlingame,
London

RUSSIAN EMPIRE
Tracy, Odessa
Tracy, Riga (for 1900)

SWITZERLAND
Hoffman, Zürich

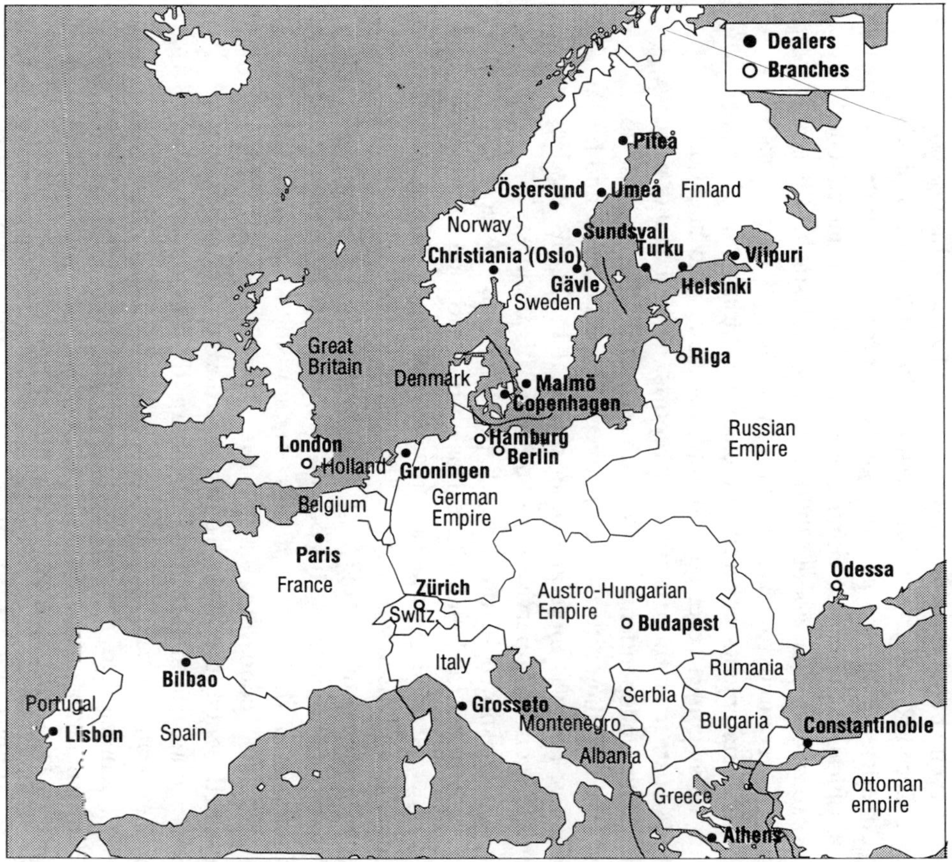

Source: Machine Records. Mss 3x, box 26.

The achievements of Couchman were also reflected in his position in the McCormick Company's organization. In 1902, McCormick's had in America 65 salaried general agents reporting to the central office at Chicago. Under the general agents worked about 12 000 local or subagents and about 2000 travelling salesmen. General agents were required to make reports every month to the main office. These statements covered, for example, hirings and firings, numbers of machines received and sold, accounts paid and notes collected. Apart from these mandatory reports, general agents were almost totally independent of Company control. There was no method of controlling the relations between a general agent and his subordinates; nor was there any other means than the dismissal of the general agent to supervise the conduct of his business.[31]

As has been shown in previous chapters, foreign business had been directly under the home office. This custom had long traditions. As a former general agent, Couchman could not accept this treatment. "I cannot see why I should handle this office in any other different manner than I would a general agency..."[32] Consequently, Couchman began to increase his powers step by step. He conducted the negotiation of the transfer of Lankester's business and made the final decisions concerning this. His standing was further strengthened when the home office transferred all transactions between agents, shipments from Hamburg to agents, and other similar European matters, entirely under his authority.[33] By these actions Couchman had reached a position similar to that of the American general agents and considerably raised the status both of the European business, and of course also of himself.

In the beginning of 1901, the McCormick brothers began to plan a new change in the Company organization and were looking for candidates to be placed in charge of the new departments. The superintendent of the reaper works was allocated an assistant, and collection, twine, purchasing, transportation, accounting, advertising and order and shipping departments were separated, each under its own manager. The legal, patent and experimental departments were merged, and most important for the foreign trade, this was finally separated as a distinct unit. From now on, the domestic department took care of the United States and Canada; the foreign department was put under W.C. Mundt, who had under his charge Mexico,

---

31 *The International Harvester Co.* 1913, 327-328.

32 *Shapiro* 1958, 4. It has not been possible to find the letter which Shapiro refers to.

33 *McCormick Co. to Couchman* 7.24.1900. Mss 1a, box 118.

South America, Cuba, Africa, Australia and New Zealand; but Couchman had overall responsibility for the business in Europe. Mundt was only his local representative at home office, and Couchman was not responsible to him, their positions being of a co-ordinate nature.[34]

Figure 5. Organization of the McCormick Harvesting Machine Company, 1901.

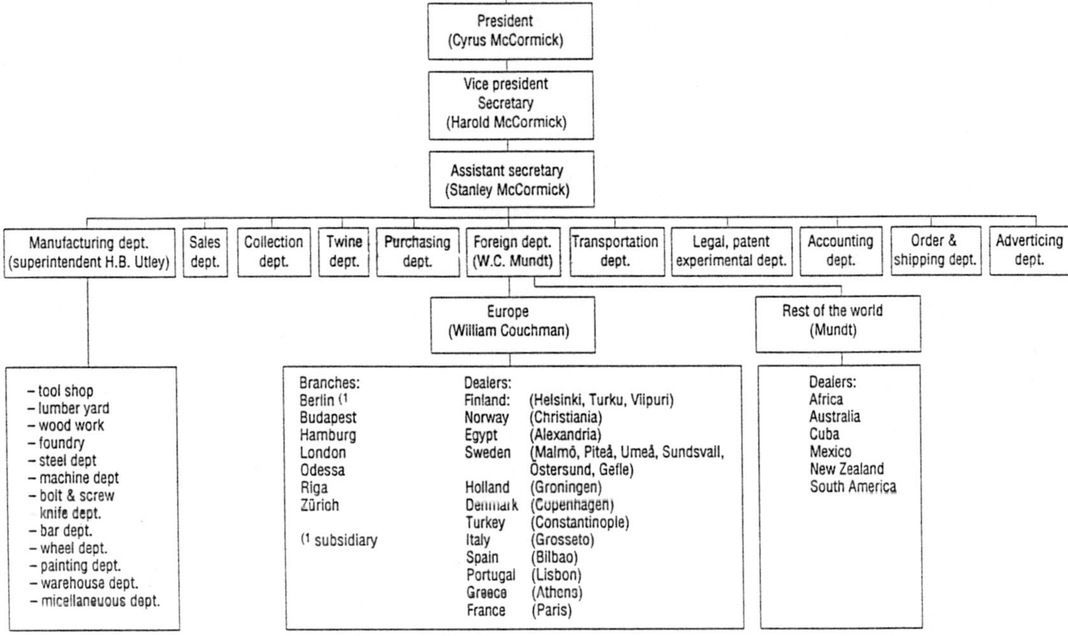

At the turn of the century, the McCormick Company's activities reached all parts of the agricultural world. It had received new inquiries for agencies from South Africa and even from Japan and India.[35] Although the two latter countries were mere curiosities, that was not the case with Australia and New Zealand.

---

34 *Cyrus Jr. to Harold McCormick* 1.19.1901. Mss 8c, box 11; *The International Harvester Co. 1913*, 330-335. This information comes from the statements of Stanley McCormick, Mr. Bentley and G.W. Perkins 6.27.1902.

35 *Mundt to Couchman* 2.27.1900. Mss 1x, LPCB 466; *Mundt to the Kan Sain Trading Co.* 4.25.1900. Mss 1x, LPCB 466 and to *F.W. Horne* 5.2.1900. Mss 1x, LPCB 466. Especially Horne from Yokohama had anticipated a large sale for agricultural machines in Japan; *McCormick Co. to Abdool Tayab* in India 8.17.1900. Mss 1a, box 118.
Charles B. Harris from the U.S. Consulate in Nagasaki let McCormick know that it was not worth while looking for trade in Japan. *Harris to McCormick Co.* 7.27.1901. Mss 1a, box 120.

South America, Cuba, Africa, Australia and New Zealand; but Couchman had overall responsibility for the business in Europe. Mundt was only his local representative at home office, and Couchman was not responsible to him, their positions being of a co-ordinate nature.[34]

Figure 5. Organization of the McCormick Harvesting Machine Company, 1901.

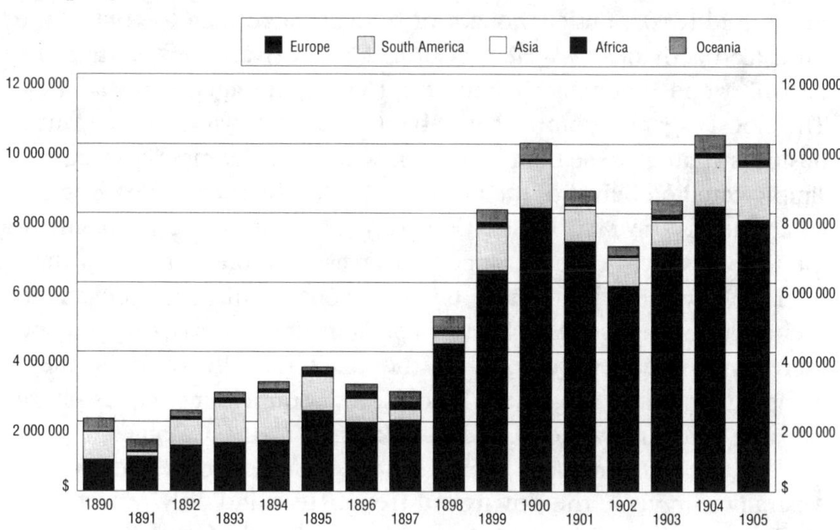

Source: Organization; McCormick Reaper Works, "Staff". Mss 2c, box 29; Officials of the McCormick Harvesting Machine Company. Mss Special Reports File, box 14; President's annual report, June 1901. Mss M/I, box 18; The International Harvester Co. 1913, 334-335.

At the turn of the century, the McCormick Company's activities reached all parts of the agricultural world. It had received new inquiries for agencies from South Africa and even from Japan and India.[35] Although the two latter countries were mere curiosities, that was not the case with Australia and New Zealand.

---

34 *Cyrus Jr. to Harold McCormick* 1.19.1901. Mss 8c, box 11; *The International Harvester Co. 1913*, 330-335. This information comes from the statements of Stanley McCormick, Mr. Bentley and G.W. Perkins 6.27.1902.

35 *Mundt to Couchman* 2.27.1900. Mss 1x, LPCB 466; *Mundt to the Kan Sain Trading Co.* 4.25.1900. Mss 1x, LPCB 466 and to *F.W. Horne* 5.2.1900. Mss 1x, LPCB 466. Especially Horne from Yokohama had anticipated a large sale for agricultural machines in Japan; *McCormick Co. to Abdool Tayab* in India 8.17.1900. Mss 1a, box 118.
Charles B. Harris from the U.S. Consulate in Nagasaki let McCormick know that it was not worth while looking for trade in Japan. *Harris to McCormick Co.* 7.27.1901. Mss 1a, box 120.

In 1892, when the McCormick Company's salaried representative Edward Ackerman left Oceania, McCormick's harvesters were the leading machines in the Colonies. Nevertheless, competitors had intensified their operations and Massey-Harris especially had invested in its marketing efforts, trying to oust its competitors from the Colonies with legal actions by preventing the sale of any open-end binder but their own. This action forced at least the McCormick and the Osborne companies to a joint effort to oppose it.[36]

When the reorganization of the European trade was in full swing in the mid-1890s, Butler thought of sending Ackerman to control the Australian activities as well. McCormick's old rivals were getting the best of it and time was ripening for a change of agents in that area. The local agents complained about the McCormick Company's business strategy to sell only for cash, whereas its competitors' agents simply acted on behalf of the makers.[37] Under the pressure of growing competition, the McCormick Company faced the same problems in the Colonies as in Europe. Since there was no representative of their own to oversee the trade on the spot, difficulties began to accumulate. Agents were short of goods in the best sales period and there were frequent defects in the machines. In spite of these complaints, Butler regarded Europe and South America as more valuable, and finally did not send Ackerman to the Colonies.[38]

In 1900 McCormick's finally sent a salaried representative to Australia to reverse the downward trend. He could only report how McCormick's machines had dropped in a decade from the lead to fourth place in the trade. In Australia the harvester companies were for the first time successful in reaching an agreement in 1902 over sales areas and prices.[39]

---

36 *Butler to Ackerman* 7.18.1892. Mss 1x, LPCB 461; *D.M. Osborne & Co. McLean Bros. & Rigg* 2.3.1896. Mss 2x, box 282; *McLean Bros. & Rigg to McCormick Co.* 10.27.1897. Mss 2x, box 298.

37 *Butler to Ackerman* 9.14.1897. Mss 1x, LPCB 463; *McLean Bros. & Rigg to McCormick Co.* 1.22.1897. Mss 2x, box 298. Strongest competitors were Massey-Harris, Wood and Hornsby.

38 *McLean Bros. & Rigg Limited to McCormick Co.* 12.17.1897. Mss 2x, box 298; *Butler to Ackerman* 3.12.1897. Mss 1x, LPCB 463.

39 W.P. Postin's promising start in the Colonies collapsed for personal reasons, and he had to return home. He was succeeded by Fred Hewetson, who had earlier visited the area. William Mundt could only regret that the Company had not sent earlier its own man to look after its interests. In 1900 Massey-Harris sent to the Colonies 13 000 tons and Deering from 1500 to 2000 tons of goods, while McCormick's shipments amounted only to 1450 tons. Mundt also asked Postin to investigate if there would be room for opening a sideline in drills, plows and other machines, if the Company decided to open its own branches in Oceania. *Mundt to Postin* 5.14.1900. Mss 1a, box 118 and 12.14.1900. Mss 1x, LPCB 467. *Mundt to Hewetson* 4.8.1901. Mss 1x, LPCB 467; *Schonberger* 1964, 170-171.

The first McCormick machines had been sent to South Africa in 1882, but since then no special efforts had been made to expand its business there. Smuts & Koch at Malmesbury remained their reliable agents from the very beginning, and not until 1894, when Edward Ackerman visited these colonies, did he appoint three more agents. After the first orders they did not renew their orders; in 1898 a more prominent dealer replaced them and McCormick's could express its satisfaction with the situation.[40] As can be seen in Table 10, trade in South Africa did not rise to considerable amounts during the span of this study.

Neither did the trade meet the expectations of the McCormick Company in South America. Agar, Cross & Co. of Glasgow, Scotland, were McCormick's main agents for Argentina. They had made use of the experts and travelers that the manufacturer had sent to help their business, but the expected large sales did not materialize, and Agar, Cross & Co. even declined to take new machines for 1898. Irritated, E.K. Butler sent them a biting letter in which he threatened to replace them: "...as the agency for our line of manufacture does not go begging in any country, we can afford to be more than fair with you, and tell you now in advance that you had better be making your plans to get something that suits you better."[41]

Argentina was the most important selling area in South America, Uruguay coming next, and McCormick's could not afford to neglect these countries. Ackerman was once again sent to oversee the business there. In spite of Butler's warnings, Agar, Cross & Co. retained its agency, and increasing demand for headers in South America forced the McCormick Company to add this to the Company's line of machines.[42] In 1899 Ackerman was able to report having sold all the machines,[43] but as can be seen in Table 10,

---

40 *McCormick Co. to Smuts & Koch* 8.5.1882. Mss 1x, LPCB 456; *Mundt to Postin* 12.7.1899. Mss 1x, LPCB 466; *Mundt to Cyrus Jr.* 9.4.1900. Mss 1a, box 118.

41 *Butler to Agar, Cross & Co.* 5.11.1897. Mss 1x, LPCB 463; *Butler to Ackerman* 6.28.1897. Mss 1x, LPCB 463. Edward Ackerman had visited Argentina, and Agar, Cross & Co. had let him make an order for 300 binders and for a couple of hundred mowers. Butler's anger was further aroused when he learned that Agar, Cross & Co. was buying twine from outside sources and was selling repair parts which were made in England after its own patterns. *Butler to Ackerman* 3.12.1897. Mss 1x, LPCB 463.

42 The problems were settled when Agar & Co. explained the situation in Argentina. They were unable to sell machines at $176, since Massey-Harris, Buckeye, Wood and Osborne sold at $140. Besides, the locus plague had harassed farmers for so many seasons that they were unwilling to invest on machines. *Agar, Cross & Co. to McCormick Co.* 4.16.1898. Mss 2x, box 290; *Butler to Ackerman* 10.22.1897. Mss 1x, LPCB 463; *Mundt to Ackerman* 8.2.1899. Mss 1x, LPCB 465; *Mundt to Cyrus Jr.* 9.4.1900. Mss 1a, box 118. Ackerman was told to keep a close watch on the machines and report on their working.

43 *Cyrus to Ackerman* 2.3.1899. Mss 1x, LPCB 464.

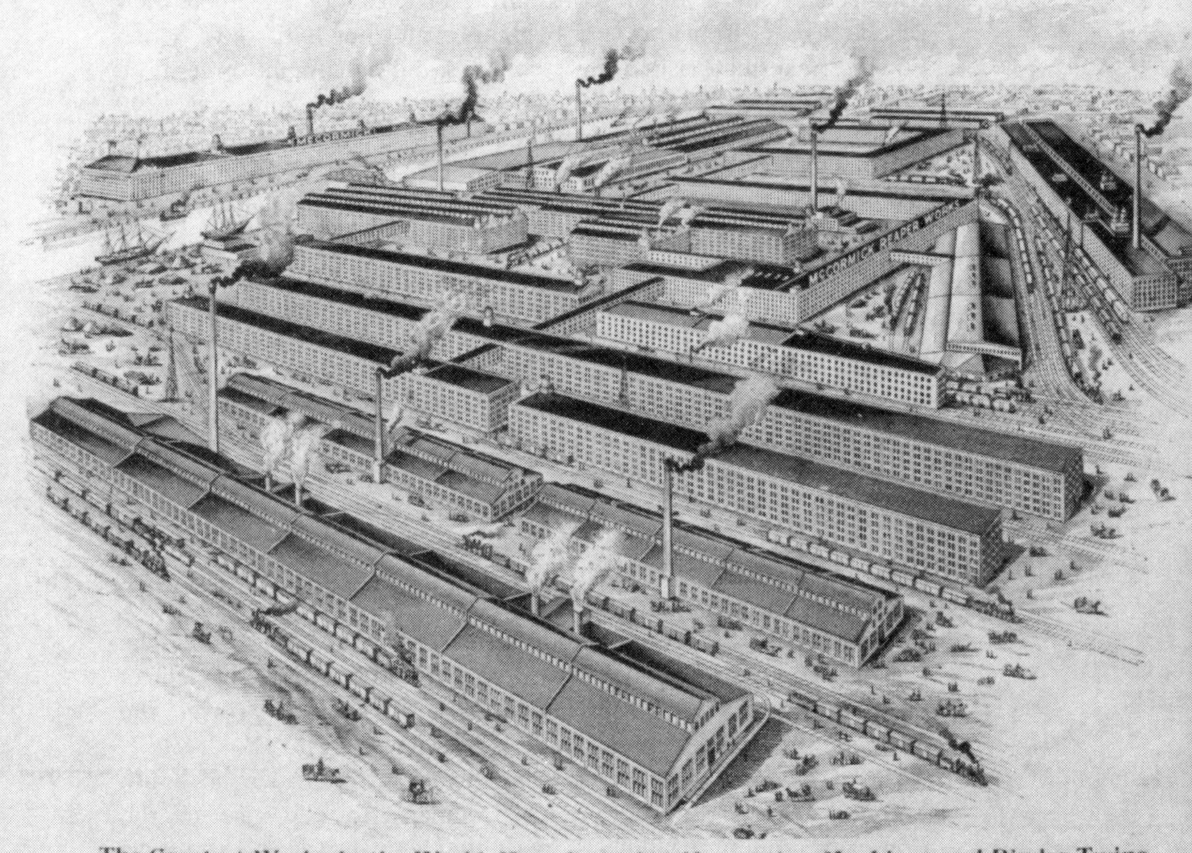

The Greatest Works in the World Manufacturing Harvesting Machines and Binder Twine.

business with South America remained at a minimal level compared with McCormick's European trade.

9.1.2. Expansion has its limits

The recession that hit the United States in 1893 turned to an upswing in 1897. Prices of farm products began to rise again and prosperity returned to rural areas. The editors of 'Farm Machinery' could rejoice that "at last the footing has been knocked from under the chronic calamity howler... the evidence of a complete return of prosperity is full, corrobative, unambiguous of tone, and unmistakable as to

By 1900 the McCormick Company was once again a family-owned and managed company which had, nevertheless, hired an army of middle managers. Together with the Deering Harvester Company it fought for the supremacy of the harvester markets. (McCormick Collection. State Historical Society of Wisconsin).

conclusion".[44]

Impacts of the boom were soon also felt in the harvester industry. From Table 11 can be seen that McCormick's sales reached an all-time peak in 1897 and continued their growth, with net profits growing 31 % from the previous year. At the same time the two market-leaders had entered into negotiations over the purchase of the Deering Company by its arch-rival. Negotiations broke off after a couple of months because of the difficulties to obtain funding for the merger. There are, however, good reasons to point also to improving sales as an explanation for the failure. The economic boom reopened attractive prospects, and perhaps the McCormicks considered it possible to crush their competitor with sales.

The McCormick Company's financial standing remained good throughout the 1890s. Its net profits sank from 1893 to 1894 by about $500 000, but the following year it could inform its shareholders of a dividend of profits of $2 500 000. As Table 11 shows, its investments increased during 1897, the first year after the recession. Simultaneously its borrowings also rose from a very minimal $150 000 in 1895 to $1 550 000 in 1896, peaking at $3 400 000 in 1899, which of course strained its economic standing. On the other hand, the Company's sales began to increase in 1897, and Butler could report an unprecedented demand for mowers.[45]

All these facts tell of increasing activity, and in fact just at this time the McCormick Company began the modernization and enlargement of its production plant. Parts of the old factory were torn down to make way for more efficient additions.[46] In 1895 a five-story paint and packing building was erected, a year later the new knife shop, in 1898 a new warehouse was erected and the older one was extended. In 1897 the old grey iron foundry was pulled down and replaced by a new one with a capacity of 20 000 tons per year. In 1899 Cyrus McCormick Jr. instituted a new program of plant expansion which involved the building of a new twine mill. It was finished in 1900 and thereafter the McCormick Company could guarantee prompt delivery of twine. Besides, a new foundry with a capacity of 30 000 tons a year was completed in 1900. Technological improvements were combined with the spirit of

---

44 *Hughes* 1987, 280-282; *Shannon* 1945, 292-295; *Farm Machinery* 4.13.1897 in *Schonberger* 1964, 127.

45 *Butler to Ackerman* 6.28.1897. Mss 1x, LPCB 463. According to Butler, McCormick's had produced 77 000 mowers, 57 500 binders and 6000 reapers; An undated statement showing the key figures of the McCormick Co. in 1895-1899. Mss 3b, box 16.

46 *Butler to Lankester* 7.20.1897. Mss 1x, LPCB 463.

getting out as large an output as possible. "The predominating motive of the entire organization... was the attainment of as great a volume of completed machines as was humanly possible as this was seen to be a great factor in the financial predominance of the McCormick interests."[47]

From the above, it becomes obvious that the McCormick Company's aim was, by increasing its output and on the other hand by cutting production costs, to raise its profits and thereby turn the situation to its benefit. Table 11 shows an uninterrupted growth of net sales, although in 1899 profits began to decline. The exceptionally good seasons of 1897 and 1898 encouraged harvester companies to increase their output to their utmost levels.[48] The modernization of the McCormick's factory was carried out at an opportune time. Nevertheless, more orders for machines came in than the Company could supply. In the midst of the worst rush, there broke out the Spanish-American War, which scared Nettie McCormick. She demanded in vain that Cyrus Jr. cut production, and then requested General Manager Butler to stop it and store all the machines if not sold for cash. For the first time Cyrus Jr. heavily attacked his mother, who had to back off. The factory was thereafter run at full speed to the end of the season. The President of the Company could claim in his report to the stockholders that unprecedented demand had exhausted the capacity of the plant.[49]

---

47 *Johnson* 1927, 28-43. Mss Special Reports File, box 12. Machinery in the whole factory was driven by a 1200 horsepower engine. Transmission of power to various sections and parts of the large complex caused many problems, since all the departments were linked with the main transmission. It was almost impossible to work overtime except in some departments when others were inactive. Also, trouble at one point of the chain meant a complete interruption of power for every point beyond. Later, the power problem was partially solved when additional power machines were introduced in various parts of the factory. Gas lights were replaced by electric lighting in 1893 and telephones were installed in the departments over the two next years. *R.W. Drake to R.O. Johnson* 10.1.1931. Mss Special Reports File, box 13; *A.A. Halverson to Johnson* 9.26.1930. Mss Special Reports File, box 13.
In addition to the new buildings and power machines the McCormick factory was equipped with the newest technology. The screw top jig significantly increased the accuracy of work. Efficiency and saving in production costs were achieved for example by a 4-spindle drill which permitted one operator to take care of several tools. Numerous other improvements followed, but perhaps most important for the success of the McCormick Company was the high precision in its products reached by intensive jigging. *M.B. Edgerton to R.O. Johnson* 10.10.1931. Mss Special Reports File, box 13. Edgerton, who was an old employee from McCormick's engineering department, states that although the firearms industry had used jigging of parts for decades, the birth of jigging in the implement industry should be located at the McCormick works; *Edgerton to Johnson* 10.13.1931. Mss Special Reports File, box 13.

48 *The Farm Implement News* 11.10.1898.

49 *President's Annual Report for the Stockholders* 7.13.1899. Mss 3b, box 21; *Schonberger* 1964, 130-132.

McCormick's competitors had also enjoyed record sales and consequently began to enlarge their factories too. In the fall of 1898 the Deering Company had six buildings under construction. The Plano Company was enlarging their old plant and had two new buildings under construction. Everybody expected the next season to be even better than the one before. In spite of some warning remarks, harvester companies expanded their volume. McCormick's raised the number on its payroll from 2471 in 1898 to 3702 in 1899.[50] Bad weather conditions during the next winter and spring spoiled all the expectations. Demand for the harvesting machines died out because of the collapse of the wheat crop. As a consequence, the McCormick Company ended the season of 1899 with over 42 000 machines in hand, though there had been a shortage of mowers.[51]

Indications of better harvests in 1900 were still promising in April. McCormick's anticipated a rush of orders, and problems in finding storage for the 50 car loads of machines that it was building daily. But instead, the prospects weakened hand in hand with deteriorating weather conditions. Finally Cyrus Jr. decided to apply the brakes on June 1st, and also began to reduce the workforce. In spite of the enlargements, the McCormick Company was soon short of storage room. It sent to its general agents a hurried demand for early orders and shipments. "Storage capacity in Chicago being limited, the importance of making shipments *early* in the season is apparent, and we think fully understood by all".[52] Deering had encountered similar experiences, and was already looking with apprehension at the increase in the cost of material for the next year. Similar ideas were circulating at McCormick's too. Material costs had increased five dollars on a binder and about one dollar and a half on a mower over the cost of 1899. Cyrus Jr. considered reducing prices, but decided to keep them as they were, since "a reduction of $5.00 would not materially increase our trade, as our agents have been trained so many years in the belief that whatever prices we fixed our competitors would go just so much lower". This comment opens

---

50 *The Farm Implement News* 8.25., 9.14. and 10.13.1899; *Schonberger* 1964, 133; *President's Annual Report to Stockholders* 7.13.1899. Mss M/I, box 18. This report has partly survived. Parts of it can be found also in Mss 3b, box 21.

51 Untitled estimate of machines manufactured, sold and carried over 11.17.1899. Mss 3b, box 21; *President's Annual Report to the Stockholders* 7.13.1899. Mss 3b, box 21.

52 *Mundt to Couchman* 4.25.1900. Mss 1a, box 118; *Cyrus Jr. to Harold McCormick* 6.4.1900. Mss 3b, box 24; *McCormick Co. to general agents* 11.14.1900. Mss M/I, box 7.

a totally new insight on the business ideology of the McCormick Company.[53]

Since McCormick's had carried over 13 920 machines from 1898, it had 42 619 unsold machines at the end of 1899. In 1900 the Company had to carry over 35 200 machines. This forced McCormick's to erect a new warehouse for 400 carloads of goods and to began planning a similar warehouse for 1901. Other manufacturers were in not much better shape. Deering constructed a five-story brick warehouse for its surplus machines and the Milwaukee Company followed somewhat later.[54] The overall assessment of the state of the harvester industry in 1901 was hopeless. Aultman, Miller & Co. was practically insolvent, the Minneapolis Company was in the hands of creditors, the Plano Company had paid no dividends for years, the Milwaukee Company was seeking a purchaser and Warder, Bushnell & Glessner faced steadily decreasing profits.[55]

Although the standing of the harvester companies had deteriorated, at least the McCormick and Deering companies nominally managed the situation well. McCormick's total profits increased continuously from 1900 to 1902 and Deering experienced a similar development except in 1901. The figures in Table 21 differ somewhat from the information in Table 11, which shows a slight decline from 1898. Gaps in the source material do not, however, allow a closer comparison and study on their reliability.

---

53 *Cyrus Jr. to Harold McCormick* 10.12.1900. Mss 1a, box 117; Unsigned letter from the home office *to Cyrus Jr.* 9.7.1900. Mss 1a, box 118; *Cyrus Jr. to Harold McCormick* 1.25.1901. Mss 2c, box 29.

54 Untitled estimate of machines manufactured, sold and carried over 11.17.1899. Mss 3b, box 21; *Schonberger* 1964, 136.

55 *The International Harvester Case.* A compilation of public statements and court documents in connection with the government's investigation of the International Harvester Company and its anti-trust suit against that company, covering the period from May 20, 1907, to May 28,1920. p.18

Table 21. Total profits, cost of product, average prices and number of machines manufactured in the McCormick Harvesting Machine Company and the Deering Harvester Company ($), 1900-1903.

	Total profits[1]		Cost of product		Average price		Number manufactured	
	McC	De	McC	De	McC	De	McC	De
1900	4 893 988	3 987 616					371 312	281 574
Harvester			47.2	53.2	104.3	104.5		
Mower			16.8	18.7	36.5	35.5		
Reaper			25.7	32.4	56.7	53.6		
1901	5 270 559	3 087 150					399 735	263 767
Harvester			46.5	50.2	99.1	98.9		
Mower			16.8	18.4	36.8	34.7		
Reaper			27.2	32.1	59.4	52.8		
1902	5 464 922	4 455 010					503 517	370 107
Harvester			49.6	48.1	99.1	99.1		
Mower			16.6	17.9	36.1	34.7		
Reaper			26.6	31.4	54.4	52.9		

[1] The statement showing total profits is from a different archival entity than the one showing net profits in Table 11. Total profits also include deprecation, sundry adjustments, and construction and selling expenses deducted.

Source: McCormick Harvester Machine Company and Deering Harvester Company. Comparative statement of net sales for seasons 1900, 1901 and 1902. Mss 2c, box 31; Comparative statement of cost of product manufactured and sold for seasons 1900, 1901 and 1902. Mss 2c, box 31; Deering Harvester Company. Comparison of profits for the three seasons 1900, 1901 and 1902 showing variations in selling prices, selling expenses and cost of production on the basis of the figures of McCormick Harvesting Machine Company. Mss 2c, box 31.

Notwithstanding the fact that the figures in Table 21 are only indicative, they come from the same entity and are in that sense usable. McCormick 's output was significantly larger than Deering's; its efficiency was higher, since it was able to keep a higher outsale price than Deering's.

This does not, nevertheless, tell the whole truth. In Table 10 can be seen how the McCormick Co. extended its investments by over $5 million from 1896 to 1898, reaching almost $22 million. This can be explained by the erection of the new buildings and renewal of its machinery. To make the investments possible, the McCormick Company had to borrow increasing amounts from outside sources. In 1900 it had to use $7 742 705 borrowed money; the next year $8 192 705; and a year later $10 592 705. These funds were for the most part borrowed from local sources in Chicago but also from banking houses in New York and Boston.[56] It is obvious that investments increased the state of indebtedness of the McCormick Company at the turn of the century.

What then was the role and significance of the foreign trade in the light of the information presented above? Mira WILKINS explains how many American companies failed during the depression, but those who survived grew mighty and expanded into giant corporations. On the other hand companies saw the foreign markets as a way to get rid of their surplus goods.[57]

Developments in the harvester industry confirm the first argument. At the end of the 1890s there remained only two large firms in the harvesting machine industry, and some smaller ones, which could not compete with their superiors. Wilkins' latter observation may have been true in the industries where identical products could be produced and sold both in America and in Europe; this was the case, for example, for the Singer Sewing Machine Company, which sold uniform goods throughout the world.[58] Products in the harvester trade were diversified according to the markets where they were sold. The standard American machines did not suit the foreign conditions, which varied from country to country. Goods had to be adjusted to local conditions, which of course reduced the possibilities of selling American-style surplus machines. Machines for the foreign trade needed costly jigs and patterns of their own. This fact alone reduced the validity of this statement.

As the American harvester trade was concentrated into two categories, competition between the remaining companies intensified but there were also signs of efforts to find a common base.[59] As has been earlier shown, there was competition; but was it so acute that it forced some of the companies to find new markets abroad?

---

56 *Comparative statement of money borrowed and sources from which borrowed.* 4.7.1902. Mss 2c, box 30.

57 *Wilkins* 1976, 71-72.

58 *Carstensen* 1984, 23-26.

59 All the main harvesting machine companies at the turn of the century held meetings to discuss, for example, the fixing of prices, production quotas, presentation in fairs, form of contracts or canvassing. *Mayer and Daniels to Harold McCormick* 8.18.1900. Mss 3b, box 24; *Daniels and Perkins to (Harold ?) McCormick* 12.22.1900. Mss M/I, box 6.

As can be seen in Table 16, McCormick's foreign business expanded from about five thousand machines in 1894 to at least about 34 000 five years later. There is a clear jump in sales after 1897. From 1897 to 1898 foreign sales grew by 71 percent, while the domestic trade showed a growth of 31 percent. On the other hand, the volume of the foreign business was so modest that even a small addition in volume produces a high percentage growth. Foreign trade increased every year from 1894 to 1902, but its share of the total trade remained all the time less than 18 percent. All the above tells only about steady growth and about the expanding interest in foreign business, but fails to give reasons for it.

Alfred D. Chandler has explained this evolution on the same grounds as in the domestic market. Big enterprises had to expand abroad to maintain their cost advantages of throughput. Loss of share to a competitor not only increased one's production costs but also decreased those of one's competitor. Besides, in the 1890s the volume of trade had grown to such an extent that small agents were no longer able to manage. Furthermore, products had become too complicated and specific for an independent agent to put up and service them properly.[60]

The Supreme Court of the United States, on the other hand, regarded the saturation of domestic markets as a central reason for foreign expansion. It stated that there had been no growth in the number of binders, reapers and mowers sold annually in the United States during the five years preceding 1902. Besides, there was not even any further growth to be expected.[61]

The McCormick Company had, nevertheless, developed its foreign business since the time Butler got it under his thumb. Sales increased steadily year after year at the same rate as the Company was ready to invest on it; and when it then really decided to expand also on that front, the decision was put into effect with the same determination as on the home field. The McCormick Company anticipated profits in the expanding foreign business, which was not as developed and as competitive as the domestic trade. If we can trust the sample year of 1899, the profits were very lucrative indeed, as Table 18 shows.

Keeping this in mind, it should actually be asked why McCormick's did not invest earlier and more intensively on foreign sales. Certainly

---

60 *Chandler* 1988, 41-42.

61 *Supreme Court of the United States. Brief for Appellants.* No. 757, 1914.

there was will and vigor for it.[62] The American harvester manufacturers fought in the Paris World Fair in 1900 as if they were for
the first time in Europe. In a way that is true. Only a couple of them
had established their own branch offices, while the majority still
worked through jobbing houses. The situation closely reflects the
attitudes toward foreign countries among the companies. This
appears, for example, from William Couchman's statement: he
considered Europe "practically virgin soil for the increase of
business". Trade was well developed only in England and partially
so in France and in Germany. Charles H. Haney, inspector of the
foreign sales of the Deering Company, had similar opinions.
European jobbers wanted to make quick returns and after the sale
of a machine their interest ceased.[63]

The American consuls reported on the situation in Europe in
largely the same light, although there were also differences. In
Austria-Hungary modern agricultural implements had been widely
introduced and their number was steadily expanding. The harvesting

---

62 The McCormick Company prepared carefully and in good time their display for
the Paris World Fair in 1900: they needed to secure at least the same amount of
awards as the Deerings. McCormick's application for a space in the agricultural
pavilion was not granted, however, whereas the Deerings obtained an outstanding
location and were commissioned to furnish the official historical agricultural exhibit.
As counterattack, the McCormicks showed an exhibit of 26 models, and in addition
they erected an additional building to house thirteen full-sized machines. Both
companies tried to influence the decisions of the jurors, brought over an
experimental auto-mower and an army of experts to assist in the field trials. In spite
of their efforts to defeat each other they were favored with the same number of
awards. *R. Wallut to Stanley McCormick* 6.23.1900 and to *Cyrus Jr.* 6.25.1900. Mss
1a, box 120; *Cyrus Jr. to Harold McCormick* 6.28. and 6.30.1900. Mss 1a, box 117;
Stanley McCormick to McCormick Co. 7.24. and 9.3.1900. Mss 1a, box 119; *Stanley
McCormick to Cyrus Jr.* 7.28.1900. Mss 1a, box 119; *Stanley McCormick to
Couchman* 11.23.1900. Mss 1a, box 118; *Benson* 1936, 15-16.
Cyrus McCormick was made a member of the Legion of Honor with the rank of
officer, as was William Deering. *McCormick Co. to Couchman* 1.24.1901. Mss 1x,
LPCB 467. The McCormick Company began an intensive campaign against the
Deerings when James Deering was appointed Chevalier de la Legion d'Honneur while
Stanley McCormick was excluded from the list. *Cyrus Jr. to Wallut* 4.22.1901. Mss
1a, box 121.
Witnesses in the International Harvester case underlined in their testimonies the
importance of the Paris World Fair in awakening the interest of the McCormick
brothers in foreign business. This is somewhat strange, however, since at least Cyrus
Jr. had made frequent visits to Europe ever since his first trip to the Derby show in
1878. He must have had a good understanding of the foreign business, for he was
in direct contact with the Company's agents throughout his career. In the District
Court of the United States for the District of Minnesota. The United States of America,
Petitioner, vs. International Harvester Company and others, Defendants. *Appendix to
Defendants' Brief. Evidence as to certain points abstracted and topically arranged.*
Vol. II. Testimony of H.F. Perkins, 31.

63 In the District court of the United States for the District of Minnesota. The United
States of America, petitioner. IH et. al., Defendants. Volume XIII. Testimony of
Witnesses for the Defendants. *Testimony of Charles H. Haney.* p. 134-136 and
testimony of William V. Couchman. p.190.

machine trade was almost entirely in the hands of the American companies. A similar situation was found in Belgium, too, where English and German makers, however, also had some share of the imports. Other agricultural machines and implements were mostly of English or German manufacture. France was considered by the consuls to be an important and well developed agricultural country in terms of machinery. As a whole the American makers dominated the French agricultural machinery trade; in fact the strongest competitor for an American manufacturer was another American maker.[64]

At the turn of the century Germany was a rapidly growing industrial power, whose strength was felt in the agricultural machine trade also. The McCormick Company had noticed this danger, and founded its first subsidiary in Berlin; but they had to react quickly before it was too late.[65] Germany was known for the quality of its products. The plows, harrows, cultivators, and drilling, threshing and potato-digging machinery in use were of domestic origin. German firms like Heinrich Lanz or H.F. Eckert were concerns with large resources which were able to turn out excellent goods, even if modeled after American patterns. According to consular reports, only the McCormick Company, because of its investments in that country, was able to compete profitably with them in the harvesting machine business.[66] On the other hand, Paul Sack of Land-maschinenfabrik Rudolf Sack made in 1893 and Heinrich Lanz as late as 1902 a trip to America to learn new production methods. According to Fritz BLAICH only the harvesting machine trade remained at the turn of the century in the hands of foreign manufacturers.[67]

Russia was understood to be a great future market for American agricultural machines. The McCormicks had noted the possibilities it offered and in Odessa they had their own office. Consequently, once again only American harvester machines could compete with German and English machines. Many other American goods like

---

64 *Markets for Agricultural Implements and Vehicles in Foreign Countries.* 1903, 1-4, 8, 10-11, 18-20.

65 In the District Court of the United States for the District of Minnesota. The United States of America, Petitioner, vs. International Harvester Company and others, Defendants. Appendix to Defendants' Brief. *Evidence as to certain points abstracted and topically arranged.* Vol. II. p. 51.

66 *Markets for Agricultural Implements and Vehicles in Foreign Countries.* 1903, 23-24.

67 *Blaich* 1984, 70.

plows were superior in quality, but due to inefficient marketing strategy had lost markets to their European rivals.[68]

Superiority of the American harvesting machines became visible also in the machine test held by the Finnish agricultural societies. In 1897 all the tested machines were American and in later tests American machines were regarded to be better suited to the Finnish conditions than the Swedish ones.[69] In this light the quality of the American machines gave them a great advantage over their European competitors.

From the above it can be deduced that in the agricultural implement trade, only the harvesting machine companies had organized their European trade in a proper way. They had sent their travelers, experts and agents to Europe; they had appointed reliable local dealers to take care of the business, which they kept under constant control. On the other hand, the European trade and agriculture were perhaps not quite as undeveloped as Couchman and Haney had found. In the 1890s, most of the machines were still sold to large estate holders; when the smaller farmers began in increasing numbers to buy machines, manufacturers were forced to extend credit to them and educate them in the use of complicated new machinery.[70]

The European experience and research partly confirm the statements of the harvester men. In Sweden the estate owners were the first to adopt new harvesting machines; interest began to arise among smallholders during the 1890s.[71] Finland experienced a similar development, but a decade later than in Sweden. In Finland mowers were the most popular agricultural machines: in 1900 there were about 20 000 of them and ten years later 57 000.[72] In Germany the interest of farmers in the harvesting machines began to grow only in the middle of the 1890s.[73]

Information on the distribution of agricultural machines has been difficult to obtain, but even the fragmentary material shows how

---

68 Ibid. 39-41, 46-47.

69 *Grotenfelt* 1911, 102-103.

70 In the District court of the United States for the District of Minnesota. The United States of America, petitioner. IH et. al., Defendants. Volume XIII. Testimony of Witnesses for the Defendants. *Testimony of Charles H. Haney.* p. 137-138 and *testimony of William V. Couchman.* p.197.

71 *J. Kuuse* 1970, 47-54.

72 Maataloustiedustelu Suomessa vuonna 1910. Edellinen osa: Maanviljelys. SVT: III: 9. Maatalous. Helsinki 1916. 89-94, *Viita* 1964, 195.

73 *Blaich* 1984, 71.

reaping and mowing machines spread widely first in England, where by the 1870s they were no loner rarities. In other parts of Europe, harvesting machines were sold on a large scale only at the turn of the century. This means that the harvester manufacturers really had a large market waiting for them in Europe, an opportunity which they also used. The simultaneous, rapidly growing demand in Europe, combined with the highly competing and saturating markets in the U.S.A., offers the most feasible explanation for harvester companies' interest in the European markets at the turn of the century.

Table 22. Stock of harvesting machines and mowers in several West European countries, 1861-1910.

Year	England	France	Germany	Norway	Finland
1861	10 000	18 000	–	–	–
1862	–	–	–	–	–
1871	40 000	–	–	–	–
1874	80 000	–	–	–	–
1875	–	–	–	1 300	–
1882	–	35 000	20 000	–	–
1890	–	–	–	12 100	–
1892	–	62 000	–	–	–
1895	–	–	35 000	–	–
1900	–	–	–	31 500	20 000
1907	–	–	301.325	49 200	–
1910	–	–	–	–	57 000

Source: Dovring 1965, Table 58, p.644; for Norway Collins 1969, Table III, p.75; for Finland Maataloustiedustelu Suomessa 1910, 12 and Viita 1964, 195; for Germany in 1907 Blaich 1984, 72.

The McCormick Company had correctly foreseen the development on the European markets. It began to exploit the growing trade, but it is obvious that competition at home prevented it from diverting as much of its energy and capital to overseas activities as it wanted to. In 1902 the total property of the McCormick Company was estimated at \$29 461 481.[74] At the same time it had to use over ten million dollars of borrowed money, representing about a third of

---

74 *The International Harvester Co.* 1913, 95-96; *Benson* 1936, 14.

the total value of its property. This points toward the explanation
that the McCormick Company was on the brink of indebtedness; its
returns were good, but through its investments it was falling into
debt at an accelerating rate at the same time as the home markets
were getting saturated with machines. This explanation is
substantiated by Alexander Legge, General Manager of the
International Harvester Company, who comments that there was in
the McCormick Company a constant pressure for further expansion
of the business, but the McCormicks were short of the necessary
funds. "They had really passed the danger line in the amount of
borrowed money they were using". That forced the McCormick
Company to consider expanding the working capital by putting out
a bond issue on the same lines as Studebaker and J.I. Case had
done.[75]

The cost of raw materials was a factor working against the
expansion of foreign business. The managers of the McCormick
Company warned the McCormick brothers of hasty decisions before
the question of raw materials was settled. The manager of the
purchasing department had found strong fluctuation in prices, and
suggested the building of their own steel mill and blast furnaces.
Otherwise, large-scale foreign business was not based on safe
ground. The project would have required at least $7 000 000, which
turned out to be too much for the McCormick Company.[76]

The transfer from business through the jobbing houses to their
own branches in Europe increased McCormick's expenditure too.
In the jobbing trade, the jobber normally bought the machines for
cash in New York against shipping documents, or in some cases on
credit on all or part of the shipment. The responsibilities of the
manufacturer ceased at that point. After the change to its own
branches, the expenses were heavy. Thereafter, McCormick's had
to pay ocean freights, import duties, storage costs and expenses of
handling and selling. That included the warehouses, numerous
experts and travelers who were needed to assist farmers to set up
and repair the machines and train them in their use. In his testimony

---

75 In the District Court of the United States for the District of Minnesota. The United
States of America, Petitioner, vs. International Harvester Company and others,
Defendants. Appendix to Defendants' Brief. *Evidence as to certain points abstracted
and topically arranged.* Vol. II. Testimony of Alex Legge, 31-32.

76 Ibid 21-22, 27. *Testimony of Herbert F. Perkins.*
Actual prices of pig iron were:
1895 $ 12.00 Chicago per gross ton
1896   10.75   "
1897   9.80   "
1898   10.30   "
1899   15.75   "

at the International Harvester case, William Couchman suggested that once the McCormick Company had made the decision to go abroad it had no other options than to continue its investments and take advantage of the extension of that business. "If you have to stop, you simply have plowed for somebody else to reap".[77]

On the other hand, the risk McCormick took was not as big as it may look at first sight. The Company knew the benefits and limits of the jobbing houses and it was well aware that it could not expand its trade without its own direct investments. Practise showed in a few months what the firm's own able management could do. Incomes began to grow, and the Company saved the agent's commission. Productivity and throughput of the foreign machine production line was with great certainty raised in line with the rest of the business. Besides, Couchman had brought new accuracy to the orders which reduced the stock in hand.

McCormick's main competitors had encountered similar experiences. Deering had no established, regular business in Europe before 1892, when it sent Charles H. Haney abroad to inspect and develop the foreign business. Deering conducted its foreign operation in a similar manner, through jobbing houses, with all the weaknesses arising therefrom, as did the McCormicks. He also confirmed Couchman's claim about the high expenses of the branch houses, which actually prevented the Deerings from expansion in that direction. In his testimony Haney also corroborated how difficult it was to sell abroad machines designed for American conditions. A whole army of experts had to follow the machines from country to country. Besides, machines had to be fitted to different animals as a means of propulsion, varying from horses, mules, oxen, buffalo, and camels to cows.[78]

The McCormick Company had begun its foreign business early on in its history but both it and the Deering Company began to invest seriously on it only in the 1890s, when they were also competing for supremacy over the American markets. The development in Table 23 indicates that,

---

77 In the District court of the United States for the District of Minnesota. The United States of America, petitioner. IH et. al., Defendants. Volume XIII. *Testimony of Witnesses for the Defendants. Testimony of William V. Couchman.* 195-198.

78 Ibid 134-141. *Testimony of Charles C. Haney.*

Table 23. Foreign sales of the McCormick and Deering
Companies, 1898-1902.

Year	McCormick		Deering	
	$ [a]	Number of[1] machines	$	Number of machines
1898	1 864 280	–	1 414 892	–
1899	2 347 249	–	2 155 709	–
1900	3 134 351	40 677	2 594 955	31 611
1901	4 222 377	48 549	2 750 634	30 525
1902	4 336 558	51 241	3 488 250	35 628

Source: In the District court of the United States for the District of Minnesota. The United States of America, petitioner. IH et al., Defendants. Volume XIII. Testimony of Witnesses for the Defendants. Testimonies of Charles C. Haney and William V. Couchman, 179-180, 199-200.
[a] The dollar values also include sales to Canada, which are omitted from the number of machines because of the accounting practises of the companies.

[1] The figures showing machine sales of the McCormick Company in Tables 11 and 23 do not coincide. Table 11 is based on the Company's own Machine Records and therefore should be considered as reliable. The information in Table 23 was produced in a hurry by Company officials at the request of the court, and does not include sales to Canada (which also explains the difference between the figures).

although the McCormick Company was also in the lead on foreign fields, this difference was not significant until after 1900. Establishment of the branch houses after 1900 increased McCormick's efficiency compared to Deering's, which held to its old jobbing-house business.

The other harvester manufacturers followed in the footsteps of their bigger competitors. The Milwaukee Harvester Company operated only through the dealers as did the Johnston Harvester Company. In 1895, on the other hand, D.M. Osborne & Co. already had branches in Vienna, Odessa, Bremen and in Buenos Aires.[79] The Massey-Harris Company, too, sold almost exclusively through its own branch houses.[80]

---

79 *The Milwaukee Harvester Co.* Catalogs 1900, 1901. Mss 4z, box 16; *The Johnston Harvester Co.* Catalog 1896. Mss 4z, box 14; *D.M. Osborne & Co. Catalog* 1895. Mss 4z, box 18. The Milwaukee Co. had 28 agencies abroad in 1900. The Johnston Co. had its European office in Paris but this was not in the list of its branches.

80 In the District court of the United States for the District of Minnesota. The United States of America, petitioner. IH et. al., Defendants. Volume XIII. Testimony of Witnesses for the Defendants. *Testimony of Thomas Findley, vice-president and assistant general manager of the Massey-Harris Company,* 185-187.

The McCormick Harvesting Machine Company had found its limits at the turn of the century. The McCormick brothers, who took charge of the Company, pushed aside the old cautious policy of the Company and enthusiastically made large investments in their attempt to overpower the Deerings. If they had expected large returns from the Company's foreign sales, they had soon to give up such hopes. The situation of the McCormick Company deteriorated further when Deering began its investment program. The manager of the purchasing department of the McCormick Company had warned the McCormicks of the necessity to guarantee a fluent supply of raw materials and possibly even to erect a steel mill of its own. Deering had become aware of the same problem; while the McCormick Company had to abandon its plans for lack of capital, the Deering Company began to integrate vertically backwards by purchase of iron ore and timber properties, coal lands and a controling interest in a steel plant in Chicago. The Deerings aimed at total self-sufficiency in raw materials, which would reduce their material costs.[81]

9.2. The great merger

At the turn of the century harvester manufacturers found themselves in a desperate situation. In an endless series of efforts to overcome each other, they had finally reached deadlock. The McCormick Company had trimmed its marketing organization both in domestic and in foreign markets to a nicely working machine which could easily fight its competitors. The Deerings, on the other hand, had anticipated the forthcoming structural changes in the industry. Their answer to the McCormick's sales force was integration backwards by acquisition of raw material sources.

This chain of events corresponds with the findings of Naomi LAMOREAUX, who noted that manufacturers who produced large quantities of homogeneous goods tried to reduce unit costs to a minimum and at the same time protect themselves against price cuts. In times of depression they tried to expand, at the expense of weaker competitors.[82]

---

81 In the District Court of the United States for the District of Minnesota. The United States of America, Petitioner, vs. International Harvester Company and others, Defendants. Appendix to Defendants' Brief. *Evidence as to certain points abstracted and topically arranged.* Vol. II., 24; *McCormick* 1931, 112-113.

82 *Lamoreaux* 1988, 28.

In 1902, the two giants were in a situation where neither was able to destroy the other. The markets in America had for some time shown signs of saturation. In spite of that, manufacturers tried to intensify their operations by cutting production costs and prices; what they could not change was the seasonal character of agriculture. It restricted the selling period to only a few months a year before the harvest. For the rest of the year the money invested in branch houses, warehouses and in machines and repair stock lay idle. Continuous problems also arose from the yearly disruption of production. Factories closed their doors when the orders were completed. Something had to be done also in that respect.

The first attempt to merge the interests of the harvester manufacturers ended up in disaster; nor did the negotiations between the McCormick and Deering companies lead to much better results. Nevertheless, the harvester companies held discussions every now and then on the possibilities to reach agreement on machine prices and production quotas. These meetings never ended in any permanent results.[83]

Beneath the surface, however, hopes for a solution to the competition survived. Various persons offered their help to arrange a consolidation of the harvester manufacturers. It had become obvious even to outsiders that business conditions were deteriorating.[84] Over 75 000 agents were needed to clear out the yearly output, and the heavy sales expenses began to cut into manufacturers' profits. Especially the McCormick Company's sales expenses had been larger in proportion to business done than those at Deering, but in both companies the relation of sales expenses to sales had steadily risen from 1898 to 1902.[85]

The preconditions for renewed discussions for a settlement of the problems of the industry were now available. At the end of 1900,

---

83 *A.E. Mayer and H.L. Daniels to Harold McCormick* 8.18.1900. Mss 3b, box 24. This meeting was attended by the Adriance, Deering, Johnston, Milwaukee, Osborne, Plano, Warder, Wood and McCormick Companies. Three questions were considered: prices, canvassing and salaried contracts. An understanding was found on special canvassers, who were not allowed to enter the field prior to March 1st, 1901. There was discussion of setting binder prices at $95.00 cash and $100 with time, but no understanding was reached; *Mayer and Daniels to Harold McCormick* 10.12.1900. Mss M/I, box 7; *Mayer and Perkins to Harold McCormick* 12.22.1900. Mss M/I, box 6.

84 See for example *R.L. Ardrey to Cyrus Jr.* 8.5.1898. Mss 2c, box 30 and *W.F. Abbot to Cyrus Jr.* 4.3.1900. Mss 2c, box 30.

85 Deering Harvester Company. *Explanatory statement of variation in profits in seasons 1898 to 1902* (inclusive). Mss 2c, box 31; McCormick Harvesting Machine Company. *Explanatory statement of variation in profits in seasons 1900, 1901 and 1902.* Mss 2c, box 31; *International Harvester Company. Exhibit L.* Mss 2c, box 31.

the Deerings and the McCormicks hold frequent negotiations over a possible merger; the Deerings suggested an equal share for each company in the new company, which the McCormick family was not ready to accept. Cyrus Jr. also discussed the situation with his mother and uncle, who proposed taking in outside capital and absorbing the Deering Company.[86]

The mutual suspicions prevented the reaching of any agreement on a combination in discussions between the McCormicks and the Deerings. Finally, through the aid of John D. Rockefeller Jr., negotiations were transferred to New York under the auspices of the house of J.P. Morgan. Under the guidance of George Perkins an understanding of the terms of the consolidation was arrived at and the contract was signed on July 28th, 1902. The stock of the International Harvester Co. was set at $120 000 000 of which 42.6 percent went to the McCormicks, 34.4 percent to the Deerings, the Plano's share was 5.2 percent, the Champion's 3.7 percent and J.P. Morgan & Co. got 14.0 percent. Financially the new giant was agreed to be sound.[87]

The formation of the International Harvester Company had drastic impacts on the harvesting machine industry. Together the five merged companies manufactured about 90 percent of the grain binders and 80 percent of the mowers built in the United States at the time of the consolidation. Within a couple of years after its organization the International Harvester Company enlarged its activities by the purchase of D.M. Osborne & Co., Aultman, Miller & Co. and the Minnie Harvester Company. At the time of the merger in 1902, International Harvester went into negotiations with numerous other firms, like Massey-Harris, the Acme Harvester Co. and the Walter A. Wood Mowing & Reaping Machine Co., for the

---

86 *Diary of Cyrus Jr.* 11.23., 11.27. and 11.30.1900. Mss 2c, box 84.

87 The International Harvester Company merged the McCormick Harvester Machine Company, the Deering Harvester Company, the Plano Manufacturing Company, the Warder, Bushnell & Glessner (Champion) Company and later on also the Milwaukee Harvester Company. Cyrus McCormick was elected President of the corporation, and James Deering, Harold McCormick, W.H. Jones (Plano) and J.J. Glessner (Champion) Vice-Presidents. In the voting trust of the new Company were Cyrus McCormick, Charles Deering and George Perkins. Of the eighteen directors ten represented the merged harvester companies, while the rest were from the banking world. *F.H. Kennedy to Cyrus Jr.* 10.23.1901. Mss 2c, box 30; *Notes and supplemental notes on sundry transactions of McCormick Harvesting Machine Company.* Mss 2c, box 31. No date. 5-9; *Opinions of the Judges in the case of the United States v. the International Harvester Co. et. al.* 1914, 5-7; *Report on the Agricultural Implement and Machinery Industry* 1938, 6-7; *Kramer* 1964, 290-297; *Marsh* 1985, 39-42; *McCormick* 1931, 113-118.

purpose of bringing them into the combination.[88]

The International Harvester Company was not only a market leader, but due to its horizontal integration, was also the party that dominated them. Through its organization, the stockholders of the Company had sought for economies of scale and reduced sales expenses. These hopes did not materialize immediately. On the contrary, the binder and mower sales of the International Harvester Company declined during the first ten years after its formation.[89] The decline can partly be explained by the diminishing of competition and partly by saturation of the domestic markets.

From the very beginning the Company leaders tried to rationalize its operations and productions. Manufacture of the harvesting machines at the Milwaukee plant was transferred to the McCormick plant. At the Milwaukee plant the production of gasoline engines, cream separators and tractors was started. Production of the Plano Company was shifted to the Deering plant while the Plano factory was thereafter used for the manufacture of manure spreaders and wagons. The purchase of the D.M. Osborne Co. had already brought in a line of tillage tools and marked a clear change in the ideology of the Company. Through horizontal integration it had achieved all the gains available in the harvester industry at home, to increase its profits any further, it had to look for new lines. Acquisition of the Weber Wagon Co. in 1904 opened the way for the manufacture of farm wagons, which was further reinforced by the contract with Bettendorf Axle & Co. to sell the output of steel gears. In addition, the International Harvester Company purchased from the Kemp Manufacturing Co. its plant for the manufacture of manure spreaders, and made contracts with the Parlin & Orendorff Co. and with the Oliver Chilled Plow Co. for selling their plows in Canada. Furthermore, it made an arrangement with the American Seeding

---

88 *Report on the agricultural implement and machinery industry* 1938, 7. The D.M. Osborne Co. was acquired by the International Harvester Co. on 1.15.1903. for $3 200 000. It also had an important line of tillage implements and a binder twine and cordage works. The Minnie Harvester Co. was a successor of the Minneapolis Harvester Co. Its stock was purchased for $945 000 and transferred to a subsidiary company of International Harvester, the International Flax Twine Co., on 9.30.1905. Aultman, Miller & Co. became bankrupt in 1903, and was thereafter reorganized as the Aultman & Miller Buckeye Co. Its plant and property were transferred to International Harvester in 1905. The value of the company was estimated at $555,928. *The International Harvester Co.* 1913, 136-140, 143-145; In the District Court of the United States for the District of Minnesota. The United States of America, Petitioner, vs. International Harvester Company, Defendants. *Defendants' answer to petition*. p. 29-35.

89 *The International Harvester Co.* 1913, 64-65, 181-182; In the District Court of the United States for the District of Minnesota. The United States of America, Petitioner, vs. International Harvester Company, Defendants. *Defendants' answer to petition*. p. 47-49.

Machine Co. for the sale of its seeding machines.[90]

By these actions the International Harvester Company had extended its operations to totally new lines, which made it a full-line company. At the same time, the Company expanded upon backward vertical integration, by acquiring iron ore and coal mines, timberlands and railroads. New operations were organized under subsidiary companies. Of the new companies, the Wisconsin Steel Company's capacity developed beyond the needs of the International Harvester Co. In addition to the net profits gained by these new acquisitions they also assured for the Company the quality and quantity of supplies as needed. The integration forward to the marketing of the products did not dramatically change the existing situation. Like its predecessors, the International Harvester Company exercised full control over the distribution of the machines down to the dealer and in the case of credit sales even to the purchaser.[91]

Technically the International Harvester period brought with it great opportunities in the production of harvesting machines. Machines were made more uniform, and the experimental department of the Company could learn from past failures. Practically all of the binders and mowers were rebuilt by 1910. The Plano mower, which was regarded as decidedly inferior was rebuilt twice, while the Champion mower was entirely redesigned; nor did the McCormick and Deering machines escape modifications. It was estimated that the rebuilding of both machines amounted to at least $100 000 each.[92]

Development in the harvesting machine industry was, nevertheless, nothing new or exceptional. On the contrary, it can be seen as a continuation of a long trend from the 1870s. As machines became more complicated, their manufacture demanded advanced production technology and consequently also larger capital than small firms could afford. The same phenomenon was also found in other fields of the agricultural machine industry: the number of producers was reduced, the remaining big companies sought economies of scale and developed into full-line giants following the example of the International Harvester Company. Deere & Co. extended its operations from the manufacture of plows to wagons,

---

90 *The International Harvester Co.* 1913, 141-145; *Eckles* 1953, 104-106.

91 *The International Harvester Co.* 1913, 148-149, 267-268; *Eckles* 1953, 107-109.

92 In the District Court of the United States for the District of Minnesota. The United States of America, Petitioner, vs. International Harvester Company, Defendants. *Appendix to defendants' brief, evidence as to certain points abstracted and topically arranged.* p. 179-186; Vol. XIII. *Testimony of Witnesses for the Defendants.* p. 317-322.

drills, manure spreaders and even to harvesting machines. The Moline Plow Co. had a very similar history: it began as a plow maker but later on added other tillage implements, drills, wagons, etc. to its lines. The International Harvester Co.'s third competitor was the Emerson-Brantingham Co., also a plow manufacturer that expanded to new lines of agricultural machines.[93]

In this way the agricultural machinery industry reflected the overall development in the American economy. As Alfred Chandler has shown, the first step on the road to a modern industrial enterprise was the emergence of professional middle managers. The process that had its birth in the 1850s and 1860s in the railroads and transportation sector spread by the end of the century to other industries. The new managerially administered enterprises developed into giant corporations either by way of horizontal integration, which meant a merger of competing firms, or by vertical integration, to incorporate raw materials and to marketing. The large mergers of the 1890s came in two waves. During the first wave, between 1890 and 1893, 51 holding companies were formed. The first large merger movement began after the depression of the 1890s had ended, with at least 212 consolidations taking place from 1898 to 1902. The International Harvester held a notable position among the big enterprises. In 1909 its relative size according to its assets entitled it to fourth place in the ranking list of the companies.[94]

In the domestic trade and manufacture of harvesting machines, the formation of the International Harvester Company marked a clear break from the previous period. It had reached a clearly dominant position in the American markets. However, during the antitrust suit brought against it by the Justice Department in 1912, the defendants maintained that the purpose of its formation was to collect capital for expansion of the foreign trade.[95] Cyrus McCormick Jr. testified that "another purpose was to help in regard to the foreign manufacture, the development of the business was such that it was evident that some capital would be required and this was the method of getting it."[96]

---

93 *The International Harvester Co.* 1913, 50-55, 188-189; In the District Court of the United States for the District of Minnesota. The United States of America, Petitioner, vs. International Harvester Company, Defendants. *Statement, brief and argument for defendants.* p. 42-43.

94 *Chandler* 1977, 86-87, 120-121, 315-316, 331-333; *Chandler* 1987a, 5, 23-25, 29-32.

95 *Kramer* 1964, 299; *Marsh* 1985, 48.

96 *The International Harvester Co.* 1913, 70.

To some extent, this statement was true. The International Harvester Co. began a similar expansion abroad as on the domestic front. Osborne & Co. m.b.H. was organized in Germany in 1903; the brand name of Osborne was probably saved because of its long-established reputation, in spite of the fact that it had already been merged with International Harvester about half a year earlier. It could also have been a strategy to keep many nominally competing lines in the eyes of the customers in order to give a picture of continued rivalry. The same phenomenon was repeated in England, where the Deering Harvester Co., McCormick Harvesting Machine Co. and Osborne-Plano Co. were organized in 1904. Two years later, however, their business was taken over by the International Harvester Co. of Great Britain.[97] The International Harvester clearly wanted to continue the establishment of branch houses and subsidiaries which McCormick had started.

What clearly separates the International Harvester Company epoch from earlier periods in terms of foreign business was its expansion to the production of machines also abroad. In 1903 it erected a large factory at Hamilton, in Canada. The International Harvester Co. of Canada was incorporated under the laws of Ontario with a capital stock of $1 000 000, to stem the growth of the Massey-Harris Company.[98]

The American manufacturers had dominated the harvesting machine sales since their emergence in Europe. Only a couple of English makers had offered noteworthy resistance, but even they were more or less copies of the American machines. In most European countries small firms, dependent on the local markets, had also appeared. In Sweden the domestic producers of agricultural machines were exceptionally active: historically there were long traditions in the mechanical industry in Sweden. Consequently, by the beginning of the 1860s numerous implement makers had emerged, such as Kockums, Överum, Munktells, Thermaenius and Åkers. In addition to these companies, towards the end of the century, there sprang up numerous specialized new companies, in such a measure that total imports of agricultural machinery were reduced to less than half of Sweden's total demand. In 1890 Sweden was already a net exporter of agricultural machines. Of the new manufacturers, Westerås mekaniska verkstad and Arvika mekaniska verkstad made mowing and harvesting machines.[99]

---

97 Ibid. 146-147, 168.

98 Ibid. 146-147

99 *Kuuse* 1970, 91-92.

In the case of Sweden it is interesting to note that simultaneously with growing exports, similar kinds of machines were imported in great numbers from America. Jan Kuuse has found that the American high-technology machines were aimed at the large estate owners; the domestic manufacturers copied the advanced American machines and adjusted them to local needs. Although the quality of the Swedish factory products was not comparable with that of the American implements, they found their markets because of their cheaper price and suitability for Swedish conditions. On the same grounds, the Swedish machines also found their way to other Scandinavian countries, and to Russia too. Consequently, Kuuse argues that the Swedish technology was a mediator of the more advanced American technology in Northern Europe, adapting it, developing it for local needs, and mediating it forward to its neighbors.[100]

It was therefore no wonder that the International Harvester Company began to show signs of concern about the situation. The continued rivalry in the top management of the Company almost paralyzed all decision-making in the International Harvester Co. before the Board of Directors authorized Cyrus McCormick to exercise supreme power in 1904. Only after that settlement was the foreign business also reorganized. All the sales in Great Britain, western and southern Europe and northern Africa were consolidated to Deering's Paris office. Northern and eastern Europe and Asia were concentrated to McCormick's Hamburg office, under William Couchman. Couchman had recognized the growing pressure that the Swedish companies offered; to stem it he organized in 1904 a separate subsidiary in Stockholm and a year later one in Denmark, but was unable to meet the competition. In the situation the International Harvester Co. had either to abandon profitable markets or to extend its activities to production. Finally in 1905 a small factory primarily producing mowers was purchased in Norrköping. This move turned out to be a great success. Two years after its acquisition Aktiebolaget International Harvester Co. returned a profit of 17.4 percent on invested capital.[101]

The Company management did not expect effective local competition in Germany, France and England. Only one German manufacturer had in 1905 shown some success. The Swedish situation was regarded as a special case, but otherwise International

---

100 Ibid. 92-99.

101 *Carstensen* 1984, 132, 139-140, 146; *The International Harvester Co.* 1913, 146-147; *Kuuse* 1974, 334; *McCormick* 1931, 132-133.

Harvester was reluctant to open new plants abroad while it had sufficient capacity in existing factories to meet demand; but the market worked against the Company. International Harvester continued McCormick's policy and replaced jobbers with its own sales force. Consequently, jobbers who were also manufacturers of agricultural machines began to produce harvesting machines as well.

Contrary to all expectations, it was in Germany and France that competition grew most alarmingly. In Germany alone, in 1908, nineteen firm were making mowers and two more reapers also. Once again the Company management was forced to make decisions under external pressure. That same year, the Board of Directors approved plans to erect plants in France, Germany and Russia. Two years later Compagnie Internationale des Machines Agricoles at Croix in France and International Harvester Co. m.b.H. at Neuss in Germany were partly completed. In 1910 the International Harvester acquired an old air-brake factory at Lubertzy near Moscow and converted it into a harvesting machine plant. When in 1909 the Company finally also extended its credit policy by accepting farmer's papers in all European markets, its foreign business had moved into a totally new epoch compared with the previous period.[102]

---

102 *The International Harvester Co.* 1913, 146-147, 150-152, 165-166, 174; *Carstensen* 1984, 153-154, 156, 163-164; *McCormick* 1931, 133.
In July 1912 the International Harvester Co. had the following foreign subsidiaries:

Company	Location of plant or business offices	Business
Compagnie Internationale des Machines Agricoles	Croix, France	Manufacturing
Compagnie Internationale des Machines Agricoles de France	Paris, France	Marketing
Deutche International Harvester Co. m.b.H.	Berlin, Germany	Marketing
International Harvester Co. m.b.H.	Neuss, Germany	Manufacturing
International Harvester Co. in Russia	Moscow, Russia	Manufacturing and marketing
Aktiebolaget International Harvester Co.	Norrköping, Sweden	do
Aktieselskabet International Harvester Co.	Christiania, Norway	Marketing
Aktieselskabet International Harvester Co.	Copenhagen, Denmark	Marketing

Table 24. Sales of International Harvester ($m), 1903-1912.

Year	Domestic + Canada		Foreign (A)		Total (B)		A as % of B	
1903		(39.8)		(12.2)		(52.0)		(23.5)
1904	33.6	(34.4)		(15.3)		(49.7)		(30.9)
1905	40.0	(36.2)	13.2	(16.9)	53.2	(53.1)	24.8	(31.8)
1906	46.7	(42.0)	15.7	(20.2)	62.4	(62.2)	25.2	(32.5)
1907	51.9	(46.4)	19.0	(24.5)	70.9	(70.9)	26.8	(34.5)
1908	47.6	(41.8)	19.0	(24.8)	66.6	(66.6)	28.5	(37.2)
1909	56.2	(50.1)	20.6	(28.1)	76.8	(78.2)	26.8	(36.0)
1910	65.2	(56.5)	24.2	(34.2)	89.4	(90.7)	27.1	(37.7)
1911	69.3	(56.9)	28.5	(42.3)	97.8	(99.2)	29.1	(42.7)
1912	77.5	(64.0)	36.2	(50.9)	113.7	(114.9)	31.8	(44.3)

Source: Kuuse 1974, 307 table 66; Figures in the parantheses are from Appeal from the District Court of the United States for the District of Minnesota. p. 1751. Defendants' Exhibit 85.

International Harvester Co.of Great Britain (Ltd.)	London, England	Marketing
International Harvester Co. Gesellschaft m.b.H.	Vienna, Austria	Marketing
International Harvester Co. A.G.	Zürich, Switzerland	Marketing
International Harvester Co. of Canada (Ltd.)	Hamilton, Canada	Manufacturing
Oliver Chilled Plow Works of Canada (Ltd.)	Hamilton, Canada	Manufacturing
Eastern Building Co. (Ltd.)	Hamilton, Canada	Building
International Harvester Co. of Australia (Ltd.)	Melbourne, Australia	Marketing
International Harvester Co. of New Zealand (Ltd.)	Christchurch, New Zealand	Marketing
Macleod & Co.	Manila, Phillipine Islands	Purchase of fiber
Salango Export Co.	Ecuador	Fiber production

Source: *The International Harvester Co.* 1913, 165-166.

This becomes obvious also in Table 24, which shows a marked increase in the volume of the International Harvester Company's foreign business and especially on the proportion of foreign sales compared with the total sales.

X

■ Concluding Analyses

In the present study Cyrus Hall McCormick and his life-work, the McCormick Harvesting Machine Company, have been taken as examples of an emerging American multinational company. On the basis of the evolution of this sample enterprise, a model has been produced in Figure 6 which describes the various stages of its development. This model is of course simplified, and shows only the main features which have been discussed in detail in the text.

The development of McCormick's foreign trade can be divided into two phases; firstly Cyrus McCormick's strictly private foreign sales, separate from the Company; secondly, the Company business phase from the beginning of the 1880s. During the first stage, which began in 1851 at the Crystal Palace World Fair, Cyrus McCormick mainly relied on European manufacturers, to whom he sold production rights, whereas in America he had only recently given these up.

Reasons for the extension of sales abroad in the first stage have been difficult to identify. However, it is obvious that they were closely bound to the person of Cyrus McCormick. He had foreseen the chances implied in the possible success at the World Fair; the triumph at Crystal Palace, which warranted him the fame of the invention of the reaper, was rapidly materialized in America, but he anticipated money in Europe, too, and responded immediately to these opportunities.

A characteristic feature of the first stage of McCormick's foreign business is the unsystematic approach to business. Cyrus McCormick let the business continue under its own weight, reacting only to external impulses but not actively directing them. His machines at first attracted great interest and dominated the market. His European manufacturers were, however, unable to produce the quality required. Besides, McCormick's competitors had also discovered the European market and soon outdid him there. McCormick's reaper was too expensive and heavy for the European farmers. In addition, the capacity and production technology of McCormick's Chicago plant was too limited to meet even American demand, and Cyrus' foreign trade with its special requirements led to continuous quarrels with his younger brother Leander. Manufacture of the European machines demanded totally new patterns and was too costly at the time of handfitting. As a result, Cyrus McCormick's foreign enterprise gradually died off after the mid-1860s. Mechanization of the American farm was still in its infancy, and new territories opening up for agriculture further increased demand at home. The time was not yet ripe for large-scale foreign business.

Especially so when the state of European agriculture is taken into consideration. In Europe, the demand for harvesting and mowing machines was strictly concentrated among the wealthy, upper-class land-owners. In the terms of Paul E. David, the threshold size for a farm where it was profitable to buy a harvesting machine was still very high. A key factor affecting the use of harvesting machines was availability of farm labor: as long as labor was abundant and cheap, there was no incentive for capital-intensive and labor-saving machines. In Britain, the first marketing area for reapers, this relation changed slowly towards the end of the 19th century, and consequently the demand for harvesting machines began to grow, although even there the scythe was at the same time only replacing the sickle. On the Continent, the situation was more complicated. Both in France and Germany there were both large holdings but also areas of small farms. Russia was one of the leading grain producers of the world, but had a surplus of labor as did France and Germany.

Figure 6. Main stages in the McCormick Company's foreign trade.

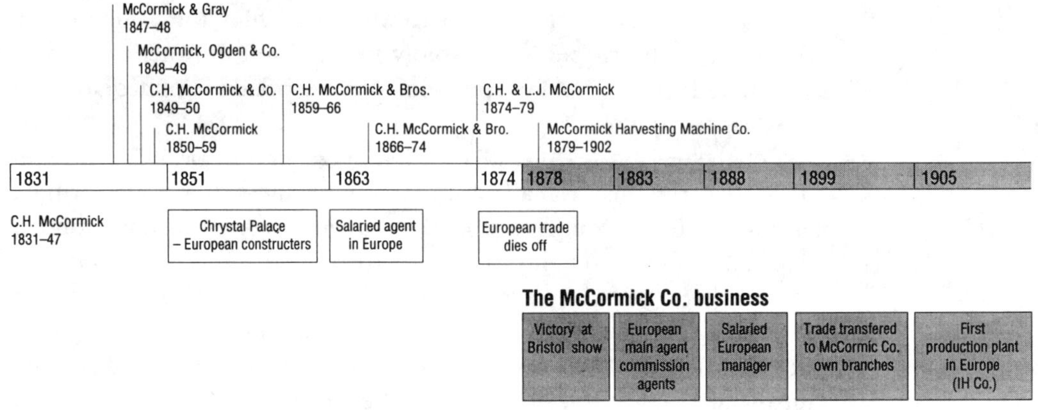

Why did Cyrus McCormick nevertheless continue his faltering business in Europe? He had reaped his greatest public victories in Europe, was decorated with high honors around Europe and consequently had found a place in European high society. In that respect business was more for his ego than profit. Besides, the business did not cause him great losses. It was handled through independent agents, who bought machines f.o.b. in London or New York. In addition, Burgess & Key and other contractors paid him royalties on every implement manufactured and sold. The aim of McCormick's first foreign enterprise was therefore more or less to keep his name in the limelight.

The next stage of McCormick's foreign enterprise was bound to and backed by many internal changes in the Company. Cyrus McCormick was among the first to put the hire-purchase system to large-scale use in America, by offering long-term credit to his customers. McCormick's reapers were sold at first through independent commission agents, but already at the end of the 1870s, jobbing houses were replaced by the Company's own branch houses. The trade had grown beyond the scope of small agents and besides, the machines were too complicated and deserved better service than agents were able to offer.

Although keeping a close eye on all the functions of his company, Cyrus McCormick's emphasis was increasingly on the structural level of the business. He was more and more engaged in organizational questions, fighting his endless patents wars or traveling with his machines from one show to another. In the long run this, combined with the totally conflicting business strategy of his business partner and brother Leander, led to an open conflict and finally to the formation of the McCormick Harvesting Machine Company in 1879. In 1880 Leander McCormick was fired as Superintendent and a professional manager was hired for the first time to oversee production. This event began a new era in the McCormick Company; a managerial approach to business. It began to use night shifts, new order and discipline was brought in and for the first time gauges, patterns and jigs appeared in the factory. Also cost accounting emerged for the first time in the business practises of the Company.[1]

It is difficult to find a clear answer to the question, why Cyrus McCormick decided to return to Europe at the end of the 1870s. The development of technology was a necessary prerequisite for the next stage in McCormick's foreign business. The new wire-binder opened a chance for a new breakthrough in Europe. However, the

---

1 *Hounshell* 1987, 179-180.

beginning of the second stage of foreign enterprise was also a result of Cyrus' personal interest. In 1877 he sent two machines to the Liverpool Agricultural Show, and appointed a new agent, but was hit by setbacks. The Show was a total loss and the new agent died. Yet something remained. His son Cyrus Jr. had accompanied the binders to Britain and had learned a hard lesson. Machines no longer sold on their own weight, but needed a careful preparation of the market.

1878 was crucial for the future of the foreign business. Cyrus McCormick sent his trusted man in good time before the Royal Agricultural Society's show to Britain and nominated a new agent. Once again, Cyrus Jr. decided to attend the show. All the necessary preparations were taken into account, and as a result the McCormick binder reaped the gold medal. But had these investments been made for the sake of the foreign trade or because of the Paris World Fair, where Cyrus Sr. was decorated with the Cross of the Legion of Honor? The answer inclines more on the side of a thirst for honors. Nothing crucial had yet changed in the production plant. The McCormick Company was not able to meet the demand, and Leander opposed extension of production. Besides, the technology was not yet ready to meet the changes necessary for machines for the European markets. Nevertheless, the name and fame of McCormick had returned to Europe.

New important decisions followed. McCormick extended his operations to Australia and New Zealand and his own salaried representatives were sent to Russia. These moves were not of a corporate action, but Cyrus McCormick's private initiative. Foreign trade was under his supervision, although correspondence and its daily functions were transferred to the Company's officers. Foreign incomes were, nevertheless, of some importance for the Company because they were carefully calculated on the revenue side, and were expected to cover some old debts. On the other hand, this example shows the haphazard nature of accounting practises, where corporate matters were mixed with Cyrus's personal accounts.

George Freudenreich can be regarded as McCormick's first salaried representative abroad. He took care of the Russian sales which were regarded as promising for the future. The whole of western Europe was left under Percy Lankester's jobbing house; but there was no visible plan for the running of foreign affairs. Only in the midst of the depression and labor problems did General Manager E.K. Butler in 1886 make an investigation trip to Europe. Thereafter, foreign business was put directly under his supervision and Lankester was nominated as the Company's salaried European main agent. In these actions McCormick's was a latecomer; its main American competitors had founded their own agencies in Europe several years before.

In a position to oversee the entire field of business, E.K. Butler developed the foreign trade as a part of overall company business. When he anticipated good potentials for earnings on a foreign field he showed the green light for extension of activities. When the domestic trade was more plausible, the foreign had to give way until the end of the 1890s. New accounting practises became visible behind his decisions. Machines for foreign markets needed fitting, and in many cases totally new patterns, which increased the cost of manufacture at a time when the Company was seeking economies of scale. As long as the foreign demand did not surpass the production costs it caused, it was more profitable to invest in domestic trade. Consequently, what appeared to be the McCormick Company's stubborn approach towards the frequent request of its foreign agents to provide machines suited for local conditions, in fact was part of a carefully calculated overall company strategy. However, Butler extended McCormick's operations rapidly to all agricultural areas of importance. New agents were appointed, but they all remained under Lankester's supervision. The Company's first investment in property abroad was its warehouse in Odessa in 1894. Butler's annual trips to Europe culminated in the decision to establish the Company's own branch houses. McCormick's old European agent, Percy Lankester, was first released from the charge of the European business and reduced to a normal jobber, and finally his business was entirely turned over to McCormick,s. Lankester was replaced by McCormick's own European manager, who was sent to Hamburg. William Couchman's actions proved this decision correct and he soon gained large autonomy for his operations. Expansion culminated in 1900 in the founding of McCormick's first subsidiary, the McCormick Harvesting Machine Co. m.b.H. in Berlin. By 1902, when the largest harvester manufacturers merged their interests, McCormick had extended its branches to all the key agricultural areas and places where competition necessitated it.

Underlying reasons for McCormick's expansion abroad during the 1880s seem to be factors connected to preservation of the market share and keeping the name in the limelight. The events in the Pacific indicate that at least in the harvester business the whole world had already at that stage become one market area where rival companies could not afford to give up their achieved share of markets. In that competition the McCormick Company did not invest on its own facilities, nor did it actively develop the trade. It only more or less passively reacted to the impulses of its competitors, or request for agencies from various countries. Jobbing houses were for it a safe way of doing business and showing the Company flag; sales f.o.b. did not bind its hands to long credits as was the case in the U.S.

Explanations for the rapid extension and change of the strategy in foreign business are in the 1890s at least as manifold as during the previous decades. In her model of the evolution of a multinational company, Mira Wilkins describes how companies first appointed independent agents to represent them abroad; this stage was followed by a salaried export manager, who was replaced in the third stage by a branch house or a distribution subsidiary, and eventually by a finishing, assembly or manufacturing plant in the final stage.[2]

When operations are examined from a general level, there is a danger that everything seems to fit into the frames of the sought model. Keeping this reservation in mind, it can be stated that the foreign business of the McCormick Company followed the outlines that Mira Wilkins has drawn; in other words, the McCormick Company was a typical representative of an American enterprise.

Wilkins' explanation describes only the various stages in the evolution of a multinational, but not the underlying reasons, nor why it succeeded to outdo its competitors on both sides of the Atlantic. When it comes to Wilkins' argument of selling the surplus abroad as a sign of a modern enterprise on its way to becoming multinational, the McCormick Co. did not in this part fit into her outlines. For a start there was surplus only in a couple of years. Secondly, harvesting machines had to be adjusted to the foreign special conditions, which required refitting and in many cases special patterns. Consequently, in this trade, economies of scale were not easily available. During the depression of the 1890s, McCormick's certainly had a large number of machines in store; but even then it did not resort to dumping them abroad at low prices.

If the McCormick Company did not send its surpluses to external markets, what then was the reason for the expansion of its business during the 1890s? Mira Wilkins offers domestic competition and the division of markets between giant corporations as an explanation for this growth. She states that they "were in a position to do more abroad".[3]

The harvesting machines companies tried to merge their interest for the first time in 1890. After the failure of this attempt, renewed competition and recession forced the smallest companies out of the field, and the two largest manufacturers, McCormick and Deering, came to dominate the market in America. Domestic markets had so far continuously expanded but in the 1890s there was no more virgin

---

2 *Wilkins* 1970, 45-46.

3 *Wilkins* 1970, 72.

land for farmers. Simultaneously, falling machine prices forced manufacturers to improve their productivity to maintain their profits. As a consequence, more machines at lower prices poured onto the saturating U.S. markets.

In that light it really looks as if foreign trade was merely an extension of domestic competition, as Wilkins argues. Only in 1892 did the Deering Company send its first traveling agent abroad; by that time McCormick's had firmly established itself in the foreign market and could easily fight its arch-rival. For smaller companies, foreign markets might at first have offered rescue from domestic competition, but in the long run they were unable to hold their position there either.

Manufacturers were forced to seek new markets, but they still had to resolve the problems in manufacturing and marketing. Foreign branches were expensive, but without after-sale service capacity and repair service there was no chance of durable business. On the other hand, competition at home curtailed possibilities for investments abroad. Besides, productivity was further curtailed by production cuts after each season when factories closed their doors and dismissed their workers. This problem was settled only after the formation of the International Harvester Company, which extended its functions to new lines.

But an eye should once again be cast on the state of agriculture in the importing areas too. The success of the harvester companies was closely connected to the state of agriculture in the receiving countries. In that respect especially the European countries differed considerably from each other. Britain was the country where agriculture first began to mechanize and was, consequently, up to the 1890s, the main marketing area for harvesting machine companies too. Especially at the turn of the century in many west and north European countries mechanization began to spread also among the peasantry. For machine manufacturers it meant growing demand. This phenomenon partly explains the eagerness of the U.S. harvesting machine manufacturers to intensify their efforts in Europe at the turn of the century. But how was it possible that the American companies conquered the European markets so easily?

American manufacturers regarded Europe as an undeveloped market with great resources. The American manufacturers had long years of bitter rivalry behind them at first in the United States but in Europe too. Only the English companies could compete with the American makers up to the 1880s, when they too were driven out of the market. Other European manufacturers had only local significance. In the highly competing harvester markets the U.S. companies, due to their large domestic markets, were able to reach, at least in some measure, economies of scale. Large production series and careful calculation of costs made it possible to lower the prices to such a level which the European or other competitors were not able to follow. Thus, English manufacturers were known for their high-quality products and use of iron in their machines. Until the 1880s, American makers, on the contrary, used wooden frames in their harvesters. When the foreign business was still limited to small amounts, English makers were able to compete, but once the American factories began to produce all-iron machines, they could benefit from their size and lower machine prices.

An integral part of the American companies' selling effort was aggressive marketing. The McCormick Company, like its competitors, tried to create demand where it did not exist. Machines were sent from one agricultural show and test to another all across Europe. The jurors of the tests were put under pressure and even bribed. Sales catalogs were translated into European languages; flyers, showcards and other advertising material was sent to European farmers. European customers were also introduced to the written testimonials of satisfied customers. Sales spheres were split into smaller entities by appointing new agents, as in Germany; agents were required to travel around in their areas in good time during the winter, and to show reluctant farmers the necessity of the new machinery; and at harvest time they had to haul machines to harvest grounds to show them in action. Besides, almost every year McCormick's developed new models which were adapted to the special conditions prevailing in various areas. In addition, it extended to Europe its after-sale service, including traveling experts.

But there was one key element in the U.S. harvester trade which was not transfered abroad during the time period under investigation. Cyrus McCormick Sr. was one of the first to introduce the hire purchase system into the American business traditions. In the United States, credits to farmers were finally extended up to three years and harvester companies were engaged in direct sales with farmers. In its foreign trade the McCormick Co. stayed strictly on f.o.b. terms and declined extending credits to farmers.

In the present study have been outlined the reasons for and development of the foreign trade of the McCormick Company. Now it is time to look at the context in which the Company operated; the emergence of American large companies and their development into multinational enterprises. Was the McCorcmick Co. a multinational company and did it follow the prevailing trends in American business life?

Alfred D. Chandler has in his works described the main features and stages of the development of the American business enterprise. With some variations, the McCormick Company fits relatively well into Chandler's model. Although the McCormick Co. never was a single unit firm operated by an individual, it was also clearly transformed in the beginning of the 1880s into a professionally managed corporation with a hierarchy of salaried executives, and with distinct operating units. Ownership and management were separated from each other during E.K. Butler's time as the General Manager. After his resignation, the corporate management was nominally in the hands of the McCormick brothers, but the professional middle managers took charge of the daily activities of the Company and took part in the corporate decision-making too.

In the chandlerian sense[4], the McCormick Co. invested in product-specific marketing and distributing, and finally its own branch houses replaced the wholesalers. In addition, the Company invested considerably in production facilities and organization of work to achieve advantages of scale. As a result, there emerged during the 1890s a new kind of competition, where the two leading companies were able to meet the domestic demand. This was one of the prerequisites for international operations in Chandler's model.[5]

This concept is closely connected to Chandler's idea of throughput. Increased throughput forced manufacturers first to replace independent agents with their own sales force and finally led to the extension of operations abroad.[6] To Chandler, foreign trade and the emergence of a multinational enterprise were an extension of domestic business. He does not give any explicit answers as to, why American companies began to operate overseas, but describes how it was accomplished.

Mira Wilkins, on the other hand, has concentrated in her works almost entirely on the multinational enterprise. She defined the four stages through which firms turned into multinationals. These phases can be identified also in the evolution of the McCormick Co. It began to sell through independent agents, appointed salaried personnel in the next and founded its own branch houses in the third. McCormick's can be said to represent new technology as well; in its final years it expanded abroad to reach new markets, it had from the 1880s on, at an accelerating rate, substituted capital for labor, and finally it also felt the pressure of the domestic economic conditions. Morover, the foreign market was only complementary to the McCormick Co. too. The Company was well aware of the prospects of the markets through its travelling agents and the

---

4 *Chandler* 1990, 29-32.

5 Ibid. 30.

6 *Chandler* 1988, 31-34, 41-42.

personal visits of its highest management in Europe. It followed keenly the political and legal development in its market areas, and invested very carefully and only in familiar conditions.

The McCormick Company is connected to the evolution of American business life at the turn of the century in yet another way. Decision-making is an art of its own. It has been, and still is, at the same time the most difficult but also the most important part of business activities. The decisions and the strategy selected depend on the personality of the manager, on the education and experience he has behind him. It was therefore no wonder, when the old General Manager of the McCormick Co., E.K. Butler resigned, that his conservative and careful strategy which aimed at the maximizing of profit per produced unit was replaced by the strategy of growth. Instead, profits were sought from economies of scale and the amount of throughput was emphasized. This phenomenon was common in other branches of American industry too.

Thus the McCormick Company seems to fit into the outlines defined as characteristic of American multinational enterprises by Alfred D. Chandler and Mira Wilkins. It certainly had its own peculiarities bound to the branch of industry, but as a whole it can be seen as the first stage of an emerging harvester multinational. In the sense of Mira Wilkins, it was the marketing stage in the evolution process. By that time the McCormick Co., by investing in branch houses and in a subsidiary of its own, had become a multinational company. Although it was still a matter of marketing organization, investments were direct foreign investments, which numerous researchers regard as the determining feature of a multinational. Besides, the Company developed many of the basic strategies and approaches so typical in the next stage, the International Harvester Company period, when activities were expanded to overseas production plants too.

The American multinational enterprise has been regarded as a unique phenomenon in history. Nevertheless, European foreign investments preceded their American counterparts by decades. Among Continental companies, the German were particularly active before World War I. Lawrence D. FRANKO even states that the total number of foreign manufacturing subsidiaries of Continental enterprises outnumbered those of American firms. European companies became multinational in a similar way as did American firms: after obtaining oligopolist advantages in technological innovation, they began by exporting. European innovations were, however, frequently first uprooted to the U.S. and commercialized there. American firms innovated new, labor-saving, highly income-elastic products. Continental companies, on the contrary, rarely pioneered in these sectors, but operated in synthetics and luxury products.[7]

Both American and European companies used similar strategies in their expansion abroad. In the first stage, European enterprises too founded their marketing organization, normally independent agents. Only after that did companies establish their own branches, and in the same way as American companies, thereafter production plants. This was the case with the German firm Siemens & Halske, that in 1900 had 42 offices in Germany, 37 in other European countries, and 38 overseas. A similar development can also be found in the German corporation AEG.[8]

Did the manoeuvres of the harvester companies described above simultaneously affect the development of agriculture? What was the role of a giant corporation like the McCormick Company? Was it a Schumpeterian innovator that changed structures in agriculture?

The reaper and its later modifications were important to the development of agriculture. Harvesting, always laborious, became considerably lighter and faster, thereby decreasing the risk of spoiling. Nevertheless, McCormick's was hardly able to increase mechanization to an important extent; the factors affecting in the opposite direction were too strong. However, it introduced and made known an important implement, and in that respect it had important long-term effects on European agriculture. Especially once it, and especially the International Harvester, had become a market leader, it could to some extent dictate what kinds of machines were available for farmers.

In the 1960 and 1970s, the mechanization of agriculture attracted innovation theorists to test their ideas. They produced models on diffusion of innovations from one country to another, from center to periphery and from large estates to smallholders. These works offered valuable new information on the mechanization of agriculture. We know now when the first threshers or reapers appeared in Britain, Sweden or in Finland; and we also know who were the first to adopt new implements.

Diffusion of innovations can be approached from the other side of the mirror, too, as has been done in the present work. Is the innovation theory relevant any more, if it is approached from the side of the innovator or entrepreneur? Can we see centers and peripheries even then? What are the early adapters like?

When in 1851 Cyrus McCormick introduced his reaper to European farmers in England, he was not thinking about foreign

---

7 *Franko* 1976, 3, 8-10, 24, 75, 77.

8 *Hertner* 1988, 146-151.

sales. The World Fair offered him new opportunities which he was ready to use. In that respect England was hardly a center area, although farming there was the most advanced in Europe. After Crystal Palace, the question was of earning plain money wherever possible. Here slowly growing demand and fast growing supply met each other. Early adapters were normally wealthy estate owners, who, for the sake of curiosity or need, purchased the first machines. On the other hand, Gould P. Coleman has shown that adaptability of an innovation depended upon the cropping system, the size of the farms, the prevailing soils and climate and accessibility to markets.[9] In the 1860s, these factors in England were favorable. In that respect the present study without doubt confirms the basic concept of diffusion of innovations.

But is this approach of any relevance? If the question is approached from the manufacturer's side, he did not see any traditional centers. He sold his products wherever he could find demand. At first Britain offered the best markets; thereafter the focus of activity moved for a while to the Pacific and then again to Europe. In this process it is impossible to define any center in the manner used in theories of diffusion of innovation.

It is self-evident that wealthy people can always acquire new things first; thereafter they rapidly become common as the price sinks below a certain level. That was the case with the harvesting machines too. What is interesting and of value, however, is to figure out the timing and evolution of this process in different countries. It explains the state of development in terms of demand in sample areas, but totally omits the supply side. As we have seen, this has been a severe shortcoming, since manufacturers were active in creating demand and consequently were an important factor affecting the diffusion of innovations. In previous works, this effect has normally been totally excluded or underestimated.[10]

9 *Coleman* 1968, 183.

10 See for example *Coleman* 1968, 175. According to Coleman the most important factors in the adoption process were experiences of neighbors. In addition, persuasion of salesmen and information in the agricultural press played a secondary role.

Figure 7. Emergence of various types of harvesting machines in Europe.

Emergence of various types of harvesting machines in Europe

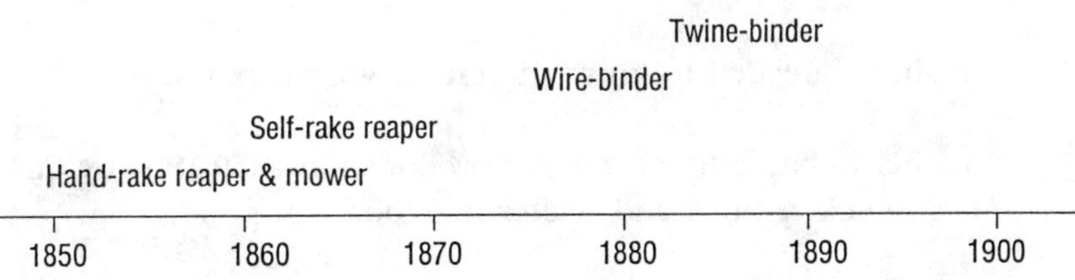

Figure 7 shows the outlines of the diffusion of various kinds of harvesting machines in Europe. In the case of Europe it is impossible to draw the kind of curves that Jan Kuuse has constructed for the United States, where old-fashioned machines were rapidly replaced with newer ones, although not all of the farmers even there were capable of such investments. In Europe adoption of new agricultural technology was not as uniform. There were big regional differences between and also within countries. Old and new implements could be used side by side, as was the case in Russia where domestic manufacturers began to produce hand-rake reapers, while their foreign competitors were already selling twine-binders. Similar views could be met also on the Finnish fields during the 1950s and even the 1960s.

■ List of Tables

■ List of Figures

■ List of Pictures

List of Maps

1. Harvesting and mowing machine manufacturers of the U.S,
    1890. Page 195.

2. Branch houses and main agents of the McCormick Company in
    Europe, 1901. Page 253.

■ List of Sources

1. ARCHIVE SOURCES

The University of Helsinki Library
-Kansalliskirjastokokoelman pienpainatteet: valmistajien ja kauppiaiden mainoskuvastot ja hinnastot (The National Library Collection: manufacturers' and dealers' catalogs and pricelists)

The University of Turku Library
-valmistajien ja kauppiaiden mainoskuvastot ja hinnastot (manufacturers' and dealers' catalogs and pricelists)

Åbo Akademi Library
-valmistajien ja kauppiaiden mainoskuvastot ja hinnastot (manufacturers' and dealers' catalogs and pricelists)

The State Historical Society of Wisconsin. Division of Archives and Manuscripts, Madison, U.S.A.
-McCormick Family Papers
Cyrus Hall McCormick (Series A)
Nettie Fowler McCormick (Series B)
Cyrus H. McCormick Jr. (Series C)
Harold F. McCormick (Series F)
Stanley R. McCormick (Series G)
-McCormick Company Papers
McCormick Harvesting Machine Company (Series X)
Catalog collection (series Z)
American Harvester Company (Series W)
-McCormick Historical Records
Special Reports File
-McCormick Estate Papers (Series M)

The State Historical Society of Wisconsin. Library.
-pamphlet collection

2. PRINTED SOURCES

In the District Court of the United States for the District of Minnesota. The United States of America, Petitioner, vs. International Harvester Company and others, Defendants. No. 624.
- Appendix to Defendants' Brief. Evidence as to certain points abstracted and topically arranged. Vol. II.
- Defendants' answer to petition.
- Hearing Before Circuit Judges Sanborn, Hook and Smith, at St. Paul, Minn., Nov. 3-5, 1913. Oral argument of Wm. D. McHugh.
- Volume V. Testimony of Witnesses for the Defendants.
- Volume XIII. Testimony of the Witnesses for the Defendants.

The Federal Trade Commission.
- Causes of High Prices of Farm Implements. Washington 1920.
- Report on the Agricultural Implement and Machinery Industry. Part I. Concentration and comparative methods. Washington 1938.

The International Harvester Case. A compilation of public statements and court documents in connection with the government's investigation of the International Harvester Company and its anti-trust suit against that company, covering the period from May 20, 1907, to May 28,1920.

International Harvester Company. Senate, 63d Congress, 2d Session. Document No.569. Opinions of the Judges in the case of the United States v. the International Harvester Co. et.al. in the District Court of the United States for the District of Minnesota. Washington 1914.

Maataloustiedustelu Suomessa vuonna 1910. Edellinen osa: Maanviljelys. SVT: III:9. Maatalous. Helsinki 1916.

Petition to the Legislature of Wisconsin: Petition against the renewal by Congress of Letters Patent granted to Cyrus H. McCormick and others, for improvements in the Reaping Machine, 1855.

Statistical Abstract of the United States 1870-1917.

Supreme Court of the United States.
-International Harvester Company of New Jersey (formerly International Harvester Company) et al., Apellants, vs. The United States. Appeal from the

District Court of the United States for the District of Minnesota. Vol. II. Transcript of record. October term, 1914. No. 757.
-International Harvester Company of New Jersey (formerly International Harvester Company) et al., Apellants, vs. The United States. Appeal from the District Court of the United States for the District of Minnesota. Brief for Appellants. No. 757, 1914.

United States Department of Agriculture. Division of Statistics.
-The course of prices of farm implements and machinery for a series of years. Washington 1901.

United States Department of Commerce. Bureau of Census.
-Manufacturers 1860, 1865.

United States Department of Commerce. Bureau of foreign and domestic commerce.
- Miscellanious series No. 81. Selling in Foreign Markets. Selected readings from published statements of business men and reports of experts on methods employed in export trade. Washington 1919.

United States Department of Commerce and Labor.
-The International Harvester Co. Washington 1913.

United States Department of Commerce and Labor. Bureau of Statistics.
-the Foreign Commerce and Navigation of the United States 1869-1915.
-Markets for Agricultural Implements and Vehicles in Foreign Countries. Special Consular Reports. Vol. XXVII. Washington 1903.

United States Department of State
-Reports from the consuls of the United States No.2/ 1880, No.3/1881, No.20/1882, No.38 and 48/1884, No.72/1886, No.123/1890, No.154/ 1893,

-Consular reports. Commerce, manufacturers, etc. No.195/1896, Vol. LV/1897 No.205, Vol. LXI/1899 No. 228, 229, 230, 231, Vol. LXIII/1900 No. 236, 237, 238, 239.

3. NEWSPAPERS

Farm Implement News 1892, 1893, 1984, 1896, 1897, 1898, 1899
Farm Machinery 1897.

Harvester World 1936.
Illustrerad Landtbrukstindnings Årsbok för år 1876.
The Implement Age 1902.
Lantmannen 1878.
Nya Pressens Landtbruksafdelning 1890.
Suomen Huoneenhallitusseuran Sanomia 1851.
Teknikern 1894.

4. BIBLIOGRAPFY

Aldcroft, Derek H.: Introduction: British Industry and Foreign Competition, 1875-1914 (Derek H. Aldcroft: The Develoment of British Industry and Foreign Competition 1875-1 914. Studies in Industrial Enterprise). London 1968.

Anttila, Veikko: Talonpojasta tuottajaksi. Suomen maatalouden uudenaikaistaminen 1800-luvun lopulla ja 1900-luvun alussa. Helsinki 1974.

Ardrey, Robert L.: American Agricultural Implements: A Review of Invention and Development in the Agricultural Implement Industry of the United States. Chicago 1894.

Argersinger, Peter - Argersinger, Jo Ann E.: The Machine Breakers: Farmworkers and Social Change in the Rural Mid-West of the 1870s. Agricultural History 58 (1984): 393-410.

Beaumont, Olga - Higgs. J.W.Y.: Agriculture: farm implements. Part I (Charles Singer - E.J. Holmyard - A.R. Hall - Trevor I. Williams: A History of Techonoly. Vol. IV. The Industrial Revolution c 1750 to c 1850). Oxford 1958.

Benson, Howard William: Organization and first years of the International Harvester Company. Unpublished MA thesis, University of Chicago. Chicago 1936.

Berend, Ivan T. - Ranki, György: The European periphery and industrialization 1870-1914. Translated by Eva Palmai. Budapest 1982.

Bidwell, Percy Wells -Falconer, John I: History of Agriculture in the Northern United States, 1620-1860. New York 1941.

Blackford, Mansel G. - Kerr K. Austin: Business Enterprise in American History. Boston 1986.

Blaich, Fritz: Amerikanische Firmen in Deutschland 1890-1918. US-Direktinvestionen im deutschen Maschinenbau. Zeitschrift für Unternehmensgeschichte 30. Wiesbaden 1984.

Bogue, Allan G.: From Prairie to Corn Belt. Farming on the Illinois and Iowa Prairies in the Nineteenth Century. Chicago 1968.

Broehl, Wayne G.: John Deere's Company. A History of Deere Company and Its Times. New York 1984.

Carstensen, Fred V.: American Enterprise in Foreign Markets. Studies of Singer and International Harvester in Imperial Russia. Chapel Hill 1984.

Casson, Herbert F.: Romance of the Reaper. New York 1908.

Casson, Herbert F.: Cyrus Hall McCormick. His life and work. Chicago 1909.

Chandler, Alfred D.: The Visible Hand. The Managerial Revolution in American Business. Cambridge, Mass. 1977.

Chandler, Alfred D.: Strategy and Structure. Chapters in the History of the Industrial Enterprise. Fifteenth printing. Cambridge, Mass. 1987. (Chandler 1987a)

Chandler, Alfred D.: Technology and the Transformation of Industrial Organization (Joel Colton - Stuart Bruchy: Technology, The Economy and Society: The Emerican Enterprise). New York 1987.

Chandler, Alfred D.: Technological and organizational underpinnings of modern industrial multinational enterprise: the dynamics of competitive advantage (Alice Teichova - Maurice Levy-Leboyer - Helga Nussbaum: Multinational enterprise in historical perspective). Cambridge 1988.

Chandler, Alfred D.: Scale and Scope. The Dynamics of Industrial Capitalism. Cambridge, Mass. 1990.

Clapham, J.H.: The economic development of France and Germany 1815-1914. Cambridge 1951.

Clark, Victor S.: History of Manufacturers in the United States. Volume II, 1860-1893. New York 1929.

Cochran, Thomas C.: Did the Civil War Retard Industrialization? Mississippi Valley Historical Review 48 (1961):197-210.

Coleman, Gould P.: Innovation and Diffusion in Agriculture. Agricultural History 3 (1968): 173-188.

Collins, E.J.T.: Labour supply and demand in European agriculture 1800-1880 (E.L. Jones and S.J. Woolf: Agrarian Change and Economic Development. The Historical Problems). Bungay 1969.

Collins, E.J.T.: The "machinery question" in English agriculture in the nineteenth century (George Grantham and Carol S. Leonard: Agrarian organization in the century of industrialization: Europe, Russia, and North America. Research in economic history. Supplement 5, 1989. Part A). London 1989.

Cooper, Martin R. - Barton, Glen T. - Brodell, Albert P.: Progress of Farm Mechanization. U.S. Department of Agriculture. Miscellaneous Publications No. 630. Washington 1947.

Cyrus H. McCormick, 1809-1884. Chicago 1925. Wisconsin Historical Library. Pam 85-1872.

Cyrus H. McCormick and the Reaper. S.L.: s.n.: S.A. Wisconsin Historical Library. Pam 81-1390.

Danhof, Clarence H.: Change in Agriculture: The Northern United States, 1820-1870. Cambridge, Mass. 1969.

Danhof, Clarence H.: The Tools and Implements of Agriculture. Agricultural History 46 (1972):81-94.

David, Paul E.: The Mechanization of Reaping in the Ante-Bellum Midwest (Henry Rosovsky: Industrialization in two Systems: Essays in the Honor of Alexander Gerschenkron by a Group of His Students). New York 1966.

David, Paul E.: The landscape and the machine: technical interrelatedness, land tenure and the mechanization of the corn harvest in Victorian Britain (Donald N. McCloskey: Essays on a Mature Economy: Britain after 1840. Papers and Proceedings of the Mathematical Social Science Board Conference on the New Economic History of Britain, 1840-1930, held at Eliot House, Harvard University, 1-3 September 1970). London 1971.

Deering Harvester Company. Official retrospective exhibition of the development of harvesting machinery for the Paris exposition of 1900. Paris 1900.

Denison, Merrill: Harvest Triumphant: The Story of Massey-Harris. New York 1949.

Dovring, Folke: The Transformation of European Agriculture (H.J. Habakkuk and M. Postan: The Cambridge Economic History of Europe. Volume VI. The Industrial Revolutions and after: incomes, population and technological change, II). Cambridge 1966.

Dunning, John D.: The multinational enterprise: the background (Dunning, John D.: The Multinational Enterprise). London 1971.

Dunning, John D. - Cantwell, John A. - Corley, T.A.B.: The Theory of International Production: Some Historical Antecedents (Peter Hertner and Geoffrey Jones (eds.): Multinationals: Theory and History). Shaftesbury 1986.

Eckles, Elvis Luverne: The Development of Oligopoly in the Farm Implement Industry. Unpublished Ph.D. thesis, University of Illinois, Urbana. Urbana 1953.

Engerman, Stanley: The Economic Impact of the Civil War. Explorations in Entrepreneurial History 3 (1966):176-199.

Eskeröd, Albert: Jordbruket under femtusen år. Redskapen och maskinerna. Borås 1973.

Facta 2001. Volume 11. Porvoo 1984.

Fagan, V.P.: Reaping Machine Development After 1831. 1931. S.L.

Foreman-Peck, James: A History of the World Economy. International Economic Relations since 1850. Brighton 1983.

Forsberg, K: Kertomus seitsemännestä yleisestä Suomen maanviljelyskokouksesta Helsingissä 30 ja 31 p:nä elokuuta sekä 1, 2, ja 3 p:nä syyskuuta w.1876. Helsinki 1877.

Forty harvest seasons. Deering 1898. S.L.

Fowler, Eldridge M.: Agricultural Machinery and Implements (Chauncey M. Depew: One Hundred Years of American Commerce. A History of American Commerce by One Hundred Americans). New York 1895.

Franko, Lawrence G.: The European Multinationals. A Renewed Challenge to American and British Big Business. London, New York 1976.

Fussel, G.E.: The Farmer's Tools. The history of british farm implements, tools and machinery before the tractor come. London 1952.

Fussel, G.E.: Farming Technique from Prehistoric to Modern Times. London 1966.

Galtung, Johan: Foreign policy opinion as a function of social position. Journal of peace research 1 (1964): 206-231.

Gates, Paul W.: Agriculture and the Civil War. New York 1965.

Grantham, George: Agrarian organization in the century of industrialization: Europe, Russia, and North America (Grantham, George - Leonard, Carol S.: Agrarian organization in the century of industrialization: Europe, Russia, and North America. Research in economic history. Supplement 5, 1989. Part A). London 1989.

Gray, R.B.: Development of Agricultural Tractor in the United States. Part I: Up to 1919 inclusive. United States Department of Agriculture. Agricultural Research Service. Beltsville 1954.

Greeno, Follet L.: Obed Hussey. Who, of all inventors, made bread cheap. Rochester 1912.

Grigg, D.B.: The Aricultural Systems of the World. An Evolutionary Approach. Cambridge 1974.

Grigg, D.B.: The Dynamics of Agricultural Change. The historical experience. Tiptree 1982.

Grotenfelt, Gösta: Jordbruksredskap och -machiner afprovade i Finland intill slutet af år 1910. Skrifter utgifna af K.F. Hushållningssällskapet, Nylands och Tavastehus läns och Österbottens svenska landtbrukssällskap n:o 1, 1911. Helsingfors 1911.

Hafstad, Margaret R.: Guide to the McCormick Collection of the State Historical Society of Wisconsin. Madison 1973.

Hanna, W.J.: Collection of Historical Material. S.L. 1885.

Hansen, Hal: From the American Manufacturing System to Haymarket: the transformation fo McCormick Harvesting Machine Co. 1880-1886. Unpublished seminar paper, University of Wisconsin. Wisconsin 1989?

Harley, Charles K.: The shift from sailing ships to steamships, 1850-1890: a study on technological change and its diffusion (Donald N. McCloskey: Essays on a Mature Economy: Britain after 1840. Papers and Proceedings of the Mathematical Social Science Board of Conference on the New Ecnomic History of Britain, 1840-1930, held at Eliot House, Harward University, 1-3 September 1970). London 1970.

Haushofer, Heinz: Die deutsche Landwirtschaft im technischen Zeitalter. Zweite, verbesserte Auflage mit 19 abbildungen und 16 Bildtafeln. Stuttgart 1972.

Heikkonen, Esko: Ulkomaisen maatalousteknologian tulo Suomeen 1800-luvun alusta ensimmäiseen maailmansotaan. Unpublished MA thesis, the University of Turku. Turku 1983.

Heikkonen, Esko: Yhdysvaltojen leikkuukoneteollisuuden kehitys ja leviäminen Eurooppaan v.1851-1902 erityisesti McCormick Harvesting Machine Companyn toiminnan valossa. Unpublished PhL thesis, the University of Turku. Turku 1989.

Hertner, Peter: Financial strategies and adaptation to foreign markets: the German electro-technical industry and its multinational activities: 1890s to 1939 (Teichova, Alice - Levy-Leboyer, Maurice - Nussbaum, Helga: Multinational enterprise in historical perspective). Cambridge 1988.

Hounshell, David A.: From the American System to Mass Production, 1800-1932. The Development of Manufacturing Technology in the United States. Baltimore 1987.

Hughes, Jonathan: American Economic History. Second Edition. Glenview, London. 1987.

Humphries, W.R. - Gray R.B.: Partial history of haying equipment. United States Department of Agriculture. Agricultural Research Administration. Bureau of Farm Industry, Soils, and Agricultural Engineering Division of Farm Machinery. Information Series No. 74, 1949. Beltsville 1949.

Hutchison, William T.: Cyrus Hall McCormick. Seed time, 1809-1856. New York 1930.

Hutchinson, William T.: The Reaper Industry as Related to the Agricultural Development of the Middle West from 1855 to 1875. Paper read at the American Historical Association meeting Urbana, Illinois. December 1933. Mss. Spec. Rep, File B.5.

Hutchinson, William T.: Cyrus Hall McCormick. Harvest, 1856-1884. New York 1935.

Hällstöm, Emil af: Berättelse öfver nionde allmänna finska landbruksmötet i Viborg 1887. Wiborg 1889.

Jewell, C. Andrew: The Impact of America on English Agriculture. Agricultural History 50 (1976): 125-136.

Johnson, R.O.: Manuscript History of McCormick Works. S.L. 1927.

Jones, Geoffrey - Schröter, Harm G.: Continental European multinationals, 1850-1902 (Jones, Geoffrey - Schröter, Harm G.: The Rise of Multinationals in Continental Europe). Bodmin 1993.

Jones, L.J.: The Early History of Mechanical Harvesting. History of Technology 4 (1979): 101-148.

Juhlin Dannfelt, H.: Kung. Landtbruksakademien 1813-1912 samt Svenska landthushållningen under nittonde århundradet. I. Stockholm 1913.

Jörgensen, Emil: De danske Redskabs- og Maskinpröver. Landsboskrifter XIII, 1902. Köbenhavn 1902.

Kasson, John M.: Civilizing the Machine. Technology and the Republican Values in America, 1776-1900. New York 1988.

Kero, Reino: Ulkomaisen teknologian patentointi Suomessa ennen ensimmäistä maailmansotaa. Historiallinen Arkisto 90 (1987): 115-214.

Kindleberger, Charles P.: Economic growth in France and Britain, 1851-1950. Cambridge, Massachusetts 1964.

Klein, Ernst: Geschichte der deutschen Landwirtschaft im Industriezeitalter. Wiesbaden 1973.

Kobayashi, Kesaji: Organizational Innovation in the McCormick and International Harvester Companies (Keiichiro Nakagawa: Strategy and structure of big business). Tokyo 1974.

Koivukangas, Olavi: Sea, Gold and Sugarcane. Attraction versus distance. Finns in Australia, 1851-1947. Turun yliopiston julkaisuja. Annales Universitatis Turkuensis. Sarja-Ser. B. Osa-tom. 173. Turku 1986.

Kramer, Helen M.: Harvester and High Finance: Formation of the International Harvester Company. Unpublished MA thesis, University of Wisconsin, revised version. Wisconsin SA.

Kramer, Helen M.: Harvester and High Finance: Formation of the International Harvester Company. Business History Review 3 (1964): 283-301.

Krzymowski, Richard: Geschichte der deutschen Landwirtschaft bis zum Ausbruch des Weltkrieges 1914 unter besonderer Berüksichtigung der technischen Entwicklung der Landwirtschaft. Stuttgart 1939.

Kuparinen, Eero: An African Alternative: Nordic Migration to South Africa, 1815-1914. Suomen Historiallinen Seura. Studia Historica 40. Migration Studies C 10. Jyväskylä 1991.

Kuuse, Jan: Från redskap till maskiner. Mekaniseringsspridning ock kommersialisering inom svenskt jordbruk 1860-1910. Meddelanden från ekonomisk-historiska institutionen vid Göteborgs universitet 20. Göteborg 1970.

Kuuse, Jan: Interaction between Agriculture and Industry. Case studies of farm mechanisation and industrialisation in Sweden and the United States 1830-1930. Meddelanden från ekonomisk-historiska institutionen vid Göteborgs universitet 34. Göteborg 1974.

Kuznets, Solomon: Agricultural Machinery Industry (Encyclopedia of the Social Sciences. Vol. I). New York 1937.

Lamoreaux, Naomi R.: The Great Merger Movement in American Business, 1895-1904. Cambridge 1988.

Landes, David S.: The Unbound Prometheus. Technological change and industrial development in Western Europe from 1750 to the present. New York 1989.

Larsen, C.C.: Det danske landbrugs historie. En kortfattet oversigt. Naermest udarbejedet til brug ved landbrugs- og folkehojskoler. Köbenhavn 1895.

Leigh, Warren W.: The Development of the Present System for the Marketing of Harvesting Machinery. Unpublished Master's thesis in Northwestern University, 1924,

Liakka, Niilo: Orisbergin ruukki ja maatila. Porvoo 1920.

Long, W. Harwood: The Development of Mechanization in English Farming. The Agricultural History Review XI (1963): 15-26.

Lähteenmäki, Olavi: Colonia Finlandesa. Uuden Suomen perustaminen Argentiinaan 1900-luvun alussa. Toim. Reino Kero. Suomen Historiallinen Seura. Historiallisia Tutkimuksia 154. Vammala 1989.

Maanviljelyshallituksen kertomus 1898. Helsinki 1899.

Marsh, Barbara: A Corporate Tragedy. The Agony of International Harvester Company. New York 1985.

Manning, Eugene A.: Foreign Business of the McCormick Harvesting Machine Company, 1885-1902. Unpublished seminar paper, University of Wisconsin. Wisconsin 1961.

McCormick Cyrus: The century of the reaper. Boston and New York 1931.

McKIBBEN, Eugene G. - GRIFFIN, Austin R.: Changes in Farm Power and Equipment. Tractors, trucks and automobiles.Phipadelphia 1938.

Miller, Merrit Finley: The Evolution of Reaping Machines. U.S. Department of Agriculture. Office of Experiment Stations. Bulletin No. 1903. Washington 1902.

Oliver, John W.: History of American Technology. New York 1956.

Olmstead, Alan L.: The Mechanization of Reaping and Mowing in American Agriculture, 1833-1870. The Journal of Economic History 35 (1975): 327-352.

Olmstead, Alan L.: The Civil War as a Catalyst of Technological Change in Agriculture (Paul Uselding: Business and Economic History. Second series. Volume five. Papers presented at the twenty-second annual meeting of the business history conference 12-13 March 1976.). Urbana 1976.

Olmstead, Alan L. - Rhode, Paul W.: Beyond the Threshold: An Analysis of the Characteristics and Behavior of Early Reaper Adopters. The Journal of Economic History 55 (1995): 27-57.

Owings, M.R.D.: New Methods and New Machines for the Farm. What the Inventor has done for Agriculture. Scientific American 114 1911): 170-174.

Ozanne, Robert: Union Wage Impact: A Nineteenth-Century Case. Industrial and Labor Relations Review 15 (1962): 350-375.

Phillips, W.G.: The Agricultural Implement Industry in Canada. A Study of Competition. Toronto 1956.

Queen, George German: The United States and the Material Advance in Russia, 1881-1906. Unpublished Ph.D. thesis. University of Illinois 1942.

Rasmussen, Wayne D.: The impact of technological change in American agriculture, 1862-1892. Journal of Economic History 22 (1962):578-591.

Rasmussen, Wayne D.: The Civil War: A Catalyst of Agricultural Revolution. Agricultural History 39 (1965):187-195.

Rogin, Leo: The Introduction of Farm Machinery in its Relation to the Productivity of Labor in the Agriculture of the United States during the Nineteenth Century. Berkeley 1931.

Rönnbäck, Ernst: Kertomus kahdeksannesta yleisestä Suomen maanviljelyskokouksesta Turusta. Turku 1883.

Saul, S.B.: The Engineering Industry (Derek H. Aldcroft: The Development of British Industry and Foreign Competition 1875-1914. Studies in Industrial Enterprise). London 1968.

Saving the World from Starvation. The Miracle of Modern Farm Machinery. Pioneered by Cyrus McCormick and Perfected Now by the Worldwide "Harvester" Organization. International Harvester Company of America. S.A. S.L.

Schlebecker, John T.: Whereby We Thrive: A History of American Farming, 1607-1972. Ames 1975.

Schonberger, Howard Bernard: The Foreign Business of the McCormick Harvesting Machine Company. Unpublished MA thesis, University of Wisconsin. Wisconsin 1964.

Schumpeter, Joseph: The Creative Response in Economic History. Journal of Economic History 2 (1947): 149-159.

Shannon, Fred A: The Farmer's Last Frontier. Agriculture, 1860-1897. Volume V. The Economic History of the United States. New York 1945.

Shapiro, Eugene: Expansion of the foreign market, 1898-1902. Inaugurating the branch house system. Unpublished seminar paper, University of Wisconsin. Wisconsin 1958.

Simon, Matthew - Novack, David: Some Dimensions of the American Commercial Invasion of Europe, 1871-1914: An Introductory Essay. Journal of Economic History 24 (1964): 591-605.

Skalweit, August: Agriculture on the Continent in modern times (Edwin R.A. Seligman and Alvin Johnson: Encyclopedia of the social sciences. Volume one). New York 1937.

Slicher van Bath, B.H.: The influence of economic conditions on the development of agricultural tools and machines in history (J.L. Meij: Mechanization in Agriculture). Amsterdam 1960.

Soininen, Arvo M.: Vanha maataloutemme. Maatalous ja maatalousväestö Suomessa perinnäisen maatalouden loppukaudella 1720-luvulta 1870-luvulle. Forssa 1975.

Spence, Clark C.: Experiments in American Steam Cultivation. Agricultural History 33 (1959):107-116.

Tallqvist, J.V.: Nordamerikas industriella öfverlägsenhet. Ekonomiska samfundet i Finland. Föredrag och förhandlingar. Fjärde bandet. Helsingfors 1906.

Taylor, George Roger: The Transportation Revolution, 1815-1860. Volume V, The Economic History of the

United States. New York 1951.

The Invention of the Reaper. International Harvester Company 1931.

Thomas, Norman F.: Minneapolis-Moline. A History of Its Formation and Operations. New York 1976.

Thwaites, Reuben Gold: Cyrus Hall McCormick and the Reaper. Proceedings of the State Historical Society of Wisconsin, 1908. Madison 1909.

Tracy, Michael: Agriculture in Western Europe. Crisis and adaptation since 1880. London 1964.

Tracy, Michael: Agriculture in Western Europe: The Great Depression 1880-1900 (Charles K. Warner: Agrarian Conditions in Modern European History). New York 1966.

Tveite, S.: Teknologisk diffusjon og oekonomisk tilpasning. Slåmaskinens gjennombrudd i norsk jordbruk. Den första ekonomisk historiker möte i Stavanger, september 1980.

van Metre, T.V.: Transportation in the United States. Chicago 1939.

Viita, Pentti: Koneet ja kalusto Suomen maataloudessa vuosina 1900-1959. Kansantaloudellisia tutkimuksia XXV. Kansantaloudellisia ongelmia. Julkaissut kansantaloudellinen yhdistys. Helsinki 1964.

Walton, J.R.: A study in the diffusion of agricultural machinery in the nineteenth century. Research papers no.5. School of Geography. University of Oxford. 1973.

Vanderlip, Frank A.: The American "Commercial Invasion" of Europe. Reprint of 1902. New York 1976.

Virtanen, Keijo: Atlantin yhteys. Tutkimus amerikkalaisesta kulttuurista, sen suhteesta ja välittymisestä Euroopaan vuosina 1776-1917. Jyväskylä 1988.

Wik, Reynol M.: Steam Power on the American Farm, 1830-1880. Agricultural History 25 (1951):181-186.

Wik, Reynold M.: Steam Power on the American Farm. Philadelphia 1953.

Wik, Reynold M.: Henry Ford's Tractors and American Agriculture. Agricultural History 38 (1964):79-86.

Wilkins, Mira: The Emergence of Multinational Enterprise: American Business Abroad from the Colonial Era to 1914. Cambridge, Mass. 1970.

Wilkins, Mira: Defining a Firm: History and Theory (Peter Hertner and Geoffrey Jones: Multinationals: Theory and History). Shaftesbury 1986

Wilkins, Mira: European and North-American Multinationals, 1870-1914. Comparisons and Contrasts (R.P.T. Davenport-Hines and Geoffrey Jones: The End of Insularity. Essays in Comparative Business History). Chippenham 1988.

Wilkins, Mira: Comparative Hosts. Business History 36 (1994): 18-50.

Wilson, Charles: The Multinational in Historical Perspective (Keiichiro Nakagawa: Strategy and structure of big business). Tokyo 1974.